About the Author

Paul W. Zitzewitz graduated from Carleton College with a B.A. in physics and received his M.A. and Ph.D. degrees from Harvard University, also in physics. After post-doctoral positions at the University of Western Ontario and Corning Glass Works, he joined the faculty at the University of Michigan—Dearborn, where he taught and did research on positrons and positronium for more than 35 years.

During his career the university awarded him distinguished faculty awards in research, service, and teaching and named him emeritus professor of physics and science education in 2009. Zitzewitz has been active in local, state, and national physics teachers organizations, received the Distinguished Service Award from the Michigan Section of the American Association of Physics Teachers, and has been honored as a Fellow of the American Physical Society for his work in physics education.

Zitzewitz is presently treasurer and member of the executive board of the American Association of Physics Teachers. He is the author of the high school physics textbook *Physics: Principles and Problems* and is a contributing author to four middle-school physical science textbooks.

Zitzewitz enjoys classical music and opera and attending plays. His hobbies are collecting stamps of scientists (especially physicists), genealogy, and computers. He and his wife live in Northville, Michigan, but enjoy their summer cottage in Traverse City, especially when their children and grandchildren visit.

Also from Visible Ink Press

The Handy Anatomy Answer Book
by James Bobick and Naomi Balaban
ISBN: 978-1-57859-190-9

The Handy Answer Book for Kids (and Parents)
by Judy Galens and Nancy Pear
ISBN: 978-1-57859-110-7

The Handy Astronomy Answer Book
by Charles Liu
ISBN: 978-1-57859-193-0

The Handy Biology Answer Book
by James Bobick, Naomi Balaban, Sandra
Bobick and Laurel Roberts
ISBN: 978-1-57859-150-3

The Handy Dinosaur Answer Book, 2nd Edition
by Patricia Barnes–Svarney and Thomas E
Svarney
ISBN: 978-1-57859-218-0

The Handy Geography Answer Book,
2nd Edition
by Paul A. Tucci and Matthew T. Rosenberg
ISBN: 978-1-57859-215-9

The Handy Geology Answer Book
by Patricia Barnes–Svarney and Thomas E
Svarney
ISBN: 978-1-57859-156-5

The Handy History Answer Book, 2nd Edition
by Rebecca Nelson Ferguson
ISBN: 978-1-57859-170-1

The Handy Law Answer Book
by David L. Hudson Jr.
ISBN: 978-1-57859-217-3

The Handy Math Answer Book
by Patricia Barnes–Svarney and Thomas E
Svarney
ISBN: 978-1-57859-171-8

The Handy Ocean Answer Book
by Patricia Barnes–Svarney and Thomas E
Svarney
ISBN: 978-1-57859-063-6

The Handy Philosophy Answer Book
by Naomi Zack
ISBN: 978-1-57859-226-5

The Handy Politics Answer Book
by Gina Misiroglu
ISBN: 978-1-57859-139-8

The Handy Psychology Answer Book
by Lisa J. Cohen
ISBN: 978-1-57859-223-4

The Handy Religion Answer Book
by John Renard
ISBN: 978-1-57859-125-1

The Handy Science Answer Book®,
Centennial Edition
by The Science and Technology
Department Carnegie Library of
Pittsburgh
ISBN: 978-1-57859-140-4

The Handy Sports Answer Book
by Kevin Hillstrom, Laurie Hillstrom and
Roger Matuz
ISBN: 978-1-57859-075-9

The Handy Supreme Court Answer Book
by David L Hudson, Jr.
ISBN: 978-1-57859-196-1

The Handy Weather Answer Book, 2nd Edition
by Kevin S. Hile
ISBN: 978-1-57859-215-9

Please visit the Handy series website at handyanswers.com

THE HANDY

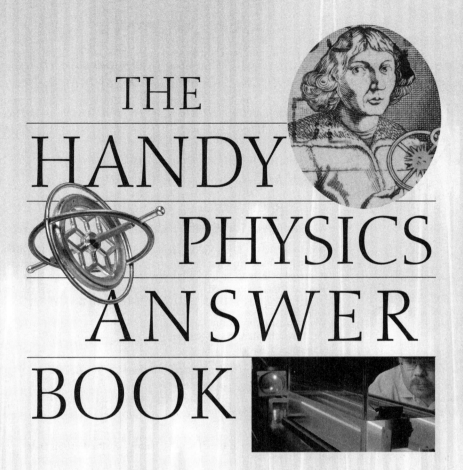

PHYSICS

ANSWER

BOOK

SECOND EDITION

Paul W. Zitzewitz, PhD

VISIBLE
INK
PRESS

Detroit

THE HANDY PHYSICS ANSWER BOOK

Visible Ink Press®
43311 Joy Rd., #414
Canton, MI 48187-2075

Visible Ink Press is a registered trademark of Visible Ink Press LLC.

Most Visible Ink Press books are available at special quantity discounts when purchased in bulk by corporations, organizations, or groups. Customized printings, special imprints, messages, and excerpts can be produced to meet your needs. For more information, contact Special Markets Director, Visible Ink Press, www.visibleink.com, or 734-667-3211.

Managing Editor: Kevin S. Hile
Art Director: Mary Claire Krzewinski
Typesetting: Marco Di Vita
Indexing: Shoshana Hurwitz
Proofreader: Sarah Hermsen
ISBN 978-1-57859-305-7

Cover images: iStock.

Library of Congress Cataloguing-in-Publication Data

Zitzewitz, Paul W.
  The handy physics answer book / Paul W. Zitzewitz.
    p. cm.
  Includes bibliographical references and index.
  ISBN 978-1-57859-305-7
  1. Physics--Miscellanea. I. Title.
  QC75.Z58 2011
  530--dc22
                                        2010047248

Printed in the United States of America

10 9 8 7 6 5 4 3 2 1

Contents

ACKNOWLEDGMENTS *vii*

INTRODUCTION *ix*

BIBLIOGRAPHY ... *323*

SYMBOLS ... *327*

GLOSSARY ... *331*

INDEX ... *359*

THE BASICS ... 1

Measurement ... Careers in Physics ... Famous Physicists ... The Nobel Prize

MOTION AND ITS CAUSES ... 25

Force and Newton's Laws of Motion

MOMENTUM AND ENERGY ... 55

Momentum ... Energy

STATICS ... 83

Center of Gravity ... Statics ... Bridges and Other "Static" Structures

FLUIDS ... 95

Water Pressure ... Blood Pressure ... Atmospheric Pressure ... Sinking and Floating: Buoyancy ... Fluid Dynamics: Hydraulics and Pneumatics ... Aerodynamics ... The Sound Barrier ... Supersonic Flight

THERMAL PHYSICS ... 117

Thermal Energy ... Temperature and Its Measurement ... Absolute Zero ... States of Matter ... Heat ... Thermodynamics

WAVES ... 137

Water Waves ... Electromagnetic Waves ... Communicating with

Electromagnetic Waves ... Putting Information on Electromagnetic Waves ... Microwaves ... The Principle of Superposition ... Resonance ... Impedance ... The Doppler Effect ... Radar ... NEXRAD Doppler Radar ... Radio Astronomy

SOUND ... 165

Speed of Sound ... Hearing ... Ultrasonics and Infrasonics ... Intensity of Sound ... Acoustics ... Musical Acoustics ... Noise Pollution

LIGHT ... 187

The Speed of Light ... Polarization of Light ... Opaque, Transparent, and Translucent Materials ... Shadows ... Reflection ... Mirrors ... Refraction ... Lenses ... Fiber Optics ... Diffraction and Interference ... Color ... Rainbows ... Eyesight ... Cameras ... Telescopes

ELECTRICITY ... 231

Leyden Jars and Capacitors ... Van de Graaf Generators ... Lightning ...

Safety Precautions ... Current Electricity ... Resistance ... Superconductors ... Ohm's Law ... Electric Power and Its Uses ... Circuits ... AC/DC ... Series/Parallel Circuits ... Electrical Outlets

MAGNETISM ... 261

Electromagnetism ... Electromagnetic Technology ... Magnetic Fields in Space

WHAT IS THE WORLD MADE OF? ... 273

AT THE HEART OF THE ATOM ... 289

UNANSWERED QUESTIONS ... 309

Beyond the Proton and Neutron ... Entanglement, Teleportation, and Quantum Computing

Acknowledgments

I want to express my thanks to a large number of others who have asked questions and challenged answers over a long career. These include students in my classes—from future elementary teachers, engineers, and physicists; members of the research group at the University of Michigan—Ann Arbor; colleagues at the University of Michigan—Dearborn in physics, the natural sciences department, and the Inquiry Institute; high school teachers in the Detroit area and the state of Michigan; and fellow members of the American Association of Physics Teachers. I owe them all a deep debt of gratitude.

Of course, the most persistent challenges have come from my children and grandchildren, who have many times asked, "But why?" My parents supported and encouraged my early interests in physics, chemistry, and electronics, and for that I am extremely grateful. More than anyone, however, I would like to thank my wife, Barb, who is my best friend and colleague. She has encouraged and supported me throughout our life together.

This second edition of the *Handy Physics Answer Book* is based on the first edition, written by P. Erik Gundersen. The new edition has adopted the structure and style of the first. Some questions and answers have not been changed, but many others have been updated and new ones have been added. Erik's work has been a tremendous help in writing this edition. I would also like to thank Roger Jänecke and Kevin Hile at Visible Ink Press for their encouragement and help during the writing of this book. While the book has been carefully researched and proofread, I take responsibility for any remaining errors.

Paul W. Zitzewitz
Northville, Michigan
November, 2010

PHOTO CREDITS

INTRODUCTION

Why don't skyscrapers sway in the wind? How does a ground-fault interrupter work? What's the ultimate fate of the universe? Who developed our understanding of the nature of the atom? Physics is full of questions. Some are about the most fundamental ideas on which the universe is based, others involve everyday applications of physics, many are just fun. Most have answers, although those answers may have been different in the past and may be different in the future.

The *Handy Physics Answer Book* is written for you to explore these and other questions and to ponder over their answers. It should lead you to ask further questions and search for other answers. Eschewing the usual mathematical explanations for physics phenomena, this approachable reference explains complicated scientific concepts in plain English that everyone can understand.

But it contains more. Physics has been developed by people over more than two thousand years. They come from diverse backgrounds from a wide range of cultures. Some made only one contribution, others made important advances over many years in several different areas. The names of some will be familiar: Einstein, Newton, Galileo, Franklin, Curie, Feynman. Others you may not have heard of: Alhazen, Goeppert-Meyer, Cornell, Heaviside. A complete list of physics Nobel Prize winners is included.

The *Handy Physics Answer Book* does not have to be read from beginning to end. Look through the index for a topic that interests you. Or, open it at random and pick a question that has always puzzled you. If a scientific term is not familiar, refer to the glossary at the end of the book. While the book describes concepts much more than equations, it does use symbols to represent physics quantities. If you're not familiar with a symbol, there is a helpful dictionary, at the end of this book, as well as a glossary of terms.

Does an answer leave you wanting more information? Look at the bibliography for a book or Website; then visit a library, bookstore, or access the Web.

But above all, enjoy your adventure!

THE BASICS

What is **physics**?

Physics is the study of the structure of the natural world. It seeks to explain natural phenomena in terms of a comprehensive theoretical structure in mathematical form. Physics depends on accurate instrumentation, precise measurements, and the expression of results in mathematical terms. It describes and explains the motion of objects that are subject to forces. Physics forms the basis of chemistry, biology, geology, and astronomy. Although these sciences involve the study of systems much more complex than those that physicists study, the fundamental aspects are all based on physics.

Physics is also applied to engineering and technology. Therefore a knowledge of physics is vital in today's technical world. For these reasons physics is often called the fundamental science.

What are the **subfields** of **physics**?

The word *physics* comes from the Greek *physis*, meaning nature. Aristotle (384–322 B.C.E.) wrote the first known book entitled *Physics*, which consisted of a set of eight books that was a detailed study of motion and its causes. The ancient Greek title of the book is best translated as *Natural Philosophy*, or writings about nature. For that reason, those who studied the workings of nature were called "Natural Philosophers." They were educated in philosophy and called themselves philosophers. One of the early modern textbooks that used physics in its title was published in 1732. It was not until the 1800s that those who studied physics were called physicists. In the nineteenth, twentieth, and twenty-first centuries physics has proven to be a very large and important field of study. Due to the huge breadth of physics, physicists today must concentrate their work in one or two of the subfields of physics. The most important of these fields are listed below.

The Greek philosopher Aristotle wrote the first known book about physics.

- **Quantum mechanics and relativity**—Both of these fields study the descriptions and explanations of the way small particles interact (quantum physics), the motion of objects moving near the speed of light (special relativity), and the causes and effects of gravity (general relativity).

- **Elementary particles and fields**—The study of the particles that are the basis of all matter. Both their properties and their interactions are included.

- **Nuclear physics**—The study of the properties of the nuclei of atoms and the protons and neutrons of which they are composed.

- **Atomic and molecular physics**—The study of single atoms and molecules that are made up of these atoms. Studies include interactions with each other and with light.

- **Condensed matter physics**—Otherwise known as solid-state physics, condensed matter is a study of the physical and electrical properties of solid materials. An exciting new study is that of nano materials, leading to nanotechnology.

- **Electromagnetism and optics**—Studies how electric and magnetic forces interact with matter. Light is a type of electromagnetic wave and so is a part of electromagnetism.

- **Thermodynamics and statistical mechanics**—Studies how temperature affects matter and how heat is transferred. Thermodynamics deals with macroscopic objects; statistical mechanics concerns the atomic and molecular motions of very large numbers of particles, including how they are affected by heat transfer.

- **Mechanics**—Deals with the effect of forces on the motion and energy of physical objects. Modern mechanics studies mostly involve fluids (fluid dynamics) and granular particles (like sand), as well as the motions of stars and galaxies.

- **Plasma physics**—Plasmas are composed of electrically charged atoms. Plasmas studied include those in fluorescent lamps, in large-screen televisions, in Earth's atmosphere, and in stars and material between stars. Plasma physicists are also working to create controlled nuclear fusion reactors to produce electricity.

- **Physics education research**—Investigates how people learn physics and how best to teach them.

Applications of Physics

- **Acoustics**—Musical acoustics studies the ways musical instruments produce sounds. Applied acoustics includes the study of how concert halls can best be designed. Ultrasound acoustics uses sound to image the interior of metals, fluids, and the human body.

- **Astrophysics**—Studies how astronomical bodies, such as planets, stars, and galaxies, interact with one another. A subfield is cosmology, which investigates the formation of the universe, galaxies, and stars.

- **Atmospheric physics**—Studies the atmosphere of Earth and other planets. Today most activity involves the causes and effects of global warming and climate change.

- **Biophysics**—Studies the physical interactions of biological molecules and the use of physics in biology.

- **Chemical physics**—Investigates the physical causes of chemical reactions between atoms and molecules and how light can be used to understand and cause these reactions.

- **Geophysics**—Geophysics is the physics of Earth. It deals with the forces and energy found within Earth itself. Geophysicists study tectonic plates, earthquakes, volcanic activity, and oceanography.

- **Medical physics**—Investigates how physical processes can be used to produce images of the inside of humans, as well as the use of radiation and high-energy particles in treating diseases such as cancer.

MEASUREMENT

Why is **measurement** so **important** for physics?

While Aristotle (384–322 B.C.E.) emphasized observation rather than measurement or experimentation, astronomy required measurements of the locations of stars and "wanderers" (now known to be planets). The study of light was another early field that began to emphasize experimentation and mathematics.

What are the **standards for measurement** in physics?

The International System of Units, officially known as *Système International* and abbreviated SI, was adopted by the eleventh General Conference on Weights and Measures in Paris in 1960. Basic units are based on the meter-kilogram-second (MKS) system, which is commonly known as the metric system.

Most of the world uses the metric system for measuring quantities such as weight. Also known as the meter-kilogram-second (MKS) system, the metric system was last refined at the eleventh General Conference on Weights and Measures in 1995.

Does the **United States** use **SI**?

Although the American scientific community uses the SI system of measurement, the general American public still uses the traditional English system of measurement. In an effort to change over to the metric system, the United States government instituted the Metric Conversion Act in 1975. Although the act committed the United States to increasing the use of the metric system, it was on a voluntary basis only. The Omnibus Trade and Competitiveness Act of 1988 required all federal agencies to adopt the metric system in their business dealings by 1992. Therefore, all companies that held government contracts had to convert to metric. Although approximately 60% of American corporations manufacture metric products, the English system still is the predominant system of measurement in the United States.

How is a **second measured**?

Atomic clocks are the most precise devices to measure time. Atomic clocks such as rubidium, hydrogen, and cesium clocks are used by scientists and engineers when computing distances with Global Positioning Systems (GPS), measuring the rotation of Earth, precisely knowing the positions of artificial satellites, and imaging stars and galaxies.

The clock that is used as the standard for the second is the cesium-133 atomic clock. The measurement of the second is defined as the time it takes for 9,192,631,770 periods of microwave radiation that result from the transfer of the cesium-133 atom between lower-energy and higher-energy states. The second is currently known to a precision of 5×10^{-16}, or one second in 60 million years!

Who **defined** or developed the **meter**?

In 1798, French scientists determined that the meter would be measured as 1/10,000,000th the distance from the North Pole to the Equator. After calculating this distance, scientists made a platinum-iridium bar with two marks precisely one meter apart. This standard was used until 1960. Today the meter is defined using the second and the speed of light. One meter is the distance light travels in 1/299,792,458 seconds.

What is the **standard unit** for **mass**?

The kilogram is the standard unit for mass in SI and the metric system. The kilogram was originally defined as the mass of 1 cubic decimeter of pure water at 4° Celsius. A platinum cylinder of the same mass as the cubic decimeter of water was the standard until 1889. A platinum-iridium cylinder with the same mass is permanently kept near Paris. Copies exist in many countries. In the United States the National Institute of Standards and Technology (NIST) houses the mass standard, as well as the atomic clocks that define the second. The kilogram is the only standard unit that is not based on atoms or molecules. Several methods are under development to define the kilogram in terms of the mass of the carbon atom. Currently one method has a precision of 35 parts per billion. That is equivalent to measuring the mass of your body and the change in mass if one hair falls off your head!

What was the **first clock**?

For thousands of years the second, and all other units used to measure time, were based on the rotation of Earth. The first method of measuring time shorter than a day dates back to 3500 B.C.E., when a device known as the gnomon was used. The gnomon was a stick placed vertically into the ground which, when struck by the sun's light, produced a distinct shadow. By measuring the relative positions of the shadow throughout the day, the length of a day was able to be measured. The gnomon was later replaced by the first hemispherical sundial in the third century B.C.E. by the astronomer Berossus (born about 340 B.C.E.).

What do some of the **metric prefixes** represent?

Prefixes in the metric system are used to denote powers of ten. The value of the exponent next to the number ten represents the number of places the decimal should be

What are the major limitations of gnomons and sundials?

This kind of clock cannot be used at night of when the sun doesn't shine. To remedy this problem, timing devices such as notched candles were created. Later, hourglasses and water clocks (clepsydra) became quite popular. The first recorded description of a water clock is from the sixth century B.C.E. In the third century B.C.E. Ctesibius of Alexandria, a Greek inventor, used gears that connected a water clock to a pointer and dial display similar to those in today's clocks. But it wasn't until 1656 when a pendulum was used with a mechanical clock that these clocks kept very accurate time.

Sundials are a very old way to tell time. While accurate, they are limited by the fact that they only work when the sun is shining.

moved to the right (if the number is positive), or to the left (if the number is negative). The following is a list of prefixes commonly used in the metric system:

femto	(f)	10^{-15}	deka	(da)	10^1
pico	(p)	10^{-12}	hecto	(h)	10^2
nano	(n)	10^{-9}	kilo	(k)	10^3
micro	(μ)	10^{-6}	mega	(M)	10^6
milli	(m)	10^{-3}	giga	(G)	10^9
centi	(c)	10^{-2}	tera	(T)	10^{12}
deci	(d)	10^{-1}	peta	(P)	10^{15}

How does "accuracy" differ from "precision"?

Both "accuracy" and "precision" are often used interchangeably in everyday conversation; however, each has a unique meaning. Accuracy defines how correct or how close to the accepted result or standard a measurement or calculation has been. Precision describes how well the results can be reproduced. For example, a person who can repeatedly hit a bull's eye with a bow and arrow is accurate and precise. If the person's arrows all fall within a small region away from the bull's eye, then she or he is precise, but not accurate. If the person's arrows are scattered all over the target and the ground behind it, the she or he is neither precise nor accurate.

CAREERS IN PHYSICS

How does one **become a physicist**?

The first requirement to be a physicist is to have an inquisitive mind. Albert Einstein (1879–1955) himself admitted, "I'm like a child. I always ask the simplest questions." It seems as though the simplest questions always appear to be the most difficult to answer.

These days, becoming a physicist requires quite a bit of schooling along with that inquisitive mind. In high school, a strong academic background including mathematics, English, and science is necessary in order to enter college with a strong knowledge base. Once you are a physics major in college you will take courses such as classical mechanics, electricity and magnetism, optics, thermodynamics, modern physics, and calculus in order to obtain a bachelor's degree.

To become a research physicist, an advanced degree is required. This means attending graduate school, performing research, writing a thesis, and eventually obtaining a Ph.D. (Doctor of Philosophy).

What does a **physicist do**?

Physicists normally do their work in one of three ways. Some are theorists who create and extend theories, or explanations of the physical world. Others are experimenters, who develop experiments to test theories to explore uses of new instruments, or to investigate new materials. The third method of doing physics is to use computers to simulate experiments, explore and extend theories, or make observations that cannot be done by the human eye.

Physicists can find employment in a variety of fields. Many research physicists work in environments where they perform basic research. These scientists typically work in research universities, government laboratories, and astronomical observatories. Physicists who find new ways to apply physics to engineering and technology are often employed by industrial laboratories. Physicists are also extremely valuable in areas such as computer science, economics and finance, medicine, communications, and publishing. Finally, many physicists who love to see young people get excited about physics become teachers in elementary, middle, or high schools, or in colleges and universities.

FAMOUS PHYSICISTS

Who were the **first physicists**?

Although physics was not considered a distinct field of science until the early nineteenth century, people have been studying the motion, energy, and forces that are at play in the universe for thousands of years. The earliest documented accounts of serious thought toward physics, specifically the motion of the planets, dates back to the years of the Chinese, Indians, Egyptians, Mesoamericans, and the Babylonians. The Greek philosophers Plato and Aristotle analyzed the motion of objects, but did not perform experiments to prove or disprove their ideas.

What contributions did **Aristotle** make?

Aristotle was a Greek philosopher and scientist who lived for sixty-two years in the fourth century B.C.E. He was a student of Plato and an accomplished scholar in the fields of biology, physics, mathematics, philosophy, astronomy, politics, religion, and education. In physics, Aristotle believed that there were five elements: earth, air, fire, water, and the fifth element, the quintessence, called aether, out of which all objects in the heavens were made. He believed that these elements moved in order to seek out each other. He stated that if all forces were removed, an object could not move. Thus motion, even with no change in speed or direction, requires a continuous force. He believed that motion was the result of the interaction between an object and the medium through which it moves.

Through the third century B.C.E. and later, experimental achievements in physics were made in such cities as Alexandria and other major cities throughout the Mediterranean. Archimedes (c. 287–c. 212 B.C.E.) measured the density of objects by measur-

Who was the founder of the scientific method?

Ibn al-Haitham (known in Europe as Alhazen or Alhacen) lived between 965 and 1038. He was born in Basra, Persia (now in Iraq) and died in Cairo, Egypt. He wrote 200 books, 55 of which have survived. They include his most important work, *Book of Optics,* as well as books on mechanics, astronomy, geometry, and number theory. He is known as the founder of the scientific method and for his contributions to philosophy and experimental psychology.

ing their displacement of water. Aristarchus of Samos is credited with measuring the ratio of the distances from Earth to the sun and to the moon, and espoused a sun-centered system. Erathosthenes determined the circumference of Earth by using shadows and trigonometry. Hipparchus discovered the precession of the equinoxes. And finally, in the first century C.E. Claudius Ptolemy proposed an order of planetary motion in which the sun, stars, and moon revolved around Earth.

After the fall of the Roman Empire, a large fraction of the books written by the early Greek scientists disappeared. In the 800s the rulers of the Islamic Caliphate collected as many of the remaining books as they could and had them translated into Arabic. Between then and about 1200 a number of scientists in the Islamic countries demonstrated the errors in Aristotelian physics. Included in this group is Alhazen, Ibm Shakir, al-Biruni, al-Khazini, and al-Baghdaadi, mainly members of the House of Wisdom in Baghdad. They foreshadowed the ideas that Copernicus, Galileo, and Newton would later develop more fully.

Despite these challenges, Aristotle's physics was dominant in European universities into the late seventeeth century.

How did the idea that the **sun** was the **center** of the **solar system** arise?

Aristotle's and Ptolemy's view that the sun, planets, and stars all revolved around Earth was accepted for almost eighteen centuries. Nicolas Copernicus (1473–1543), a Polish astronomer and cleric, was the first person to publish a book arguing that the solar system is a heliocentric (sun-centered) system instead of a geocentric (Earth-centered) system. In the same year as his death, he published *On the Revolutions of the Celestial Spheres*. His book was dedicated to Pope Paul III. The first page of his book contained a preface stating that a heliocentric system is useful for calculations, but may not be the truth. This preface was written by Andreas Osiander without Copernicus' knowledge. It took three years before the book was denounced as being in contradiction with the Bible, and it was banned by the Roman Catholic Church in 1616. The ban wasn't lifted until 1835.

What famous **scientist** was placed under **house arrest** for agreeing with Copernicus?

Galileo Galilei (1564–1642) was responsible for bringing the Copernican system more recognition. In 1632, Galileo published his book *Dialogue Concerning the Two Chief World Systems*. The book was written in Italian and featured a witty debate among three people: one supporting Aristotle's system, the second a supporter of Copernicus, and the third an intelligent layman. The Copernican easily won the debate. The book was approved for publication in Florence but was banned a year later. Pope Urban VIII, a long-time friend of Galileo, believed that Galileo had made a fool of him in the book. Galileo was tried by the Inquisition and placed under house arrest for the rest of his life. All of his writings were banned.

Galileo was also famous for his work on motion; he is probably best known for a thought experiment using the Leaning Tower of Pisa. He argued that a heavy rock and a light rock dropped from the tower would hit the ground at the same time. His arguments were based on extensive experiments on balls rolling down inclined ramps. Many scientists believe that Galileo's work is the beginning of true physics.

Galileo Galilei's *Dialogue Concerning the Two Chief World Systems* (1632) argued for the Copernican system of the solar system with the sun at the center and the planets circling the sun.

Who is considered **one of** the **most influential scientists** of all time?

Many scientists and historians consider Isaac Newton (1643–1727) one of the most influential people of all time. It was Newton who discovered the laws of motion and universal gravitation, made huge breakthroughs in light and optics, built the first reflecting telescope, and developed calculus. His discoveries published in *Philosophiæ Naturalis Principia Mathematica*, or *The Principia,* and in *Optiks* are unparalleled and formed the basis for mechanics and optics. Both these books were written in Latin and published only when friends demanded that he publish, many years after Newton had completed his work.

Where did **Newton study**?

Newton was encouraged by his mother to become a farmer, but his uncle saw the talent Newton had for science and math and helped him enroll in Trinity College in Cambridge. Newton spent four years there, but he returned to his hometown of Woolsthorpe to flee the spread of the Black Plague in 1665. During the two years that he spent studying in Woolsthorpe, Newton made his most notable developments of calculus, gravitation, and optics.

What **official titles** did **Newton** receive?

Newton was extremely well respected in his time. Although he was known for being nasty and rude to his contemporaries, Newton became Lucasian Professor of Mathematics at Cambridge in the late 1660s, president of the Royal Society of London in 1703, and the first scientist ever knighted, in 1705. He was famous as the Master of the Mint where he introduced coins that had defined edges so that people couldn't cut off small pieces of the silver from which the coins were made. He is buried in Westminster Abbey in London.

Sir Isaac Newton, one of the most famous scientists of all time, discovered the laws of motion, developed calculus, and built the first reflecting telescope, among many other accomplishments.

Who would become the **most influential scientist** of the **twentieth century**?

On March 14, 1879, Albert Einstein was born in Ulm, Germany. No one knew that this little boy would one day grow up and change the way people viewed the laws of the universe. Albert was a top student in elementary school where he built models and toys and studied Euclid's geometry and Kant's philosophy. In high school, however, he hated the regimented style and rote learning. At age sixteen he left school to be with his parents in Italy. He took, but failed, the entrance exam for the Polytechnic University in Zurich. After a year of study in Aarau, Switzerland, he was admitted to the University. Four years later, 1900, he was graduated.

He spent two years searching for a job and finally became a patent clerk in Bern, Switzerland. During the next three years while working at the Patent Office he developed his ideas about electromagnetism, time and motion, and statistical physics. In

Albert Einstein is most often remembered for his famous formula $E = mc^2$, but his Nobel Prize in physics was awarded for his explanation of the photoelectric effect.

1905, his so-called *annus mirabilis* or miracle year, he published four extraordinary papers. One was on the photoelectric effect, in which Einstein introduced light quanta, later called photons. The second was about Brownian motion, which helped support the idea that all matter is composed of atoms. The third was on special relativity, which revolutionized the way physicists understand both motion at very high speeds and electromagnetism. The fourth developed the famous equation $E = mc^2$. While these papers completed his Ph.D. requirements, it was two years before he was appointed a professor at the German University in Prague.

What did **Einstein do** to win **worldwide fame**?

By 1914 Einstein's accomplishments were well accepted by physicists and he was appointed professor at the University of Berlin and made a member of the Prussian Academy of Sciences. Einstein published the General Theory of Relativity in 1916. Among its predictions was that light from a star would not always travel in a straight line, but would bend if it passed close to a massive body like the sun. He predicted a bending twice as large as Newton's theory predicted. During a 1919 solar eclipse these theories were tested and Einstein's prediction was shown to be correct. The result was publicized by the most important newspapers in England and the United States and Einstein became a world figure. In 1921 he won the Nobel Prize in physics as a result of his work on the photoelectric effect.

Why was Einstein more than just a world-renowned physicist?

Einstein supported unpopular causes. The year he moved from Switzerland to Germany, he joined a group of people opposing Germany's entry into World Was I. He joined both socialist and pacifist causes. He opposed the Nazis, and when Adolf Hitler (1889–1945) came to power, Einstein moved to the United States. He took a position at the Institute for Advanced Study in Princeton, New Jersey. Some years later he became a citizen of the United States. After being urged by other physicists, Einstein signed a letter to President Franklin D. Roosevelt (1882–1945) pointing out the danger posed by Germany's work on uranium that could lead to a dangerous new kind of bomb. The letter helped to launch the Manhattan Project that lead to the development of the atomic bomb.

Although Einstein did not actually work on the bomb, after the defeat of Germany, and knowing the death and destruction that dropping the bomb would cause, he sent another letter to the President urging him not to use the bomb. The letter was never forwarded to President Harry Truman (1884–1972). After the war Einstein spent time lobbying for atomic disarmament. At one point he was even asked to head the new Jewish state of Israel. Einstein, both for his scientific works and his social and political views, became an international icon.

Why did **Einstein** win a **Nobel Prize** for the photoelectric effect, but not for relativity?

Einstein was a controversial person. He was Jewish and a strong supporter of pacifist causes. In addition, his approach to theoretical physics was very different from physicists of that time. He was repeatedly nominated for the Nobel Prize, but members of the Prize committee, despite his public fame, refused to grant him the Prize, most likely for political reasons. The 1921 prize was not awarded. In 1922 the committee found a way to compromise. Einstein was awarded the 1921 prize for the photoelectric effect because of the way it could be tested experimentally.

THE NOBEL PRIZE

What is the **Nobel Prize**?

The Nobel Prize is one of the most prestigious awards in the world. It was named after Alfred B. Nobel (1833–1896), the inventor of dynamite; he left $9,000,000 in trust, of which the interest was to be awarded to the person who made the most significant contribution to their particular field that year. The awards, given in the fields of physics, chemistry, physiology and medicine, literature, peace, and economics, are worth over $1,400,000, and a great deal of recognition.

Who are the other **Nobel Prize winners** in **physics**?

The table below lists the prize winners. In some cases, the award was split between winners.

Year	Recipient	Awarded For
2010	Andre Geim and Konstantin Novoselov	For groundbreaking experiments regarding the two-dimensional material graphene
2009	Charles K. Kao	For groundbreaking achievements concerning the transmission of light in fibers for optical communication
	Willard S. Boyle and George E. Smith	For the invention of an imaging semiconductor circuit—the CCD sensor
2008	Yoichiro Nambu	For the discovery of the mechanism of spontaneous broken symmetry in subatomic physics
	Makoto Kobayashi and Toshihide Maskawa	For the discovery of the origin of the broken symmetry which predicts the existence of at least three families of quarks in nature
2007	Albert Fert and Peter Grünberg	For the discovery of Giant magnetoresistance
2006	John C. Mather and George C. Smoot	For their discovery of the blackbody form and anisotropy of the cosmic microwave background radiation
2005	Roy J. Glauber	For his contribution to the quantum theory of optical coherence
	John L. Hall and Theodor W. Hänsch	For their contributions to the development of laser-based precision spectroscopy, including the optical frequency comb technique
2004	David J. Gross, Frank Wilczek H. David Politzer,	For the discovery of asymptotic freedom in the theory of the strong interaction
2003	Alexei A. Abrikosov, Vitaly L. Ginzburg, Anthony J. Leggett	For pioneering contributions to the theory of superconductors and superfluids
2002	Raymond Davis Jr. and Masatoshi Koshiba	For pioneering contributions to astrophysics, in particular for the detection of cosmic neutrinos

Year	Recipient	Awarded For
2002	Riccardo Giacconi	For pioneering contributions to astrophysics, which have led to the discovery of cosmic X-ray sources
2001	Eric A. Cornell, Wolfgang Ketterle, Carl E. Wieman	For the achievement of Bose-Einstein condensation in dilute gases of alkali atoms, and for early fundamental studies of the properties of the condensates
2000	Zhores I. Alferov and Herbert Kroemer	For developing semiconductor heterostructures used in high-speed- and opto-electronics
	Jack St. Clair Kilby	For his part in the invention of the integrated circuit
1999	Gerardus T Hooft and Martinus J.G. Veltman	For elucidating the quantum structure of electroweak interactions in physics
1998	Robert B. Laughlin, Horst L. Stormer, Daniel C. Tsui	For their discovery of a new form of quantum fluid with fractionally charged excitations
1997	Steven Chu, Claude Cohen-Tannoudji, William D. Phillips	For development of methods to cool and trap atoms with laser light
1996	David M. Lee, Douglas D. Osheroff, Robert C. Richardson	For their discovery of superfluidity in helium-3
1995	Martin L. Perl	For the discovery of the tau lepton
	Frederick Reines	For the detection of the neutrino
1994	Bertram N. Brockhouse	For the development of neutron spectroscopy
	Clifford G. Shull	For the development of the neutron diffraction technique
1993	Russell A. Hulse and Joseph H. Taylor Jr.	For the discovery of a new type of pulsar, a discovery that has opened up new possibilities for the study of gravitation
1992	Georges Charpak	For his invention and development of particle detectors, in particular the multiwire proportional chamber
1991	Pierre-Gilles de Gennes	For discovering that methods developed for studying order phenomena in simple systems can be generalized to more complex forms of matter, in particular to liquid crystals and polymers

Year	Recipient	Awarded For
1990	Jerome I. Friedman, Henry W. Kendall, Richard E. Taylor	For their pioneering investigations concerning deep inelastic scattering of electrons on protons and bound neutrons, which have been of essential importance for the development of the quark model in particle physics
1989	Norman F. Ramsey	For the invention of the separated oscillatory fields method and its use in the hydrogen maser and other atomic clocks
	Hans G. Dehmelt and Wolfgang Paul	For the development of the ion trap technique
1988	Leon M. Lederman, Melvin Schwartz, Jack Steinberger	For the neutrino beam method and the demonstration of the doublet structure of the leptons through the discovery of the muon neutrino
1987	J. Georg Bednorz and K. Alexander Müller	For their important breakthrough in the discovery of superconductivity in ceramic materials
1986	Ernst Ruska	For his fundamental work in electron optics, and for the design of the first electron microscope
	Gerd Binnig and Heinrich Rohrer	For their design of the scanning tunneling microscope
1985	Klaus von Klitzing	For the discovery of the quantized Hall effect
1984	Carlo Rubbia and Simon van der Meer	For their decisive contributions to the large project, which led to the discovery of the field particles W and Z, communicators of weak interaction
1983	Subramanyan Chandrasekhar	For his theoretical studies of the physical processes of importance to the structure and evolution of the stars
	William A. Fowler	For his theoretical and experimental studies of the nuclear reactions of importance in the formation of the chemical elements in the universe
1982	Kenneth G. Wilson	For his theory for critical phenomena in connection with phase transitions
1981	Nicolaas Bloembergen and Arthur L. Schawlow	For their contribution to the development of laser spectroscopy
	Kai M. Siegbahn	For his contribution to the development of high-resolution electron spectroscopy
1980	James W. Cronin and Val L. Fitch	For the discovery of violations of fundamental symmetry principles in the decay of neutral K-mesons

Year	Recipient	Awarded For
1979	Sheldon L. Glashow, Abdus Salam, Steven Weinberg	For their contributions to the theory of the unified weak and electromagnetic interaction between elementary particles, including inter alia the prediction of the weak neutral current
1978	Pyotr Leonidovich Kapitsa	For his basic inventions and discoveries in the area of low-temperature physics
	Arno A. Penzias and Robert W. Wilson	For their discovery of cosmic microwave background radiation
1977	Philip W. Anderson, Sir Nevill F. Mott, John H. van Vleck	For their fundamental theoretical investigations of the electronic structure of magnetic and disordered systems
1976	Burton Richter and Samuel C.C. Ting	For their pioneering work in the discovery of a heavy elementary particle of a new kind
1975	Aage Bohr, Ben Mottelson, James Rainwater	For the discovery of the connection between collective motion and particle motion in atomic nuclei and the development of the theory of the structure of the atomic nucleus based on this connection
1974	Sir Martin Ryle and Antony Hewish	For their pioneering research in radio astrophysics; Ryle for his observations and inventions, in particular of the aperture synthesis technique, and Hewish for his decisive role in the discovery of pulsars
1973	Leo Esaki and Ivar Giaever	For their experimental discoveries regarding tunneling phenomena in semiconductors and superconductors, respectively
	Brian D. Josephson	For his theoretical predictions of the properties of a supercurrent through a tunnel barrier, in particular those phenomena which are generally known as the Josephson effects
1972	John Bardeen, Leon N. Cooper, J. Robert Schrieffer	For their jointly developed theory of super-conductivity, usually called the BCS-theory
1971	Dennis Gabor	For his invention and development of the holographic method
1970	Hannes Alfvén	For fundamental work and discoveries in magneto-hydrodynamics with fruitful applications in different parts of plasma physics

Year	Recipient	Awarded For
1970	Louis Néel	For fundamental work and discoveries concerning antiferromagnetism and ferrimagnetism which have led to important applications in solid state physics
1969	Murray Gell-Mann	For his contributions and discoveries concerning the classification of elementary particles and their interactions
1968	Luis W. Alvarez	For his decisive contributions to elementary particle physics, in particular the discovery of a large number of resonance states, made possible through his development of the technique of using hydrogen bubble chamber and data analysis
1967	Hans Albrecht Bethe	For his contributions to the theory of nuclear reactions, especially his discoveries concerning the energy production in stars
1966	Alfred Kastler	For the discovery and development of optical methods for studying hertzian resonances in atoms
1965	Sin-Itiro Tomonaga, Julian Schwinger, Richard P. Feynman	For their fundamental work in quantum electrodynamics, with deep-ploughing consequences for the physics of elementary particles
1964	Charles H. Townes and jointly to Nicolay Gennadiyevich Basov and Aleksandr Mikhailovich Prokhorov	For fundamental work in the field of quantum electronics, which has led to the construction of oscillators and amplifiers based on the maser-laser principle
1963	Eugene P. Wigner	For his contributions to the theory of the atomic nucleus and the elementary particles, particularly through the discovery and application of fundamental symmetry principles
	Maria Goeppert-Mayer and J. Hans D. Jensen	For their discoveries concerning nuclear shell structure
1962	Lev Davidovich Landau	For his pioneering theories for condensed matter, especially liquid helium
1961	Robert Hofstadter	For his pioneering studies of electron scattering in atomic nuclei and for his thereby achieved discoveries concerning the structure of the nucleons
	Rudolf Ludwig Mössbauer	For his researches concerning the resonance absorption of gamma radiation and his discovery in this connection of the effect which bears his name
1960	Donald A. Glaser	For the invention of the bubble chamber

Year	Recipient	Awarded For
1959	Emilio Gino Segrè and Owen Chamberlain	For their discovery of the antiproton
1958	Pavel Alexseyevich Cherenkov, Il'ja Mikhailovich Frank, Igor Yevgenyevich Tamm	For the discovery and the interpretation of the Cherenkov effect
1957	Chen Ning Yang and Tsung-Dao Lee	For their penetrating investigation of the so-called parity laws which has led to important discoveries regarding the elementary particles
1956	William Shockley, John Bardeen, Walter Houser Brattain	For their researches on semiconductors and their discovery of the transistor effect
1955	Willis Eugene Lamb	For his discoveries concerning the fine structure of the hydrogen spectrum
	Polykarp Kusch	For his precision determination of the magnetic moment of the electron
1954	Max Born	For his fundamental research in quantum mechanics, especially for his statistical interpretation of the wavefunction
	Walther Bothe	For the coincidence method and his discoveries made therewith
1953	Frits (Frederik) Zernike	For his demonstration of the phase contrast method, especially for his invention of the phase contrast microscope
1952	Felix Bloch and Edward Mills Purcell	For their development of new methods for nuclear magnetic precision measurements and discoveries in connection therewith
1951	Sir John Douglas Cockcroft and Ernest Thomas Sinton Walton	For their pioneer work on the transmutation of atomic nuclei by artificially accelerated atomic particles
1950	Cecil Frank Powell	For his development of the photographic method of studying nuclear processes and his discoveries regarding mesons made with this method
1949	Hideki Yukawa	For his prediction of the existence of mesons on the basis of theoretical work on nuclear forces
1948	Lord Patrick Maynard Stuart Blackett	For his development of the Wilson cloud chamber method, and his discoveries therewith in the fields of nuclear physics and cosmic radiation

Year	Recipient	Awarded For
1947	Sir Edward Victor Appleton	For his investigations of the physics of the upper atmosphere especially for the discovery of the so-called Appleton layer
1946	Percy Williams Bridgman	For the invention of an apparatus to produce extremely high pressures, and for the discoveries he made therewith in the field of high pressure physics
1945	Wolfgang Pauli	For the discovery of the Exclusion Principle, also called the Pauli Principle
1944	Isidor Isaac Rabi	For his resonance method for recording the magnetic properties of atomic nuclei
1943	Otto Stern	For his contribution to the development of the molecular ray method and his discovery of the magnetic moment of the proton
1940–42		No prizes awarded because of World War II
1939	Ernest Orlando Lawrence	For the invention and development of the cyclotron and for results obtained with it, especially with regard to artificial radioactive elements
1938	Enrico Fermi	For his demonstrations of the existence of new radioactive elements produced by neutron irradiation, and for his related discovery of nuclear reactions brought about by slow neutrons
1937	Clinton Joseph Davisson and Sir George Paget Thomson	For their experimental discovery of the diffraction of electrons by crystals
1936	Victor Franz Hess	For his discovery of cosmic radiation
	Carl David Anderson	For his discovery of the positron
1935	Sir James Chadwick	For the discovery of the neutron
1934		No prize awarded
1933	Erwin Schrödinger and Paul Adrien Maurice Dirac	For the discovery of new productive forms of atomic theory
1932	Werner Heisenberg	For the creation of quantum mechanics, the application of which has, inter alia, led to the discovery of the allotropic forms of hydrogen
1930	Sir Chandrasekhara Venkataraman	For his work on the scattering of light and for the discovery of the effect named after him
1929	Prince Louis-Victor de Broglie	For his discovery of the wave nature of electrons

Year	Recipient	Awarded For
1928	Sir Owen Willans Richardson	For his work on the thermionic phenomenon and especially for the discovery of the law named after him
1927	Arthur Holly Compton	For his discovery of the effect named after him
	Charles Thomson Rees Wilson	For his method of making the paths of electrically charged particles visible by condensation of vapor
1926	Jean Baptiste Perrin	For his work on the discontinuous structure of matter, and especially for his discovery of sedimentation equilibrium
1925	James Franck and Gustav Hertz	For their discovery of the laws governing the impact of an electron upon an atom
1924	Karl Manne Georg Siegbahn	For his discoveries and research in the field of X-ray spectroscopy
1923	Robert Andrews Millikan	For his work on the elementary charge of electricity and on the photoelectric effect
1922	Niels Bohr	For his services in the investigation of the structure of atoms and of the radiation emanating from them
1921	Albert Einstein	For his services to theoretical physics, and especially for his discovery of the law of the photoelectric effect
1920	Charles Edouard Guillaume	In recognition of the service he has rendered to precision measurements in physics by his discovery of anomalies in nickel steel alloys
1919	Johannes Stark	For his discovery of the Doppler effect in canal rays and the splitting of spectral lines in electric fields
1918	Max Karl Ernst Ludwig Planck	In recognition of the services he rendered to the advancement of physics by his discovery of energy quanta
1917	Charles Glover Barkla	For his discovery of the characteristic Röntgen radiation of the elements
1916		No prize awarded
1915	Sir William Henry Bragg and Sir William Lawrence Bragg	For their services in the analysis of crystal structure by means of X rays
1914	Max von Laue	For his discovery of the diffraction of X rays by crystals
1913	Heike Kamerlingh-Onnes	For his investigations on the properties of matter at low temperatures which led, inter alia, to the production of liquid helium

Year	Recipient	Awarded For
1912	Nils Gustaf Dalén	For his invention of automatic regulators for use in conjunction with gas accumulators for illuminating lighthouses and buoys
1911	Wilhelm Wien	For his discoveries regarding the laws governing the radiation of heat
1910	Johannes Diderik van der Waals	For his work on the equation of state for gases and liquids
1909	Guglielmo Marconi and Carl Ferdinand Braun	In recognition of their contributions to the development of wireless telegraphy
1908	Gabriel Lippmann	For his method of reproducing colors photographically based on the phenomenon of interference
1907	Albert Abraham Michelson	For his optical precision instruments and the spectroscopic and metrological investigations carried out with their aid
1906	Sir Joseph John Thomson	In recognition of the great merits of his theoretical and experimental investigations on the conduction of electricity by gases
1905	Philipp Eduard Anton Lenard	For his work on cathode rays
1904	Lord John William Strutt Rayleigh	For his investigations of the densities of the most important gases and for his discovery of argon in connection with these studies
1903	Antoine Henri Becquerel	In recognition of the extraordinary services he has rendered by his discovery of spontaneous radioactivity
	Pierre Curie and Marie Curie	In recognition of the extraordinary services they have rendered by their joint researches on the radiation phenomena discovered by Professor Henri Becquerel
1902	Hendrik Antoon Lorentz and Pieter Zeeman	In recognition of the extraordinary service they rendered by their researches into the influence of magnetism upon radiation phenomena
1901	Wilhelm Conrad Röntgen	In recognition of the extraordinary services he has rendered by the discovery of the remarkable rays subsequently named after him (X rays)

> ## What country has produced the most winners of the Nobel Prize in physics?
>
> Since 1901, when the Nobel Prize was first awarded, the United States has had more nobelists in physics than any other country, although initially it took six years before a U.S. citizen won a Nobel Prize in physics.

Who was the **first American** to win the **Nobel Prize in physics**?

In 1907, for the development of extremely precise measurements for the velocity of light and his work on optical instruments, German-born Albert A. Michelson—a naturalized U.S. citizen—won the Nobel Prize in physics.

Who were the **two women** to win the **Nobel Prizes** in physics?

In 1903, Marie Curie was the first woman to win the Nobel Prize in physics. She was awarded the prize with her husband, Pierre, and with Antoine Becquerel for their discovery of over forty radioactive elements and other breakthroughs in the field of radioactivity.

In 1963, Maria Goeppert-Mayer became the second woman and the first and only American woman to win the Nobel Prize in physics for her discovery of the shell model of the nucleus.

MOTION AND ITS CAUSES

What is my **position**?

Physicists define an object's location as position. How would you define your present position? Are you reading in a chair 10 feet from the door of your room? Perhaps your room is 20 feet from the front door of the house? Or, your house is on Main Street 160 feet from the corner of 1st Avenue? Notice that each of these descriptions requires a reference location. The separation between your position and the reference is called the distance.

What is **displacement** and how does it **differ** from **distance**?

The examples above involved only distance, not direction from the reference location. Distance has only a magnitude, or size. In the example of a house, the magnitude of the distance of the house with respect to 1st Avenue is 160 feet. Displacement has both a magnitude and direction, so the displacement of the house from 1st Avenue is 160 feet west. Or, you define west as the positive direction (because house numbers are increasing when you go west). Then the house's displacement from the reference location, 1st Avenue and Main Street, could be written as +160 feet. A quantity like this that has both a magnitude and a direction is called a vector.

How can you **represent** a **vector quantity** such as displacement?

A convenient way to represent a vector is to draw an arrow. The length of the arrow represents the magnitude of the vector; its direction represents the direction of the arrow. For example, you might create a drawing where 1 inch on the drawing represents 100 feet, and west points toward the left edge of the paper. Then the displacement of the house from 1st Avenue would be represented as an arrow 1.6" long pointing toward the left.

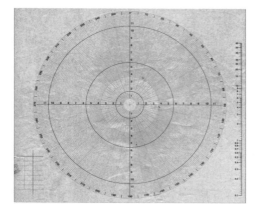

This diagram is similar to what navigators once used to calculate sea and air travel motion.

Can **displacement** be defined in **more than one dimension**?

More often than not you have to define a displacement in two or three dimensions. As an example, suppose you want to locate a house that is 160 feet west of 1st Avenue and 200 feet north of Main Street. The displacement is a combination of 160 feet west and 200 feet north. But how are they combined? You can't simply add them, because they have different directions. Go back to the drawing with the arrow. Define north as the direction toward the top of the page. Then add a second arrow 2.0" long in the upward direction. The tails of the two arrows are at the same place, representing the intersection of Main Street and 1st Avenue.

The two arrows are half of a rectangle 1.6" wide and 2.0" high. Draw lines completing the rectangle. The location of the house would be at the upper right-hand corner of the rectangle. Draw a third arrow, with the tail at the intersection of the other two vectors and the heat at the upper right-hand corner. The length of the arrow can be measured on your drawing, or calculated using the Pythagorean Theorem: the square of the length (the hypotenuse of a right triangle) is equal to the sum of the squares of the other two sides. In this case: $1.6^2 + 2.0^2 = 6.56$. Then length is the square root of that, or 2.56". So in real life the displacement would have a magnitude of 256 feet.

How is **GPS used**?

One very important use of GPS is to send time signals that allow clocks to be calibrated to within 200 ns (200 billionth of a second). Why would you need to know the time this accurately? Businesses that use computers in many parts of the world can synchronize their computers so that the precise time that transactions occurred are known.

GPS also provides navigation information. Hikers use GPS to replace maps, which are often outdated, compasses, and lists of landmarks. Automobile and truck drivers use GPS to replace paper maps and to find local businesses, such as banks, restaurants, and gas stations. Farmers use GPS to map precise locations in their fields. Not only does this information improve planting, it can be used to mark the locations of areas that need additional insect control chemicals or fertilizers.

Engineers are now working on GPS systems to improve the flow of automobile traffic and reduce crashes. Each car would have a GPS receiver that would then broadcast its position. This information could be used to change red traffic lights to green if

How can you use GPS to find your location?

GPS, or the Global Positioning System, was developed by the Department of Defense and was made operational in 1993. It consists of three parts. The first part is 24 satellites in 12-hour orbits that broadcast their location and the time the signal was sent. The second part is the control system that keeps the satellites in their correct orbits, sends correction signals for their clocks as well as updates to their navigation systems. The third component is the receiver. Some receivers are designed to be mounted in autos or trucks and display a map of the region around the receiver. Some are used by boaters to monitor their locations either in rivers or lakes or the open sea. Some are hand-held and can be used in the field by hikers and campers. Others are so small that they can be built into cell phones.

there were no opposing traffic. If two cars were equipped this way, the system could determine the distance between them and their relative speeds. If they were on a collision course the system would apply the brakes to avoid a crash.

How could you define your **position on Earth**?

If you use a GPS device, you might find that your location is given as a latitude and longitude. For example it might give you latitude: 40° 26' 28.43"N and longitude 80° 00' 34.49"W. Note that these are angles, not distances. The reference for latitude is Earth's equator. The reference for longitude is the "Prime Meridian" that runs through Greenwich, England (a suburb of London).

How can you **convert latitude** and **longitude** to a **distance** measurement?

A precise conversion is difficult because Earth is not a perfect sphere. Latitude is easier to convert. The circumference of Earth taken over the poles is 24,859.82 miles, which is equivalent to 360° of latitude. Therefore one degree is equal to 69 miles. So, the north-south distance between two cities 5° of latitude apart would be 345 miles.

Longitude is more complicated. At the equator 360° is Earth's circumference, 24,901.55 miles. But at the poles it is zero! So the conversion of longitude to distances depends on the latitude. If you use trigonometry to find the distance, you will find that the circumference at a latitude of θ degrees is the circumference at the equator times the cosine of the angle θ. So, at the latitude 40°, the circumference is 19,076 miles, and one degree of longitude is 53 miles. This result is only approximate because Earth is not a perfect sphere. For more accurate conversions consult a website such as http://www.nhc.noaa.gov/gccalc.shtml.

27

What is **speed**?

If something moves from position to position then speed is a measure of how fast it moved. Speed is defined as the distance moved divided by the time needed to move. Both change in position or distance and time are measured quantities. Frequently speed is called the time rate of change of distance. For example, if you drive 240 miles in 4 hours then your speed is 60 miles per hour (abbreviated mph). It is unlikely that you drove the whole trip at a constant 60 mph; this example calculated your *average speed*. If you were pulled over for speeding and were told you were going 80 mph, you wouldn't be able to avoid a ticket by saying "But officer, my average speed is only 60!"

The vertical lines you see on a map or globe are lines of longitude and the horizontal lines are lines of latitude.

What **units** are used to **describe speed**?

In the English units used in the United States, speeds are usually given in feet per second or miles per hour. In the metric system meters per second or kilometers per hour are more common. Here are some typical speeds in the four systems of measurement.

Description	feet/sec	miles/hour	meters/sec	kilometers/hour
Walking	5.9	4.0	1.8	6.4
Sprinting	33	22	10	36
Car speed limit	103	70	31	113
Pitched baseball	139	95	42	153
Speed of sound	1,100	750	335	1,207
Speed of space shuttle orbit	25,667	17,500	7,823	28,164
Speed of light	98,425,197	67,108,089	30,000,000	108,000,000

What is **instantaneous speed** and how is it measured?

If you reduce the time interval between measurements of position both the distance moved and the time required are reduced. If the speed is constant, then the ratio of the two does not change. Instantaneous speed is defined as the limit of distance divided by time

How was motion perceived in the ancient world?

To the ancient Greeks motion was either natural or violent. The four elements sought their natural locations. Earth (including metals) fell down because it had a property called gravity. Fire (including smoke) went up because it had a property called levity. Water was between Earth and air. Heavenly objects, made of aether, moved in circles.

Arrows or other objects that were thrown were said to move because they were given violent motion. What was violent motion? The bow transferred force to the arrow; the thrower transferred force to the rock. Once in air (or water) the medium pushed the object along. When the force ran out the medium now opposed the motion and the object fell to Earth.

In the sixth century, the commentator Philoponus doubted Aristotle's view of the role of the medium in motion. Avempace, whose Arabic name was Ibn Bajja, was a Spanish Arab who died in 1138. He also discussed the role of the medium. While Aristotle claimed that motion in a vacuum would be impossible Avempace stated that motion in a vacuum would continue forever because nothing opposed it.

It wasn't until 1330 that the possibility that motion could vary was suggested. In that period philosophers at Merton College, part of Oxford University, developed their ideas of instantaneous speed and acceleration. Scholars at the University of Paris contributed greatly to these definitions that made the modern measurements of motion possible.

interval when the time interval is reduced to zero. In practice you can't reach the limit, but it is possible to measure positions every thousandths of a second. There are indirect methods of measuring instantaneous speed. For example, police use the Doppler shift (that will be discussed later in this book). That is, the change in frequency that occurs when the radio or light wave is reflected from a moving object. Automobile speedometers often use the turning force (torque) on an aluminum disk produced by a magnet that is rotated by the turning car axle. This force will also be discussed later in the book.

What is the **difference** between **speed** and **velocity**?

Just as you can add direction to a change in position and end up with displacement, you can also specify the direction of motion. The combination of speed with direction is called velocity. Velocity is the displacement divided by the time required to make the change, or the time rate of change of displacement. Velocity is a vector quantity, like displacement.

You might walk at 4 mph north, or a balloon might move at 5 feet per second up. If you assign the variable x to represent north/south, y to east/west, and z to up/down

If you can find a place free of light polution, on a clear night you can see the stunning view of our Milky Way galaxy. The earth not only rotates around the sun, but also around the entire Milky Way. A Cosmic Year is one rotation of the Milky Way, which takes about 225 million years.

position, then both the change in position and speed would be positive for movement to the north, east, or up. Your walking velocity would be $+4x$ mph. The balloon's would be $+5z$ ft/s.

Average speed is almost always more useful than average velocity. For example, in a NASCAR race the starting and finish lines are in the same position. So, no matter how fast the cars go, their average velocity is zero because the beginning and ending positions are the same. Instantaneous velocity, however, is more useful than instantaneous speed, as will be shown shortly.

Does **velocity affect distance** and **time**?

Einstein's Special Theory of Relativity shows that both distance and time change with velocity. Einstein reached this conclusion by noting that one must define methods of measuring both distance and time. The result is that as objects move near the speed of light their length (in the direction of motion) shrinks and their internal clocks that measure time run slower. The amount of change is described by a factor called γ (gamma), which is always larger than one. Thus the time given by a moving clock is γt, where t is the time shown by a fixed clock, and length is given by $l\gamma$, where l is the length of the fixed object. The table below shows γ for a variety of velocities (note that c is the speed of light):

How can you be standing still and yet moving?

Suppose you are reading *The Handy Physics Answer Book* while sitting on a moving bus? To another bus passenger your velocity would be zero, but to a person watching you while standing on a sidewalk your velocity would be equal to that of the bus. If you were walking forward on the bus, then the standing observer could show that your velocity was the sum of that of the bus plus your walking speed. Similarly, if you were walking toward the back of the bus, your velocity would be the velocity of the bus less your walking velocity.

Earth itself is in motion. It is rotating on its axis, revolving around the sun, and moving with the entire solar system around the center of the Milky Way galaxy. If is therefore important that the frame of reference for motion be specified. Usually Earth's surface is used as the frame of reference, which means that its velocity is zero.

Velocity	γ
550 mph	$1 + 3 \times 10^{-11}$
17,500 mph	$1 + 3 \times 10^{-8}$
$0.5\,c$	1.2
$0.9\,c$	2.3
$0.99\,c$	7.1
$0.995\,c$	10.0

How much does a **moving clock** slow down?

A clock on a jet plane (v = 550 mph) would lose 0.9 milliseconds per year, while one in the space shuttle (v = 17,500 mph) would lose 0.9 s/yr. In 1971 atomic clocks were placed on planes, one of which flew around the world eastbound, the other westbound. The changes in time were measured and agreed with relativity theory. Clocks on GPS satellites must be adjusted for the loss of time. More conclusive tests have been done with very fast moving ($0.995c$) muons. Muons, when at rest, decay in 2.2 μs (microseconds). The number of muons, produced high in Earth's atmosphere by cosmic rays, were measured at the peak and base of a high mountain. The ratio of numbers at the two heights showed that the muons lived 22 μs, which agreed with their measured speed of $0.995c$. From the viewpoint of the muons, they decayed in 2.2 μs, but the height of the mountain was 10 times shorter than that measured by observers on Earth. Thus the predictions of slower clocks and shorter distances have been tested and agree with Einstein's predictions.

31

Do **all** relative **velocities add?**

Suppose you were riding on a spaceship moving at half the speed of light. If you were to point a laser in the direction the ship was moving, the person on the spaceship would be able to determine that the speed of the laser light was the speed of light, 300,000,000 meters per second. What speed would a stationary observer measure? Surprisingly, she would find that the light was traveling at the speed of light, not the sum of the speed of the spaceship and the speed of the laser light as measured by the traveling person. That is, the speed of light is the same in all frames of reference. This is another of the results of Einstein's Special Theory of Relativity. It has been tested, not with spaceships, but with gamma rays emitted by subatomic particles moving near the speed of light.

What is **acceleration?**

Speed and velocity are seldom constant. Usually they vary, and acceleration is a description of how that variation occurs. Acceleration is called the time rate of change of velocity. That is, it is the change in velocity divided by the time over which the change occurs. For example, a car might accelerate from 0 to 60 miles per hour. A sports car might do this in five seconds, while an economy car might require nine seconds. Which car has the larger acceleration? Both have the same change in speed, but the sports car requires less time to make the change, so it has the larger acceleration.

Like velocity, acceleration is a vector. That is, it includes both magnitude and direction. In speeding up, in which the change in velocity is positive, acceleration is a positive quantity. If the object is slowing down then the acceleration is negative. If, however, the object is going in the negative direction—for example, a car going in reverse—then speeding up is a negative quantity because the final velocity is more negative than the initial velocity.

Physicists do not use the term deceleration. Whether positive or negative, speeding up or slowing down, the term acceleration is always used.

The table below shows some typical accelerations in a variety of units. The rightmost column, labeled "g," compares the acceleration to that of an object falling in Earth's gravitational field. This ratio is frequently used in both everyday and scientific writing.

Description	feet/sec²	Miles per hour/sec	Meters/sec²	Kilometers per hour/sec	g
Auto from 0 to 10 mph	29.3	20.0	8.9	32.2	0.9
Sprinter first 1 s of 100m dash	18	12	5	19	0.5
Boeing 757 (takeoff)	10	7	3	11	0.3
Auto stopping from 60 mph	−30	−20	−10	−33	−0.9
Auto crash from 35 mph	−1,585	−1,080	−483	−1,739	−49
Object falling due to gravity	32.2	21.9	9.8	35.3	1.0

FORCE AND NEWTON'S LAWS OF MOTION

What **causes motion**?

Between 500 B.C.E. and 1600 C.E. there were many other ideas developed about the causes of motion. Some said a stone fell because of the "weight of a stone," supposing that weight is a property of the stone. Others, saying "the apple is attracted to Earth" supposed, like Aristotle (384–322 B.C.E.), that there is a desire of an object to move toward Earth. Others suggested that force is something that is transferred from one object to another, as discussed above. Still others, starting with Leonardo da Vinci (1452–1519), wrote that force was an external agent that exerted a push or a pull on a body.

What can **exert a force**?

The first answer that might come to mind is that a person can exert a force. He or she can throw a ball, pull the string on a bow to shoot an arrow, push a chair across the floor, or pull a wagon up a hill. Many animals can do the same thing, so one might say that living organisms can exert forces.

What if you place a rock on a table? Does the table exert a force on the rock, or does it just block the rock's natural motion toward the floor? If you put a heavy rock in your hand, it would sag downward because you would have trouble exerting the upward force on the rock. The same happens with a table. If the table is made of thin wood, it will also sag. What would happen if the table were replaced with a sheet of paper? The paper might hold small stones, but with a heavy rock it would tear because it could not exert a strong enough force. The heavier the rock, the greater the force the table or floor exerts. In summary, inanimate objects can also exert forces.

Forces like these, exerted by humans or tables that touch the object, are called contact forces. What are other contact forces? You might think of rubber bands on sling shots, roads on the wheels of your car, or ropes pulling carts. Water and air can also exert a force. Think of a stick moving down a stream, or what you feel when you stick your hand out the window of a fast-moving car.

In what **units** is **force measured**?

In *SI* (the metric system) force is measured in newtons, abbreviated N. In the English system, force is measured in pounds, abbreviated lbs.

How are **acceleration** and **force related**?

If you have a miniature toy car or even a smooth hard ball that can roll on a smooth level surface you can explore the effects of force on motion. When the toy car or ball is motionless, give it a gentle tap with your finger. Note how it moves. Now, while it is moving give it a second, then a third or even a fourth gentle tap. What happened?

You saw that when the toy car was at rest and a force was exerted on it, it started to move in the direction of the force. When a force was applied in the direction of its motion, it sped up. Each additional tap caused it to speed up more. What do you think would happen if you were able to exert a constant force on it while it was moving? It's difficult to do, but try it.

You can conclude from this exercise that a force applied in the direction of motion causes it to speed up, or accelerate in the direction of the force. If the direction of motion is called the positive direction, then both the force and acceleration would also be positive.

Now start the toy car moving and give it a gentle tap in the direction opposite its motion. Don't tap it so hard that it stops or changes direction. See if you can tap it two or three times, again without stopping the car. What did you observe?

You should be able to conclude that a force applied in the direction opposite its motion causes it to slow down, or accelerate in the direction of the force. If you had defined the direction of motion as positive, then the force and acceleration would both be negative.

What happens when no force is applied? You saw in the beginning that when the toy car started at rest it remained at rest until you exerted a force on it. What happened while it was moving? It probably slowed down some, with the amount depending on the condition of the toy car and the hardness of the surface. But, the amount it slowed was certainly much less than it was when you exerted a force in the opposite direction.

How does **force affect acceleration**?

From your explorations you could draw the conclusion that when there is only one force exerted on an object, the larger the force, the larger the acceleration.

How does **mass affect acceleration**?

The best way to find out is to have two toy cars. Then add some mass to one of them. For example, you could tape coins to the car. Then line them up side-by-side and use a pencil or ruler to apply the same force to both of them. Again, just give them a tap. Which one goes faster? You probably found that the one without the added mass sped up more.

That is, for the same force, the greater the mass, the less the acceleration.

What is **inertia**?

Inertia is the property of matter by which it resists acceleration. That is, an object that has inertia will remain at rest or will move in a uniform motion in a straight line unless acted on by an external force.

In this simple pendulum game called Newton's Cradle, lifting one hanging ball and allowing it to drop and hit the second ball exerts a force on the next ball. The force on the ball at the other end causes it to swing out and back, according to Newton's laws.

What are **Newton's Laws of Motion**?

You have just explored how force and mass affect acceleration. Sir Isaac Newton (1642–1727) summarized these results in what is called *Newton's Second Law of Motion*. For a single force, like you used, it can be written as: Acceleration is equal to the force applied divided by the mass, or $a = F/m$.

That is, the acceleration (a) varies directly with the force applied (F). The stronger the force, the greater the acceleration. And it varies inversely with the mass (m). The larger the mass, the smaller the acceleration.

What happens in there is no force? The equation says that there is no acceleration. That is, if the velocity was zero it remains zero. If the object was moving, it continues to move with the same velocity. These statements are called *Newton's First Law of Motion*.

What happens if **more than one force acts** on an **object**?

You can explore this question with your toy car or ball. Try exerting two forces, like two finger taps in the same direction. You can see that the car or ball moves faster. The forces add. What happens if the two forces are in opposite directions? You can try pushing each end of a motionless toy car. What happens if both forces are equal? If one is stronger than the other? If they're both equal, the car will remain motionless. That is, it will act as if there is no force on it. If one is stronger than the other, then it will move in the direction of the smaller force, but it will accelerate less. That is, the

35

Your toy car or ball almost certainly slowed down, even when you didn't tap it. Do moving things, by their nature, slow down, or is there some force causing them to slow down? You might say friction slows things down. Perhaps you think that friction is just there all the time, or is something that has no direction. But Newton would say that it must be a force that acts in the direction opposite motion.

results will be the difference between the two forces. We'll explore forces that act in different directions later.

Physicists say that the combination, addition, or subtraction of forces produces a *net force* and it is the net force that affects the acceleration. Thus Newton's laws should be written as follows.

Newton's First Law: If there is no net force on an object, then if it was at rest it will remain at rest. If it was moving, it will continue to move at the same speed and in the same direction.

Newton's Second Law: If a net force acts on an object, it will accelerate in the direction of the force. The acceleration will be directly proportional to the net force and inversely proportional to the mass.

That is, $a = F_{net}/m$. Or, $F_{net} = ma$.

What are the **properties** of **friction**?

If you push a couch across a room, it will feel like someone is on the other end pushing back. If you pull on the couch, it will feel like someone on the other end is also pulling. If the couch is heavier, the opposing force will be larger. The amount of force opposing you will depend on the surface the couch is on. A carpet will have a stronger opposing force than a smooth surface, like wood or tile. These observations summarize the properties of friction between two objects that are in contact. Friction is always in the direction opposite the motion. Friction is greater if the moving object is heavier. Friction is greater if the surface is rougher. But, friction does not increase if the speed of motion increases.

How is the **friction between two surfaces** characterized?

In most cases the frictional force is proportional to the force pushing the surfaces together (N). This proportionality is usually written as $F_{friction} = \mu N$. The Greek letter μ (mu) is called the coefficient of friction. It is important to remember that the two forces, $F_{friction}$ and N are not in the same direction.

Do smoother surfaces always have less friction?

Surprisingly, if two metal surfaces are polished until they are extremely smooth, the coefficient of friction will actually increase. Recent experiments show that friction doesn't depend on surface roughness at the atomic scale! Friction is usually caused by chemical bonds between atoms on the two surfaces, but the fundamental causes of surface friction are still a problem that is not totally solved.

You probably noticed that it took more force to start the couch moving than it was to keep it moving. Physicists say that the coefficient of static friction is larger than the coefficient of kinetic (moving) friction.

Surfaces	Static friction	Kinetic friction
Teflon on teflon	0.04	0.04
Oak on oak (parallel to grain)	0.62	0.48
Steel on steel	0.78	0.42
Glass on glass	0.94	0.4

How can you **reduce** the **coefficient** of **friction**?

One method is to use surfaces that have less ability to form chemical bonds. Teflon is one such surface. Another is to have a thin film of oil between the surfaces. The oil will prevent bonding of the atoms in one surface with those of the other. Oil or other lubricants can reduce the coefficient of friction to about 0.1 to 0.2.

Do **rolling objects** experience contact **friction**?

Whether it is a bowling ball rolling down the alley or a wheel rolling on the road, there is no sliding between the ball or wheel and the flat surface, so there can be no contact friction. What is the cause of rolling friction? Rolling friction is the result of deformations of either the rolling object or the surface. Think about how much faster a rolling ball stops on grass or sand than on a hard surface. In those cases the ball has to push down the surface in front of it, which acts the same as contact friction. Or, remember when you rode a bicycle with soft tires. In this case the tire is squeezed when it contacts the street, which also acts like friction.

Does a gas or a **liquid** exert a **friction-like force**?

If you stick your hand out of the window of a moving car you can explore the properties of "air drag." The faster you go, the stronger the force. If your palm is facing up or down the drag is much smaller than when your palm faces forward or backward, showing that the shape of the object matters. A smaller hand experiences less drag than a larger one. So, drag depends on velocity, area, shape, and the density of the air. These properties are different from other contact friction forces, but the force is still exerted in the direction opposite motion, so it slows down the object. Liquids exert similar, but stronger, drag forces.

Aren't there three Laws of Motion? What is **Newton's Third Law**?

Forces are interactions between objects. You need more than one object to have a force. Therefore, if two objects interact, each exerts a force on the other. If you push on a friend's hand, his or her hand pushes back on you. If you stand on a floor, you exert a force on the floor and the floor exerts a force on you.

Newton's Third Law of Motion states that these two forces are equal in magnitude but opposite in direction. If you exert an 800-newton force downward on the floor, the floor exerts an 800-newton force upward on you.

Note that Newton's Second and Third Laws are different. The Third Law describes forces on two different objects. Those forces can then be used with the Second Law to find out how the motions of the objects are changed.

What are some **applications** of the **Third Law**?

How do you accelerate your car? You press on a pedal called the accelerator. Does that cause the car to speed up? It cannot because the net force that causes the acceleration must be exerted on the car from outside it.

What does the car interact with? When it is not moving, it is touching the road, and thus interacting with it. When you press on the accelerator the engine causes the wheels to rotate in a direction that, because of friction between the tires and the road, pushes backward on the road. By Newton's Third Law, then, the road pushes forward on the tires and the car accelerates in its forward direction. What happens if you car is on ice? Often then the friction between the tires and the ice is so small that the wheels can's exert enough backward force on the ice for the ice to exert the force needed to accelerate the car.

Note that friction, instead of being a bad thing, as suggested earlier, is needed in this case. How else is friction useful in accelerating things? How do you walk or run? Your feet are also interacting with the ground. As long as there is sufficient friction, when you push your feet backwards, the ground pushes you forward.

How does **gravity act** on **us**? Is it a contact force?

Newton recognized that the gravitational force of Earth acts not only on objects close to Earth, like the famous apple, but also on objects as far away as the moon. But, how exactly does a celestial object such as the sun reach out the approximately 93 million miles and hold Earth in its orbit? An early idea was that it was an "action at a distance" force, for example, that the Sun attracted all objects, like Earth, without anything between the two.

In the middle of the nineteenth century, physicist Michael Faraday (1791–1867) proposed that one magnet creates a field of force around it, and another magnet interacts with the field at its location. Based on the field idea, Earth then creates a gravitational field. The apple and the moon have a force exerted on them, not directly by Earth, but by the gravitational field that exists at the location of the apple and the moon.

Many scientists now believe that gravity is the result of mass warping space-time. While the effects of gravity have long been studied and its laws defined, what the force of gravity actually is still is a matter of debate.

What would happen to Earth if the sun suddenly disappeared? How soon would Earth recognize that the sun's gravitational field was gone? It couldn't happen instantaneously, because Einstein's Special Theory of Relativity says that no information can travel faster than the speed of light. So, it would take about eight minutes before Earth would both experience the lack of sunlight and the lack of gravitational force.

On what does the **gravitational field**, like that of Earth, **depend**?

Newton demonstrated that the gravitational force on one object caused by another is proportional to the product of the masses of the two objects divided by the square of the distance between them. The gravitational field of an object would then be the force divided by the mass of the object on which the force is exerted. The symbol used for the gravitational field is g, and G is the so-called universal gravitational constant. It is called universal because it is the same for objects made of any material and of any mass—from a 1-kg apple to a galaxy.

The equation that describes the gravitational field at a specific location is $g = GM/r^2$ where M is the mass of the attracting object and r is the distance from the center of this mass to the location in space. The gravitational field is a vector quantity. Its direction is toward the center of the attracting object. We'll consider its magnitude shortly.

What's the **strength** of the **gravitational field** of **Earth**?

Earth's mass is 5.9736×10^{24} kg and the gravitational constant is 6.673×10^{-11} N m^2 kg^{-2}. At the surface of Earth, 6.4×10^6 m from the center, then $g = 9.8$ N/kg. Thus a

> ## Earth is huge! Why does the gravitational field depend on the distance from its center?
>
> This is not an easy question to answer. Newton recognized the problem and had to develop a new mathematics, the integral calculus, to solve it. His argument, simplified, can be illustrated with a wooden button. Hold it at arms length in front of a window so you can see a tree. Note how much area the button covers on the tree. According to Newton, assuming that the button and tree were made of the same material, the force of attraction of the button on you is exactly the same as that of the part of the tree the button covers. The reason is that the effect of the button being closer is balanced by the much larger mass of the tree. Can you show that the area the button covers is r^2 times the area of the button? The sum of the two forces is that of the sum of the masses of the button and the tree located at a distance halfway between the two.

kilogram of meat, for example, experiences a gravitational force of 9.8 N toward Earth's center.

The International Space Station orbits at about 320 km above Earth's surface. How large is the gravitational field at that altitude? Its distance from Earth's center is about 6.7×10^6 m, so $g = 8.9$ N/kg. That's not much different than at Earth's surface.

The moon is 384×10^6 m from the center of Earth. At that distance $g = 0.0027$ N/kg. So the force of Earth's gravity on the same kilogram of meat would be only three thousandths of a Newton! How does this very small gravitational field keep the moon in its orbit? The answer, of course, is that the moon has a large mass, 7.2×10^{22} kg, and so the force on it is 2.0×10^{20} N. When we explore orbits later we'll use this result.

How is the **gravitational field** related to **force**?

As was described above, the force of gravity on an object is equal to the object's mass times the gravitational field strength, expressed as $F = mg$. Thus if you have a mass of 70 kilograms (154 pounds), the force of gravity on you is 686 newtons.

Mass has been **defined** both in terms of **acceleration** and **gravitational force**. Are these the **same**?

Mass defined as $m = F_{net}/a$ is called inertial mass. Mass defined as $m = F_{gravitation}/g$ is called *gravitational mass*. Many physicists, starting with the Hungarian Baron von Eötvös, have done experiments to determine if the two kinds of mass are equal and if they are the same for all materials. Recent experiments have shown that if they are different, the difference is only one part in 10^{15}! Furthermore, Einstein's General Theory of Relativity explains that they are identical, and calls this fact the "Principle of Equivalence" between gravitation and acceleration. In other words, the laws of physics in an accelerating reference frame or a gravitating frame are indistinguishable. You can't tell the difference between falling or being acted on by gravity.

What's the **strength** of **gravitational fields** of other **astronomical objects**?

This table shows the properties of the sun, planets, and Earth's moon. It also shows the strength of the sun's gravitational field

Jupiter, the largest planet in our solar system, also has the strongest gravitational pull.

at each planet and the strength of the planets' gravitational fields at their surfaces. The sun and Jupiter, Saturn, Uranus, and Neptune are gaseous and have no solid surfaces.

Which planet has the strongest gravitational field? Is it also the planet with the largest mass? Why do you think that the sun, which has a mass 1,000 times that of Jupiter, has a surface gravitational field only about 10 times that of Jupiter? As a hint, look at the equation that defines g and see what properties other than mass are involved.

Name	Distance from Sun (10^3 km)	Radius (km)	Mass (kg)	g (Sun) (N/kg)	g (surface) (N/kg)
Sun	0	695,000	1.99×10^{30}		274.66
Mercury	57,910	2,439.7	3.30×10^{23}	3.96×10^{-2}	3.70
Venus	108,200	6,051.8	4.87×10^{24}	1.13×10^{-2}	8.87
Earth	149,600	6,378.14	5.98×10^{24}	5.93×10^{-3}	9.80
Moon	384.4 (from Earth)	1,737.4	7.35×10^{22}		1.62
Mars	227,940	3,397.2	6.42×10^{23}	2.55×10^{-3}	3.71
Jupiter	778,330	71,492	1.90×10^{27}	2.19×10^{-4}	24.80
Saturday	1,429,400	60,268	5.69×10^{26}	6.49×10^{-5}	10.45
Uranus	2,870,990	25,559	8.69×10^{25}	1.61×10^{-5}	8.87
Neptune	4,504,300	24,746	1.02×10^{26}	6.54×10^{-6}	11.15

Does the **gravitational force** also **obey Newton's Third Law**? How?

All objects have gravitational fields surrounding them. So, just as the moon has a force exerted on it due to Earth's gravitational field, Earth has a force on it due to the moon's field. As a result, the moon does not orbit around Earth's center, but the moon and Earth orbit around a point roughly 2,900 kilometers from Earth's center. The sun and other planets also exert gravitational forces on Earth, and Earth does on them.

The gravitational fields of the planets, in particular Jupiter, cause the sun to orbit around a point near its surface. Thus, if an observer on another planetary system studied the motion of the sun carefully he or she could determine that planets were orbiting the sun. This method has been used to detect well over 200 planets (called exoplanets) or planetary systems and other stars.

How did **Einstein describe** the **gravitational field**?

Einstein in his General Theory of Relativity showed that the gravitational field was actually a distortion of space-time caused by the mass of the object. Because space-time has four dimensions that are very difficult to visualize, the distortion is best seen with a two-dimensional model. Often the model consists of a rubber sheet in which a bowling ball is placed (see illustration on page 39). The sheet is pulled down by the ball, which represents the sun. Earth is a tiny ball that is placed on the sheet and given a push perpendicular to the direction of the sun. This ball "orbits" the sun a few times until friction causes it to speed up and spiral into the sun. In the words of physicist John Wheeler:

Spacetime grips mass, telling it how to move.
Mass grips spacetime, telling it how to curve.

Has **Einstein's theory** been **tested**?

It has been tested in many ways. It is being used every day in adjusting the clocks in GPS satellites. They run faster because they are at high altitudes where the distortion of space-time is smaller. Both the effects of special relativity (clocks running slower) and general relativity (clocks running faster) must be used to keep the clocks running accurately.

How does **gravity cause tides**?

As someone who lives near an ocean will know, there are two high tides and two low tides each day. They're not at the same time each day, but depend on the phase of the moon. The heights of the tides vary over the seasons as well. What causes the tides?

The moon's gravitational field exerts a force on the water. Because the field and force depend on distance, the force on the water closest to the moon is strongest. The force on Earth is weaker, and the force on the water on the far side of Earth is smallest. So the water nearest the moon is pulled toward it, and Earth is pulled toward the moon more strongly than the water farthest from the moon. For that reason there is a tidal bulge in the water near the moon, and another on the far side. The bulges, which are the high tides, are not directly under the moon because Earth's rotation drags the water along with it. Because the day is shorter than the lunar month, the high tides actually come about two hours before the moon is overhead.

Tides also vary greatly with location. The largest tidal variations occur in the Bay of Fundy between the Canadian provinces of Nova Scotia and New Brunswick, where the largest recorded range was 17 meters, or almost 56 feet. There have been many proposals to put a dam across the bay and use the tides to generate electricity, but environmental concerns have blocked construction in that bay.

The sun also affects the tides, but much less than the moon. When the sun, Earth, and the moon are aligned, which happens at both full and new moons, the tides are especially high.

How can you **describe the motion** of an object in a **gravitational field**?

As long as the force exerted by a gravitational field, such as Earth's, is the only force on an object, then Newton's Second Law can be used in the form $a = F/m_{inertial}$. But the force due to the field is given by $F = m_{gravitational}g$. And, as has been tested by experiment and explained by Einstein's theory, $m_{inertial} = m_{gravitational}$, so $a = g$.

There is one more thing to question. The acceleration a is measured in meters per second squared, while the gravitational field strength is measured in newtons per kilogram. How can these two quantities be equal? The answer can be found by looking at Newton's Second Law again. In the form $F_{net} = ma$, you can see that the units of force, newtons, must be equal to the units in which m times a are measured. The mass, m, is measured in kilograms and the acceleration in meters per second squared. Therefore a newton must be a kilogram times a meter per second squared. Thus a newton per kilogram (N/kg) is a meter per second squared (m/s^2).

How do the **speed** and **position** of a **dropped object vary** with **time**?

As long as the only force is the gravitational force, then the acceleration is the acceleration due to gravity, g. At Earth's surface the value is 9.8 m/s^2. (In the English system $g = 32.2$ ft/s^2.) If the object is dropped from rest at time $t = 0$, then the velocity at a future time t is simply $v = gt$. That is, the velocity increases by 9.8 m/s each second. If we measure the distance fallen from the position where it was dropped, then at time t

it has fallen a distance $d = 1/2\, gt^2$. The following table shows velocity and distance fallen for some selected times.

Time	Velocity	Distance	Velocity	Distance
0.10 s	0.98 m/s	4.9 cm	3.2 ft/s	1.9 in.
0.15 s	1.42 m/s	11 cm	4.8 ft/s	4.3 in.
0.20 s	1.96 m/s	20 cm	6.4 ft/s	7.7 in.
0.50 s	4.90 m/s	1.2 m	16.1 ft/s	4.0 ft.
1.0 s	9.8 m/s	4.9 m	32.2 ft/s	16.1 ft.
2.0 s	19.6 m/s	19.6 m	64.4 ft/s (44 mph)	64.4 ft.

Why are those times chosen to calculate velocity and distance fallen?

The first three times are chosen so that you can explore your reaction time. Have another person hold a ruler vertically by one end. Place your finger and thumb at the lower end of the ruler, but don't let them touch the ruler. Have the other person drop the ruler and you use your finger and thumb to grab it. Note the distance the ruler has dropped in the time it takes you to react to it being dropped. Compare the distance with the first three distances to see if your reaction time is between 0.1 and 0.2 seconds.

What happens if an object is not dropped, but is thrown up or down?

Physicists would say that the ball has been given an initial velocity. But this initial velocity doesn't affect the force of gravity on the ball, so the ball would still gain a downward velocity of 9.8 m/s each second it is in flight.

Suppose the ball is thrown down with a velocity of 2.0 m/s. Then at time $t = 0$, the time it was thrown down, it would already have a velocity of 2.0 m/s downward. If you use the table above you see that 0.10 s later it would have a velocity of 2 + 0.98 m/s = 2.98 m/s. Half a second after launch its downward velocity would be 2 + 4.9 m/s = 6.9 m/s. That is, the initial velocity is just added to the increasing velocity caused by the gravitational force.

Now suppose the ball is thrown upward with the same 2.0 m/s speed. Now we have to be very careful with the sign of the velocity. We have chosen the downward direction as positive. So the initial velocity of the ball is actually –2.0 m/s.

The gravitational force acts in the downward direction. By Newton's Second Law that means that the ball will slow down at a rate of 9.8 m/s each second. After 0.1 s it would be going –2.0 + 0.98 = –1.02 m/s. After 0.20 s it would be going –2.0 + 1.96 = –0.04 m/s. It is almost at rest! But the gravitational force is still acting on it, so it continues to accelerate downward. After 0.30 s it would be going –2.0 + 2.94 = +0.94 m/s. That is, it would now be moving downward.

How can the acceleration be independent of mass?
Doesn't a bowling ball fall faster than a soccer ball?

Which weighs more, the bowling ball or soccer ball? You can use your hands to hold first one ball and then the other. You'll notice that you have to exert a much larger upward force to hold the bowling ball at rest. From Newton's First Law, if an object is at rest then the net force on it is zero. That is, the downward force of gravity has the same strength as the upward force of your hands. So, you can conclude that the force of gravity, the weight, of the bowling ball is greater than that of the soccer ball.

Galileo was said to have dropped a pressing iron and a wooden ball from the Tower of Pisa and found that they arrived at the bottom at the same time. He also argued that if you tied two wooden balls together, thus doubling the mass, they would obviously fall at the same rate as they would have not tied together.

How can we understand this result? When you drop the ball, the force of gravity is the only force on it, so $a = F_{gravity}/m$. An object with a larger force on it will accelerate faster than one with a smaller force, but only as long as the masses are equal. If the forces are equal but the masses different, then the object with greater mass will accelerate more slowly. As described above, because the force of gravity is proportional to the mass, the two effects cancel, and $a = g$ for all masses.

When it is at its maximum height does it stop? At an instant in time it is indeed at rest, but it doesn't stay motionless for any time interval because the gravitational force keeps acting on it.

How does **air drag affect** the **motion** of dropped objects?

Air drag adds a second force exerted on the object that is in the direction opposite its motion. The net force is the difference between the downward force of gravity and the upward force of air. As was discussed above, the force of the air increases as the speed increases. Therefore, as the object falls and gains speed, the upward force increases. At some time it will equal the downward force and there will be no net force on the object. So, according to Newton's First Law, the object's speed will now become constant. This constant velocity is called the terminal velocity.

On what does the **terminal velocity depend**?

Air drag depends on the density of the air, the size, velocity, and the shape of the object. Therefore the terminal velocity will depend on all of these, plus the mass. You can test to see if the air drag depends on the velocity or the velocity squared by doing a

The force of the air on this person's parachute—the air drag—increases as speed increases, and eventually the force of gravity and the upward force of the air equalize, resulting in a constant speed called the terminal velocity.

simple experiment. You'll need only five filters for a drip coffee maker. The filters must be cup-shaped. If the drag depends on velocity, then the terminal velocity will be given by $kv = mg$, where k is some constant. Thus v will be proportional to m. On the other hand, if the drag depends on the square of the velocity then the terminal velocity will be given by $kv^2 = mg$, so v will be proportional to the square root of the mass.

The experiment depends on the fact that you can compare two velocities by dropping two objects at the same time, allowing them to fall different distances, and observing whether they hit the ground at the same time. If the drag depends on velocity, then doubling the mass will double the velocity. Stack two filters and hold them above your head. Hold a single filter in the other hand half that distance above the ground. Drop them at the same time and see if they hit the ground together, as they should if two filters fall twice as far in the same time. What did you observe? You probably found that they didn't hit the ground at the same time. Now add two more filters so you have a total of four. The square root of four is two, so if the drag depends on the square of the velocity, the four filters will now fall twice as fast as the single filter. Try it. Do they hit the ground at the same time?

Air drag is more complicated than this simple experiment suggests. At slow speeds the drag constant, k, is larger than at higher speeds. Many studies have been done on the air drag of tennis balls, baseballs, and soccer balls. The fuzz, stitches, and panels have a strong effect on air drag, as anyone who has tried to hit a knuckle ball will testify.

How do sky divers and **parachutists** use **air drag** to control their speed?

Sky divers can change their body shape, and thus their terminal velocity, by extending their arms, or separating their legs. Parachutists can tilt the chutes to control the direction in which they're falling.

What is **weightlessness**?

Weight is defined as the force of gravity. Strictly, then the only way one can be weightless is to be so far from any massive object, like a planet or star, that the gravitational field is zero! No human has achieved this state. In a satellite the gravitational field is about 80% as large as it is on Earth's surface, so the force of gravity on an astronaut in the satellite is not much smaller than it is on the astronaut on Earth.

An astronaut orbiting Earth actually experiences a gravitational force almost as large as on Earth's surface!

How do you measure weight? If you stand on a scale, the scale measures the upward force of the scale on your feet. According to Newton's Third Law, that force is equal to the downward force of your feet on the scale. (Actually, there is a slight difference because of the rotation of Earth.) If you stand on a scale in an elevator, you'll see that your "weight" changes as the elevator accelerates. When it is going up and increases its speed, the scale will record a larger weight. The same will happen when you are going down and come to a stop. When it slows while going up or speeds up while going down the scale will show a smaller weight. Newton's Second Law can explain these changes using the fact that a net force is needed to accelerate you along with the elevator. (Note that you don't really need a scale to sense these accelerations. You can feel them in your stomach!)

What would happen if the cable holding the elevator would break? You and the elevator would both be in free-fall, accelerating at g. There would be no weight shown on the scale. You and the elevator would both be in the state of "apparent weightlessness." We'll meet this term again when we explore the motion of satellites. The apparent weightlessness of astronauts will be explained there.

Are there **non-contact field-type forces** other than gravity?

Both the electric force between charged objects and the magnetic force between magnetized objects can be described as the force caused by an electric or magnetic field caused by one object on the charge or magnetic moment of the second object.

47

What is the **difference** between **force** and **pressure**?

Why does the stiletto heel of a woman's shoe sink deep into the ground while if she wears a running shoe the heel does not? The force on the two heels is the same. What is different is the force divided by the area of the heel. The quotient of force divided by area is much larger on the stiletto heel. That quantity is called the pressure.

As an example, consider the case of a 110-lb (50-kg or 490-N) woman who first wears shoes with heels 1.5 cm by 1.5 cm. The area of each heel is about 2.2 cm^2. Assuming that she puts all her weight on the two heels the pressure will be 500 N / 4.4 cm^2 or 4.5 N/cm^2. Pressure is usually expressed in pascals (Pa) or N/m^2. In this case the pressure would be 1.1 million pascals or 1100 kPa. If she were to wear running shoes with heels about 7.5 by 7.5 cm then the pressure would be only 45 kPa.

The pressure of air is about 101 kPa. Why don't we feel this atmospheric pressure? First, it is exerted uniformly on all parts of our bodies. Second, it is balanced by pressure from the insides of our bodies. We can feel the pressure if you try to breathe in but close your mouth and pinch your nose closed. You can tell that the external pressure on your lungs exceeds the internal pressure.

What about **pins, needles,** and **beds of nails**?

A pin or needle has a very small area at its sharpened tip, so even a small force on it will create a large pressure at the tip, enough to penetrate cloth or your skin.

Have you ever seen the bed of nails demonstration? A large number of nails are pounded into a board that is about the size of a person's back. The person very carefully lies on the bed. Sometimes a second bed is placed on his chest. A concrete block is put on this bed and a sledgehammer is used to shatter the block.

The person is not harmed. How can that be? Even though the nails have very tiny tips and would certainly penetrate the skin if there were only one, because there are dozens of nails in the bed, the person's weight is distributed over many nails, and the force of each nail is quite small. The pressure is not enough to penetrate the skin.

Anyone doing this demonstration has to take precautions. The parts of the body not on the lower bed have to be supported at the height of the nails. If the concrete block is broken the person must wear safety goggles, and a sheet of wood should be held to keep any concrete from flying into the person's face.

How can you **describe** the **path** of a **thrown ball**?

You know what happens when a ball is dropped. It accelerates downward, gaining speed at a rate of 9.8 m/s each second. What would happen if the same ball rolled off a table?

To answer this question you should define a coordinate system. One axis points down, the other, perpendicular to the first, points in the horizontal direction of the

How can atmospheric pressure collapse a steel drum?

A spectacular demonstration is often used to show the effect of atmospheric pressure. A 55-gallon steel drum is placed on a stand over a propane burner. A quart of water is poured into the drum. The burner will eventually boil the water, filling the drum with steam. When steam has displaced the air, the cap on the drum is tightened and the burner turns off. As the drum cools the steam condenses back to water. The volume of steam is about 1,000 times the volume of water, so when it has condensed it no longer exerts an outward pressure on the drum. But, the atmospheric pressure of 101,000 newtons on each square meter of the drum's surface is still there. The drum collapses with a thunderous noise.

This demonstration can be done on a much smaller scale with an empty soda can and a bowl of water. Put about 1/4" of water in the bottom of the soda can and heat the can on a stove or hot plate. When steam has come out of the hole of the can for several minutes, using a protective glove, quickly turn the can over and put the top into the water. Once again, the steam will condense into water, but because of the small opening, only a small amount of water from the bowl will be able to get into the can. What do you think will happen?

rolling ball. The force of gravity acts in the downward direction, causing the ball to accelerate downward, but there is no force in the horizontal direction, so the ball's horizontal speed would not change. Because only the downward acceleration affects its fall, it would hit the ground at the same time that a dropped ball would. The path the ball takes is a parabola.

Galileo's Principle of Relativity, not Einstein's, can help you understand this result. Galileo imagined a sailor dropping a ball from a high mast on a moving sailboat. The sailor would see the ball drop straight down, but an observer on the shore would see the ball having a horizontal velocity equal to the velocity of the boat. Therefore this observer would see the ball's parabolic path. But both would agree when the ball hit the deck of the boat at the same time. Galileo said that the laws of physics are independent of relative motion. This statement is called the principle of relativity.

What happens if the **ball** is **launched at an angle**?

The ball would now have both an initial horizontal velocity and an initial vertical velocity. Again, the horizontal velocity would be constant because there is no force in that direction. Its vertical velocity would be the same as it was when the ball was thrown either down or up. An initial upward velocity is much more interesting, so let's consider that.

We can specify the initial velocity two ways. First, the way we did before by choosing the horizontal and vertical velocities separately. The second, and more useful way, is to specify the velocity and direction. Suppose a batter hit a baseball at a speed of 90 mph. This speed is 132 ft/s or 40 m/s. The angle could be anything from 90°, a vertical pop-up, down to an angle between 10° and 30° that might be called a line drive to 0° or even a negative angle that would be a ground ball.

The distance the ball travels before hitting the ground depends on both the speed and the angle. If air drag is very small, then the distance is maximum for an angle of 45°. Air drag causes the angle for maximum distance to drop to around 35°. During World War II extensive tables were calculated so that gunners could find the angle for the gun to achieve the desired distance.

How can you get **circular motion**?

If an object moves in a circle, either a planet, a satellite, or even a ball on a string twirled around, then there is no change in speed as it moves around the circle, so there can be no force in direction of motion. There must be a force, because the direction of the ball is changing, so its velocity is changing. The force must be perpendicular to the motion.

Try to make a ball move in a circle. This works best with a ball the size of a soccer ball or basketball on a smooth (wood or tile) floor. Start the ball moving, then kick it gently in the direction perpendicular to its motion. Try another kick or two. Note that its direction changes in the direction of the kick—the momentary force you placed on the ball. If you could exert a constant force that is always perpendicular to the motion the ball would move in a circle. Note that the direction of the force is always toward the center of the circle. Such a force is called "centripetal," or center seeking.

The force required to keep an object moving in a circle depends on three quantities: the mass of the object, the speed of the object, and radius of the circle. The force must be larger for larger mass, greater speed, and smaller radius of the circle.

What supplies this **centripetal force**?

Centripetal force must be supplied by something or someone exerting the force on the object. In a rotating drum ride at a carnival you stand with your back on the drum. As the drum speeds up you can feel the force of the drum pushing on your back, toward the center of the circle. Often the floor drops down and only the force of friction between your back and the drum keeps you from dropping down with the floor.

When a car makes a turn what supplies the centripetal force on it? The road supplies this force. The road is in contact with the tires, and friction between the tire and road is necessary for the force of the road to be exerted on the car. If the road is covered with ice, the friction often isn't large enough and the car goes straight, rather than along the curve. A racetrack is often banked so the tilt of the track can supply at least part of this inward force, reducing the need for friction.

If you're sitting in the car there must be a sidewise force on you so that you stay with the car as it makes the turn. Usually the friction between you and the seat is sufficient, but sometimes you have to hold onto the door handle to exert more force.

Why do you feel an **outward force** on you when you are in **circular motion**?

When we introduced Newton's Laws we neglected to state an important warning. These laws work only in an "inertial reference frame," that is a reference frame where there is no acceleration. A car when it is speeding up, slowing down, or changing direction is accelerating. Therefore, a person in the car experiences other "forces." When the car is speeding up, you feel pushed back into the seat. When the car is slowing down you feel yourself being pushed forward. When rounding a curve, you feel pushed outward. These are not real forces, but are often called "inertial," "fictitious," or "pseudo" forces. For objects in circular motion there are two such forces, the centrifugal and Coriolis forces. The centrifugal force is the fictitious outward force you feel when your car is rounding a curve.

What are the effects of the **Coriolis force** on Earth?

A reference frame, or coordinate system fixed on Earth, is not an inertial frame. The Coriolis force affects the rotation of winds around high and low pressure areas in Earth's atmosphere. Air flows into a low pressure center. The Coriolis force in the northern hemisphere causes all flows to be deflected to the right, creating a counterclockwise flow around the center. In the southern hemisphere the deflection and rotation are in the opposite direction. On the other hand, air flows away from a high-pressure center, and so the deflection causes a clockwise rotation in the northern hemisphere and the opposite in the southern.

How does the **gravitational force cause** the **motion** of **planets**?

In antiquity the planets (the name comes from the fact that they seem to wander across the sky) were assumed to have circular orbits. In order to account for the observations, their motion was thought to involve circles attached to other circles. In between 1600 and 1605 Johannes Kepler (1571–1630) made careful studies of the observations of Mars made by Tycho Brahe (1546–1601). He found that

Johannes Kepler studied the orbit of Mars using the measurements made by Tycho Brahe. He concluded that Mars's orbit is actually an ellipse and not a circle. The elliptical shape of the orbits of the planets in our solar system is called Kepler's First Law.

a circular path required Tycho's observations to be wrong by 2 minutes of arc (four times the apparent size of the moon), but he knew that Tycho's work was better than that. After some 40 failed attempts he finally discovered that the orbit could be described as an ellipse. We now know an ellipse fits the orbits of all planets, comets, and other bodies about the sun, as well as satellites about planets. The shape of the orbit is called Kepler's First Law.

An ellipse is not a circle, so the gravitational force of the sun is not always perpendicular to the motion. Therefore the planet's speed changes as it moves around its orbit. Kepler's Second Law says that in equal times planets sweep out equal areas. Thus if the planet is closer to the sun, it will move faster than it does when it is farther away.

Kepler's Third Law was actually obtained in 1595 and was based primarily on philosophical and theological arguments. It stated that the relative sizes of the orbits of planets around the sun could be obtained by nesting the five Platonic solids: the cube, tetrahedron, dodecahedron, icosahedron, and octahedron. The planets' orbits were in the spheres that circumscribed each solid. Today we state this law as the square of the period of the planet is proportional to the cube of the radius of the orbit. The proportionality constant depends on the mass of the object about which the orbit occurs and the universal gravitational constant. The law summarizes two of the properties of orbits of planets or satellites about a central star or planet.

What is meant by a **law** as opposed to a **theory**?

Laws summarize observations. They describe phenomena. A theory explains a large number of observations. Einstein's theory of gravity explains Newton's law of gravitation, which can be used to derive Kepler's three laws. A theory cannot become a law because they are essentially different things.

What does **Kepler's Third Law** tell us about the motion of **satellites around Earth**?

Kepler's Third Law relates the period of a satellite to its distance to the center of Earth. The table below shows the mean radius of the orbit, the altitude above Earth's surface, and the period for some typical satellites.

Satellite	Type of Orbit	Altitude (km)	Mean Radius of Orbit (km)	Period
International Space Station	Equatorial	278-460	6,723	91.4 min
Hubble Space Telescope	Equatorial	570	6,942	95.9 min
Weather Satellite (NOAA 19)	Polar	860	7,234	102 min
GPS Satellite	Equatorial	20,200	25,561	718 min
Communications Satellite	Equatorial	36,000	42,105	1,436 min
Moon	Ecliptic	364,397-406,731	384,748	27.3 days

Weather satellites have polar orbits so that the view of their cameras sweeps across the entire surface of Earth many times each day. The moon's orbit is aligned with Earth's orbit around the sun, not Earth's equator.

Kepler's Third Law holds for satellites about other planets and for planets revolving around the sun, but you need either Newton's laws or Einstein's relativity theory to explain these results.

MOMENTUM
AND ENERGY

MOMENTUM

Why do we need to introduce **momentum and energy**? Aren't Newton's Laws good enough?

Newton's laws could be used, but when you analyze the motion of more than a few interacting objects, Newton's laws are just too complicated. When you use momentum and energy you concentrate on the beginning and end of an interaction, not on the details of the interaction.

A sports team can be said to have momentum. What does **momentum mean** in physics?

Momentum is defined as the product of mass and velocity. Momentum is a vector quantity—it has both magnitude and direction. Consider a football player—a lineman. Linemen are massive. If they can run fast, then the product of their mass and velocity is their momentum (mv). The momentum is in the direction the player is moving.

How are modern **cars designed** to **decrease** the chance of **injury** in a car crash?

If your car hits a barrier or another car, it will slow or even stop. Modern cars are designed so that the front end collapses, extending the time that the forces of the barrier or other car act, thus reducing the force needed to stop the car.

Cars also have airbags that don't act on the car, but on the passenger. When a sudden very large acceleration of the car is detected the airbags are deployed. A chemical

55

reaction within the bag rapidly fills the bag with gas. The front surface of the airbag speeds toward the passenger at speeds up to 180 mph! But the momentum of the passenger is decreased slowly because he or she compresses the airbag. In addition, because the airbag has a large area, the force isn't concentrated, but spread out. This reduces the pressure on the body, reducing the chance of injury.

Most recent cars also have side airbags to protect passengers from side collisions. But, airbags have caused injuries to smaller persons. Safer airbags inflate less, reducing the force on the passengers, who, because they have less mass, require less impulse to be stopped.

What are other examples of **impulse and change** of momentum?

If you catch a baseball or softball, especially without a mitt, you know that you move your hand in the direction the ball is moving. You certainly don't move it toward the moving ball. In stopping the ball its change in momentum is the same, no matter how you catch it. Thus the impulse your hand gives the ball is also the same. What changes? When you move your hand in the direction of the ball you increase the time the force of your hand acts on the ball. That means that the force is less.

If two **objects interact** what happens to the **momentum** of the other object?

According to Newton's Third Law if your hand exerts a force on the ball, the ball exerts a force on your hand. Thus there must be an impulse on your hand. Does the momentum of your hand change? Of course your hand is just a part of your body, and if you are standing firmly on the ground, your body is unlikely to move. Let's pick a simpler situation. Suppose you are sitting in a desk chair with wheels on a smooth floor and you catch a heavy ball thrown by someone. In this case you and the chair will roll backward. In other words, the momentum of you and the chair will increase as the momentum of the ball decreases.

What happens to the **momentum** of the ball and you and the chair **all together**?

As long as external forces are zero or small, then in any interaction the momentum of the system, in this case the ball, you, and the chair will be constant. This result is called the conservation of momentum. The sum of the decrease in momentum of the ball and the increase in momentum of you and the chair will be zero.

What are other **examples** of the **conservation of momentum**?

If there is only one object in the system, then with no external forces Newton's Second Law says that its velocity will not change. Conservation of momentum also says that its momentum won't change. If the momentum was zero, it will remain zero.

If you shoot a rifle or shotgun you are often told to hold the gun tightly against your shoulder. What's the physics explanation for this admonition? When the gun is

> ## How do you change the momentum of a lineman?
>
> If you're trying to stop a lineman, you need to exert a force in the direction opposite the direction of his motion. The greater the force and the longer you exert it, the greater the momentum change. The product of force and the time the force is applied is called the impulse. Because force is a vector, so is impulse. Thus the direction of the impulse that stops the lineman is opposite his motion. The larger the impulse the greater the change in momentum.

fired the bullet's momentum changes. Its new momentum is in the forward direction. So, according to the law of conservation of momentum the gun must gain momentum in the opposite direction. It will recoil. If the gun isn't held tightly to your shoulder its mass is relatively small, and so its recoil velocity will be large. When it hits your shoulder it could cause injury. If, on the other hand, the rifle is tight against your shoulder, then the mass is the mass of the rifle and your body. The recoil velocity will be much smaller.

How do **rockets accelerate** in space?

When a rocket's motor is fired it expels gas at a high velocity backward. Thus the gas, originally at rest in the rocket, is given a large momentum backward. With no external forces on the rocket-gas system, the rocket's momentum must increase in the forward direction. It will speed up.

When the rocket is on the launch pad there is an external force on it, the force of gravity. How can the rocket take off? Now the momentum of the gas and rocket isn't conserved, but still the impulse the rocket gives the gas in pushing it backward is equal and opposite to the impulse the gas gives the rocket. Thus the rocket rises, just more slowly than it would if there were no gravity.

How does Einstein's **Special Theory of Relativity** affect **momentum**?

The same factor, γ, that affected distance and time, affects momentum in the same way, that is relativistic momentum is γmv. Gamma is 1 at slow speeds and becomes large only when the velocity is near the speed of light.

Does **momentum** apply to objects that **rotate**?

Quantities that describe rotation are similar to, but different than, those that describe straight-line motion. Position is replaced by angle, velocity by angular rotation, accel-

57

With nothing to technically "push against" in space, it seems odd that rockets could accelerate in space. The explanation is that the force of the combusting rocket fuel itself pushes the rocket, accelerating it forward.

eration by angular acceleration. Force is replaced by torque.

What plays the role of **force** in **rotation**?

You have had experiences that illustrate how torque works. Suppose you want to push open a door that rotates about its hinges. You know that the speed with which the door opens depends on how hard you push. It also depends on how far from the hinges you push—the farther the faster. It also depends on the angle at which you push. Pushing at a right angle to the door is much more effective than pushing at a smaller or larger angle. If you push at a right angle, then torque equals the force times the distance from the axis of rotation.

What plays the role of **mass** in **rotation**?

Mass is defined as the net force on an object divided by its acceleration. By analogy, then, the property that takes the place of mass should be the torque divided by angular acceleration. The property is called rotational inertia or the moment of inertia. It depends not only on mass, but on how far the mass is from the axis of rotation. The further the mass is from the axis, the larger the moment of inertia. If you sit on a swiveling stool or chair while holding heavy weights, the further you extend your arms, the more difficult it is for someone to start you rotating. That is, it will require more torque to achieve the same angular acceleration.

What plays the **role of momentum** in **rotational motion**?

The angular momentum of a rotating object is proportional to the product of its moment of inertia and its angular velocity. If there are no external torques on the object, then its angular momentum does not change.

An object with linear momentum that has no external forces on it cannot change its mass, so its velocity is constant. But a rotating object can change its moment of inertia, so, even without external torques, its rotational speed can be changed.

How can **athletes** use **angular momentum**?

Let's explore two different sports. In platform diving a person pushes off the tower, and thus the platform exerts a force on her (Newton's Third Law). But, if she isn't standing straight up, the force also exerts a torque on her, and starts her rotating. If she pulls her arms and legs in, then her mass is closer to her axis of rotation, and her speed of rotation increases. To slow this rotation, she can extend her arms and legs. With good timing, she can hit the water with a bare minimum of rotation.

Put a toy gyroscope on a stand and spin the wheel. Gravity pulls down on the center of gravity of the gyroscope, creating a torque on the axis of rotation, causing it to rotate downward.

Consider a figure skater. She can start spinning on the point of one skate by pushing on the ice with the second skate. Again, the force of the ice exerts a torque, and so her rotational speed increases. She can extend her arms to slow the rotation, or pull them in as close as possible to attain the highest spin rate.

Can the **axis of rotation change**?

A toy gyroscope contains a rotating wheel. If you put it on a stand the axis moves in a circle. Why? The gravitational force pulls down on the center of mass of the wheel. Thus the gyroscope begins to rotate downward. The effect of this new torque is to cause the axis to change direction; to precess.

Precession is also important for bicycles and motorcycles. If the cycle starts to tip to the right, then the rotating front wheel's axis will rotate, and the wheel will turn to the right, helping to keep the cycle from tipping over.

ENERGY

What is **energy**?

An object with energy can change itself or its environment. That's a pretty abstract definition. Let's explore some of the many ways an object can have energy and what changes it can cause.

A speeding car has energy—think what damage it can do if it hits a wall or another car. The energy of motion is called kinetic energy. A rotating wheel also has energy—if you try to stop a spinning bicycle wheel with your hand, it may hurt you. This kind of energy is called rotational kinetic energy.

A compressed spring or a stretched rubber band can cause a stone to move. The energy in the squeezed spring or stretched band is called elastic energy. There are a variety of other forms of energy that are stored in a material. The random motion of the atoms that make up the material means that the atoms have kinetic energy. A measure of the amount of the kinetic energy in the random motion of the atoms is called temperature; the more energy, the higher the temperature. Kinetic energy in the random motion of atoms in a material is called thermal energy. If you charge or discharge a battery, like the one in your cell phone, you change the chemical composition of the battery materials. When you charge it you increase its chemical energy. You can also increase the chemical energy of your body by eating. Even mass has stored energy—splitting the nucleus of a uranium atom results in elements that have smaller mass but a large amount of kinetic energy.

How can **energy** be **transferred**?

Any energy transfer involves a source, whose energy is reduced; a means of transferring the energy; and an energy receiver, whose energy is increased. It's convenient to use a diagram to keep track of the source, the transfer, and the receiver (see pp. 61–62).

For example, if a moving pool ball collides with another ball it can transfer all or part of its kinetic energy to the other ball. Transfer of energy by this kind of mechanical interaction is called *work*. The moving ball does work on the stationary ball and its kinetic energy is reduced. The kinetic energy of the stationary ball is increased by the work done on it.

When a slingshot does work on the stone, its stored elastic energy decreases and the kinetic energy of the stone increases.

When you throw a ball your stored chemical energy is reduced, work is done on the ball, and the ball's kinetic energy is increased.

Other methods of transferring energy that do not involve work will be discussed later.

What **energy transfers** are involved when a **ball is tossed**?

If you lift an object like a ball you increase the energy in Earth's gravitational field. Energy is transferred from you to the field, resulting in a decrease in your stored chemical energy (see p. 62). Suppose you toss the ball up. When you toss it you do work on the ball, transferring energy from your body to the ball, increasing its kinetic energy as well as the gravitational field energy. Once you let go of the ball it continues to rise, but its velocity decreases as the gravitational field energy increases and the ball's kinetic energy decreases. It reaches its maximum height; at that instant the kinetic energy is zero—all energy is in the field. On its way back down it speeds up, so its kinetic energy

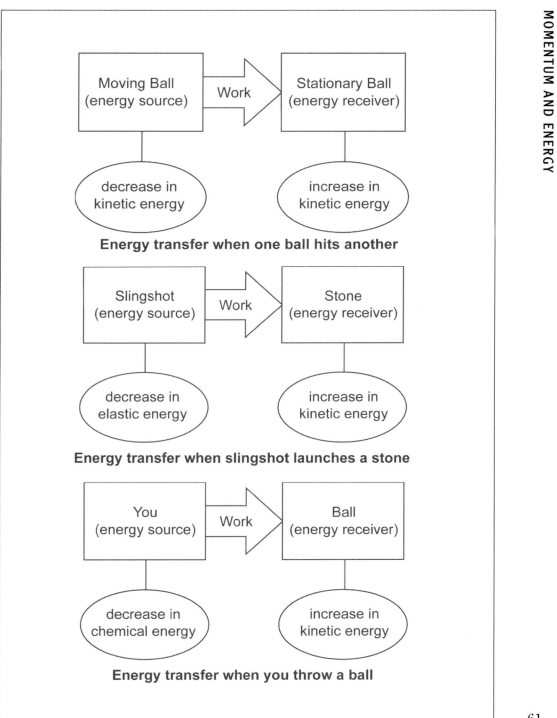

Energy transfer when one ball hits another

Energy transfer when slingshot launches a stone

Energy transfer when you throw a ball

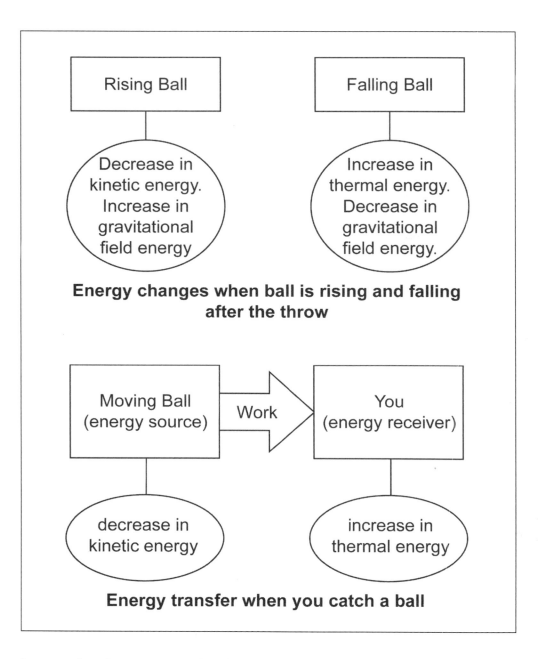

Energy changes when ball is rising and falling after the throw

Energy transfer when you catch a ball

increases but the energy stored in the gravitational field decreases. While you are not touching it the sum of the kinetic energy of the ball and the gravitational field acting on it is a constant. Energy changes from one form to the other and back again.

As you catch it, stopping its motion and thus reducing its kinetic energy to zero. The ball does work on you. But, your stored chemical energy does not increase.

When a moving pool ball collides with another one, all or part of its kinetic energy will be transferred to that ball, which is what makes this fun game possible.

Does the **total energy** of a system ever **change**?

No. Think about a block of wood on a table. You push it, doing work on the block and transferring energy from your body to the block. The block starts moving, but quickly slows and comes to a stop. Where did its kinetic energy go? What was the effect of the friction between the block and table?

To explore friction, rub a pencil eraser on the palm of your hand. Then quickly put the eraser against your cheek. You probably found that both the eraser and your hand became warmer. The friction between the block and table had the same effect, but the temperature change was probably too small to detect. If the temperature increases, then the thermal energy in the object has increased. Thus the decrease in kinetic energy of the block was accompanied by increased thermal energy in both the block and the table. The energy just changed forms.

Scientists have made careful measurements of energy in a variety of forms and have always found that energy is neither created nor destroyed. In other words, the energy put into a system always equals the energy change in the system plus the energy leaving the system.

These measurements have led to a law: the Conservation of Energy. As long as no objects are added to or removed from a system, and as long as there are no interactions between the system and the rest of the world, then the energy of the system does not change.

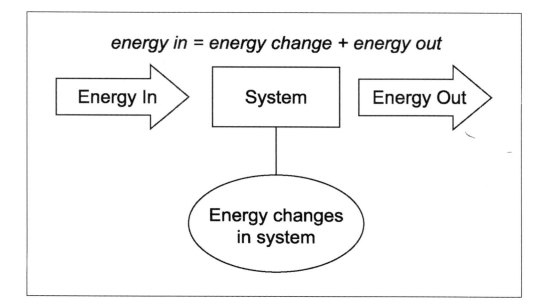

energy in = energy change + energy out

Energy In

System

Energy Out

Energy changes in system

How are **conservation of momentum** and **conservation of energy** used in **everyday life**?

These two laws are most often used when two objects collide. Momentum is conserved if there are no external forces, and it changes very little if the forces during the collision are much greater than the external forces. For example, if two cars collide, the two cars are the system. The forces between them are much larger than the forces on the wheels that come from outside the system, so momentum is conserved. Is energy conserved? While total energy is conserved, the kinetic energy before the collision is much greater than the energy afterward. Much of the energy goes into bending metal and breaking glass and plastic. Automobile crash reconstruction is a way of using conservation of momentum to figure out the speeds of one or both of the cars before the collision.

Kinetic energy is not always converted to other forms of energy in collisions. The toy "Newton's Cradle" has a set of steel balls that swing on strings. In these collisions kinetic energy is almost totally conserved. The incoming ball stops when it hits a second ball, and the second ball moves away with the same speed, and thus kinetic energy, as the incoming ball had. Collisions of pool or billiard balls is another case where kinetic energy is conserved. This kind of collision is called "elastic" while the case of cars colliding is called "inelastic."

A hot object has **thermal energy**. How can it **transfer** this **energy**?

What happened to the increased thermal energy of the eraser and your hand, or to the block and the table? A short time later all will have cooled. They have transferred their

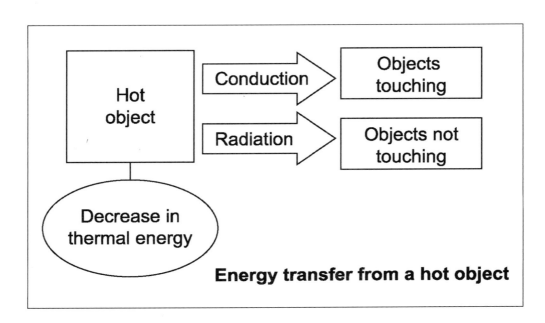

Energy transfer from a hot object

thermal energy to the cooler surroundings. Energy transfer that results from a difference in temperature is called *heat*. Heat always flows from the hotter (energy sources) to the cooler objects (energy receivers).

Thermal energy can be transferred in three ways: conduction, convection, and radiation. Conduction occurs when two objects are in contact, like when you put your hand in hot water. Convection is the motion of a fluid, usually air or water. The fluid is heated by the hotter object, then moves until it contacts a colder object where it heats that object. You can think of convection as two instances of conduction. Radiation is infrared waves that are emitted by hotter objects and absorbed by colder ones. You can feel radiation if you bring your hands near a hot electric burner on a stove. The sun heats Earth by radiation.

What happens to the thermal energy in the surroundings? They get warmer, so their thermal energy increases. Then they transfer this energy to a colder object. Earth is warmer than the space around it, and so it radiates energy into space. When that energy reaches another planet or star, those objects are warmed, and so on through the entire universe.

Is there a **difference** between **work and energy?** Between **heat and thermal energy?**

Energy, whether kinetic, stored, or thermal, is a property of an object. Gravitational field energy is a property of the gravitational field. Work and heat are means of energy transfer. Work is transfer by mechanical means. Heat is transfer between two objects

65

with different temperatures. Examples of work are you throwing a ball, a slingshot launching a stone, a ball being caught in a mitt. Examples of heat are your hand being warmed by putting it in hot water, a bottle of soda being cooled in a refrigerator, and Earth's surface being warmed by sunlight.

Who developed the ideas of conservation of momentum and conservation of energy?

Isaac Newton (1642–1727), considering collisions, first described momentum as the product of mass and velocity, but he called it "the quantity of motion." Energy took 150 years from the first statement of principles until the terminology was worked out. Collisions also inspired the Dutch physicist Christiaan Huygens (1629–1695), who wrote that in the collision of two perfectly elastic spheres the sum of what we today call kinetic energy would not be changed by the collision. The German scientist Gottfried Wilhelm Leibniz (1646–1716) gave the name *vis viva* in 1695 to kinetic energy. But how could the conservation of *vis viva* be extended beyond elastic collisions? Finding the answer to this question took over 150 years!

An important contribution was made by Benjamin Thompson (1753–1814). Thompson was born in Massachusetts, but because he opposed the American Revolution he left for England and was knighted by King George III and given the title Count Rumford. While in America he spied for the British. While in England he spied for the French and was a counter spy for the British. He moved to Bavaria, now part of Germany, and became Minister of War, among other duties. Because he ran an orphanage and wanted to save money he studied heat and invented many items, like an efficient stove and a coffee percolator. A long series of experiments led him to conclude in 1798 that thermal energy was nothing more than the vibratory motion of what we know today as the atoms that make up the material.

About twenty years before Rumford's work the French scientists Antoine-Laurent de Lavoisier (1743–1794) and Pierre-Simon Laplace (1749–1827) showed that heat produced by a guinea pig after eating was very close to the heat produced when the food was burned. The development of steam engines by James Watt and others stimulated studies of the relationship between work done and heat produced and how to make engines more efficient. Around 1807 the word "energy" was used with its modern meaning.

In 1842 a German physician, Julius Robert von Mayer (1814–1878), proposed that all forms of energy are equivalent and that the sum of all forms is conserved. He wrote in general, qualitative terms, although in later essays he included quantitative evidence based on the work done when a gas was heated. But his work resulted in little recognition until the end of his life.

About the same time, a British amateur of science, James Prescott Joule (1818–1889), began a series of experiments designed to determine the relationship between work done and thermal energy increase that resulted in heat transmitted to the outside. He explored electric generators, the compression of gasses, and stirring water. His experiments lasted eighteen years. As he continued to publish his results they were taken more and more seriously.

The German physicist and physiologist Hermann von Helmholtz (1821–1894) developed a mathematical description published in 1847 that showed precisely how energy was conserved in many fields including mechanics, thermal energy and heat, electricity and magnetism, chemistry, and astronomy. With his results the scientific community recognized the great achievement of Rumford, Mayer, Joule, and others and fully accepted energy conservation.

What does **energy efficiency** mean?

In most cases thermal energy is not a useful form of energy for a system. You want the energy content in the gasoline in your auto to give it kinetic energy, not to make it hotter. The cooling system uses a water-antifreeze mixture to cool the engine and warm the radiator, where air flowing through it is heated, thus cooling the fluid. The thermal energy in the heated air is often called rejected or waste energy. An auto is about 20% efficient. That is, only 1/5 of the energy in the gasoline is converted into the kinetic energy of the auto. In addition to the hot air from the engine, tires get warm from flexing, and the brakes get hot when they are applied. All this thermal energy is rejected or waste energy.

Your home furnace converts the chemical energy in oil or gas or electrical energy into thermal energy, either of air or water, depending on whether you have forced-air heat or hot water radiators. But not all the energy goes into heating the house; some leaves through the chimney as rejected or waste energy. Heating systems used to be about 60% efficient. Newer systems can be as much as 95% efficient.

Means of increasing the efficiency of auto and home appliances is an active area of research as nations try to conserve as much of the produced energy as they can.

Your body also uses only a fraction, again about 20%, of the food energy to move your limbs when you walk or run. Your body is cooled by contact with the air, or by evaporating liquid—either perspiration or the humid air expelled by your lungs.

How is **energy measured**?

The SI unit for energy is the joule (J), but energy is often measured in other units. The calorie (cal) and the kilo-calorie (kcal) or food calorie are used both for chemical energy stored in foods and to measure heat. The British Thermal Unit (BTU) is most often used to measure heat in homes and industries. The kilowatt-hour (kWh) is used to measure electrical energy. A completely unofficial, but useful unit is the jelly doughnut (JD), the energy in a medium-sized jelly-filled doughnut. It helps relate all these units to a tasty treat. The table below illustrates conversions among the units. To use it read across. For example, 1 J = 0.239 cal, 1 cal = 4.186 J, 1 JD = 239 kcal.

	joule (J)	calorie (cal)	food calorie (kcal)	British thermal unit (BTU)	kilowatt-hour (kWh)	jelly doughnut (JD)
J	1	0.239	0.000239	0.000949	0.000000278	0.0000001
cal	4.186	1	0.001	0.00396	0.00000116	0.00418
kcal	4,186	1,000	1	3.96	0.00116	4.18
BTU	1,055	253	0.253	1	0.293	0.001055
kWh	3,600,000	859,000	859	3.41	1	3.6
JD	1,000,000	239,000	239	949	0.278	1

How much **energy** is there in **commonly used fuels**?

In comparing fuels used to heat a home (natural gas, electricity, fuel oil, and wood) more has to be considered than just the cost of the fuel per MJ of energy. The furnaces that distribute the heated air or water have quite varied efficiencies themselves.

Fuel	MJ/liter	MJ/kg	Common Units	Cost (2009)	$/MJ	MJ/$
Gasoline	35	47	121 MJ/gal	$2.60/gal	0.021	47
E85 (ethanol)	22	38	76 MJ/gal	$2.50/gal	0.033	30
Diesel	39	48	135 MJ/gal	$2.70/gal	0.020	50
Natural gas	0.039		11,000 MJ/mcf (1,000 ft³)	$11/mcf	0.001	1,000
Electricity			0.278 MJ/kWh	$0.12/kWh	0.43	2.3
Fuel oil	38		148 MJ/gal	$2.75/gal	0.019	54
Wood		18	36,000 MJ/cord	$250/cord	0.007	144
Coal		27	12,000 MJ/ton	$40/ton	0.003	300
Uranium (not enriched)		54,000	240,000 MJ/ton			
Candy bar		2.1	0.12 MJ/1 oz. bar	$1/bar	8.4	0.12

What's a **watt**?

Suppose you climb the stairs to the second floor. Whether you run or walk, because you have gone up the same distance the increase in the gravitational field energy will be the same. The difference is the rate at which the energy has changed. The rate, the change in energy divided by the time taken is called *power*. Power is measured in the

Scottish inventor James Watt came up with the term "horsepower," which is equal to 746 watts, the amount of energy it takes an average horse to pull 33,000 pounds of coal one foot in one minute.

unit called the watt. One watt (W) is one joule (J) per second (s). A kilowatt is 1,000 watts or 1,000 joules per second.

Automobiles can accelerate from 0 to 60 miles per hour, but the more powerful ones can do it in six seconds or less while ones with less powerful engines may take more than 10 seconds.

Where did the **term horsepower** originate?

The term horsepower came from Scottish inventor James Watt. The value for a unit of horsepower was determined after Watt made an extensive study of horses pulling coal. He originally determined that the average horse was able to lift 33,000 pounds of coal one foot in one minute. The conversion between watts and horsepower (hp) is that 1 hp = 746 W = 0.746 kW.

In the United States automobile engines are rated in horsepower while in the rest of the world kilowatts are used. In a hybrid car the power of the gasoline engine is usually measured in hp while the electric motor is measured in kW.

What are the **sources** and **uses of energy** in the **United States**?

The tables below explain how much energy the United States uses per source and where the energy is used. The United States used 99.2 quads of energy in 2009, where

69

each quad equals one thousand billion BTUs. Electrical generation (39.97 quads) is very inefficient. Only 31.5 percent of the energy from the source (coal, natural gas, oil, nuclear) is transformed into electrical energy. The transportation industry is also wasteful in terms of energy, with 75 percent of the energy (mostly petroleum) it uses being wasted as heat.

Energy Source	Percentage of Use in U.S.
Petroleum	37.13%
Natural Gas	23.84%
Coal	22.42%
Nuclear	8.45%
Biomass	3.88%
Hydroelectric	2.45%
Solar	0.9%
Wind	0.51%
Geothermal	0.35%

Energy Used By	Quads Used	Quads Wasted	Percentage Wasted
Transportation	27.86	20.90	75%
Industry	23.94	4.78	20%
Residential	11.48	2.29	20%
Commercial	8.57	1.71	20%

What are some **typical power outputs**?

The following table was adapted from Wikipedia's entry on "Orders of magnitude (power)" retrieved on November 13, 2009.

Unit	Example*
femtowatt (10^{-15} watt)	
	10 fW — approximate lower limit of power reception of digital cell phones
picowatt (10^{-12} watt)	
	1 pW — average power consumption of a human cell
microwatt (10^{-6} watt)	
	1 µW — approximate consumption of a quartz wristwatch
milliwatt (10^{-3} watt)	
	5–10 mW — laser in a DVD player
watt	
	20–40 W — approximate power consumption of the human brain

**What might be the consequences of automobiles
powered by electricity rather than gasoline?**

The need for petroleum would be reduced, but unless the fuels for electrical
generation are changed, the need for coal would increase.

70–100 W — approximate basal metabolic rate used by the human body

5–253 W — per capita average power use of the world in 2001

500 W — power output of a person working hard physically

909 W — peak output power of a healthy human (non-athlete) during a
30-second cycle sprint

kilowatt (10^3 watts)

1.366 kW — power received from the sun at Earth's orbit by one square meter

up to 2 kW — approximate short-time power output of sprinting professional
cyclists

1 kW to 2 kW — rate of heat output of a domestic electric tea kettle

11.4 kW — average power consumption per person in the United States as of
2009

40 kW to 200 kW — approximate range of power output of typical automobiles

megawatt (10^6 watts)

1.5 MW — peak power output of a wind turbine

2.5 MW — peak power output of a blue whale

3 MW — mechanical power output of a diesel locomotive

16 MW — rate at which a typical gasoline pump transfers chemical energy to a
vehicle

140 MW — average power consumption of a Boeing 747 jumbo jet

200–500 MW — electrical power output of a typical nuclear power plant

gigawatt (10^9 watts)

2.074 GW — peak power generation of Hoover Dam

4.116 GW — installed capacity the world's largest coal-fired power plant

18.3 GW — current electrical power generation of China's Three Gorges Dam,
the world's largest hydroelectric power plant

terawatt (10^{12} watts)

3.34 TW — average total power consumption of the United States in 2005

50 to 200 TW — rate of heat energy release by a hurricane

71

petawatt (10^{15} watts)

    4 PW — estimated total heat flux transported by Earth's atmosphere and ocean
        away from the equator towards the poles

    174.0 PW — total power received by Earth from the sun

yottawatt (10^{24} watts)

    384.6 YW — luminosity of the sun

Higher

    5×10^{36}W — approximate luminosity of the Milky Way galaxy

    1×10^{40}W — approximate luminosity of a quasar

    1×10^{45}W — approximate luminosity of a gamma-ray burst

What does **sustainable energy** mean?

The amount of fossil fuels is limited. There have been major advances in discovering oil and extracting more from existing reservoirs, as well as recent advances in obtaining natural gas and oil from shale. But these sources, as well as coal and uranium, are not being replaced. Sustainable energy sources, primarily wind, water, and solar energy, ultimately receive their energy from the sun, and therefore will be available for billions of years. The present use of these sources is minimal. There are many difficulties in increasing their use. Wind power is highly variable. Water energy from traditional dams and reservoirs causes environmental problems. Energy from waves and tides has yet to be developed widely. Solar energy can be directly converted to electricity using photovoltaic cells. But these cells, at least at present, are inefficient and costly. Large solar "farms" exist, at which solar energy heats a fluid so it can boil water to use with steam turbines driving electrical generators. An additional problem in increasing the use of many of these sources is that they require the use of very rare materials, which are both costly and not easily obtained. Nevertheless, recent analyses suggest that a combination of nuclear, wind, water, and solar energy could replace most of the use of coal and oil for electric energy production.

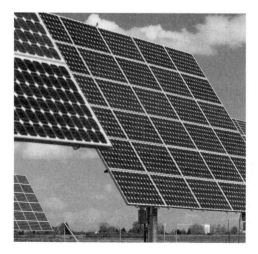

Because we will eventually run out of fossil fuels, we need to explore other, more sustainable forms of energy, such as solar power.

What are **simple machines**?

Simple machines are devices that match human capabilities to do work to tasks that need to be done. They can reduce the force required or reduce the distance or direction an object must be moved.

Suppose you want to lift a heavy object. If you use a machine you will use your stored chemical energy to do work on the machine. The machine, in turn, does work on the object, which increases its energy. If the force is constant and in the direction of motion, then work is the product of force and distance moved ($W = Fd$). The mechanical advantage (MA) of a simple machine is the output force divided by the input force ($F_{output} / F_{input} = MA$). To make the output force larger than the input force you must choose a machine that has a mechanical advantage larger than one. The drawing below illustrates the process. Note that you exert a small input force over a large distance. The machine exerts a larger output force over a smaller distance.

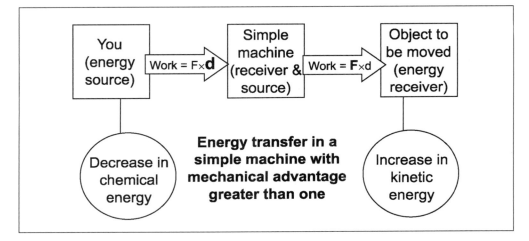

What are the **limitations** on **simple machines**?

Simple machines must, like everything else, obey the law of conservation of energy. That means that the work done on the machine equals the work the machine does plus the heat the machine puts out because friction has increased its thermal energy. Some machines are highly efficient, meaning input and output work are almost the same, while others put out only a fraction of the work put in.

The drawing on page 74 shows a simple machine where the machine has warmed up. That means its thermal energy has increased. It is now hotter than its surroundings, so it transmits heat to its surroundings (which are not shown on this diagram). The output work of the machine is smaller than the work put into it. That means that for the same input force, the output force is reduced by an inefficient machine.

How can a **machine** with a **mechanical advantage less than or equal to one** be useful?

If a simple machine has a mechanical advantage less than one then you exert a larger force on it than it puts on whatever object it contacts. The output force is less than the

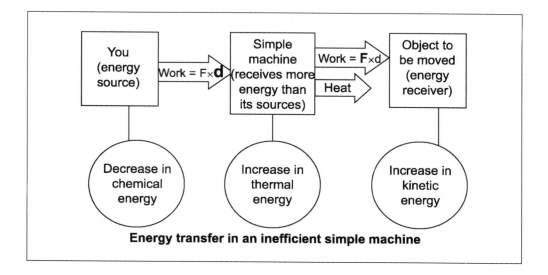

Energy transfer in an inefficient simple machine

input force. How can that be useful? It's because the distance moved (output distance) is larger than what you move (input distance). And, therefore, the output speed is also greater. A baseball bat, tennis racket, and golf club can be considered simple machines in which high speed of the end of the implement is desired.

Are machines with mechanical advantage equal to one useful? They are because they change the direction the force is exerted, which often makes it easier for the person to exert that force.

What are the **types** of **simple machines**?

There are four major groups of simple machines: levers (including wheels and axles), pulleys (including gears), and the inclined plane (including wedges and screws).

What is an **inclined plane**?

A ramp is an example of an inclined plane. Instead of lifting an object to the height at the end of the ramp, you move it a much longer distance on the surface of the ramp, but it requires much less force. So, the ramp has a larger output force than an input force; it has a mechanical advantage greater than one. That is, $MA = F_{output} / F_{input} =$ (length of ramp) / (height of ramp).

According to the Americans with Disabilities Act a wheelchair ramp should have a maximum increase in height of 1' for every 12" length of the ramp. The Act says that the maximum rise should be 2–1/2', so the ramp must be 30' long. The input work is $F_{input} \times L$, where L is the length of the ramp. The output work is $F_{output} \times d$, where d is the rise and F_{output} is the weight of the person plus wheelchair. If there is no sliding or rolling friction, then $F_{input} \times L = F_{output} \times d$. The mechanical advantage is L/d. So $F_{input} = F_{output}/MA$ and the force needed to push a wheelchair up the ramp is given by $F_{input} = F_{output} / (L/d)$. If the weight of the person plus the wheelchair is 200 lbs., then the force needed to push the person up the ramp is F_{input} = 200lbs / (30'/2.5') = 17 lbs.

Look at the cutting edge of a scissors. It's an inclined plane where the input is the force and motion of the closing blades and the output is the outward movement of the paper after it is cut.

Why is a **wedge** like an **inclined plane**?

A knife is one example of a wedge. Look carefully at the sharp edge of a kitchen knife. It looks like two inclined planes put back-to-back. As the knife blade moves down through food, the wedge pushes the pieces apart. A hatchet or axe is another example of a wedge. Wedges are also used to split wood. In this case the flat end of an axe is often used to drive the wedge into the end of a log, which is then forced apart. Wedges are inefficient machines because there is usually a large amount of friction between the wedge and the material, which leads to increased temperature of both the material and wedge, and thus heat transfer.

What type of simple machine is a **screw**?

You can think of a screw as an inclined plane wrapped around an axle. The ancient Greeks used a screw to lift water. Screws and bolts are also used to fasten two pieces of wood, plastic, or metal together.

Screws have a large mechanical advantage, and thus the distance the screw thread moves in its circular motion is much larger than the distance the screw moves into the material. Therefore the force the screw or bolt exerts to hold the material together is large in comparison to the torque used to turn the screw. Screws and bolts are also inefficient machines because of the friction between the screw and material, or between the metal of the bolt and nut.

What is a **lever**?

A lever is a bar that rotates around a fulcrum or pivot. The locations where the input and output forces are exerted relative to the location of the pivot determines the class of the lever, as shown on page 76.

The width of the arrows illustrates the force while the length shows the distance moved.

How **efficient** are **levers**?

The only place where friction can occur with a lever is at the pivot point. So, as long as friction is minimal there, the lever can approach 100% efficiency!

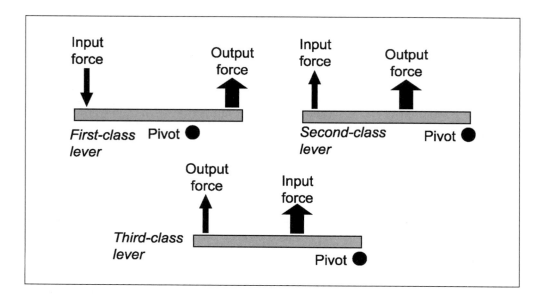

What are **examples of levers**?

The first-class lever shown above has a mechanical advantage greater than one. How could you arrange the location of the pivot to make the mechanical advantage less than one? Look closely at a pair of scissors. Note that you can adjust the mechanical advantage by moving the region of the scissors you are using to cut. Where would you put the material to be cut if more force is needed to make the cut? You would put it, nearer the pivot. On the other hand, if the material is easily cut, cutting near the tip of the blades provides enough force and speeds up the cutting. What class lever is a can or bottle opener? Locate the pivot point and compare to the three drawings above.

Your forearm is a lever, with the elbow joint being the pivot. Where does the bicep muscle attach? The attachment is close to the elbow, so the forearm is a third-class lever. Are other muscles and bones in your body so easily characterized? Most are not because tendons that transmit the force from the muscle to the bone are long and go through several bends.

Consider sports equipment like baseball bats, tennis racquets, and golf clubs. They are often used as extensions of your arms, so the person plus the bat or club has to be examined together. But note that in every case the system is a third-class lever, where a large distance moved (and therefore greater speed) is favored over an increased force.

How is a **wheel** and **axle similar** to a **lever**?

What is a wheel and axle? It consists of a disk (the wheel) attached to a thin rod (the axle) so that the two rotate together. Typically the input force is applied to the outer edge of the wheel, and the output force is exerted on something, like a rope, attached to the outer edge of the axle. If the two forces are exerted in the same direction, then it is like a third-class lever. If the two are in the opposite direction (for example, a person pushing down on one side of the wheel while a rope is pulled up on the other side), then it is like a first-class lever. One may also have the input force exerted on the axle, in which case it is like a second-class lever.

Both the lever and the wheel and axle are really torque, not force, multipliers. Recall that if the force is at right angles to the line from the axis of rotation to the point where the force is applied, then the torque is given by Fr. Therefore, if the radius of the axel is a and the radius of the wheel is w, then if the input force is applied to the wheel, the output force is given by $F_{output} = F_{input}\ (w/a)$.

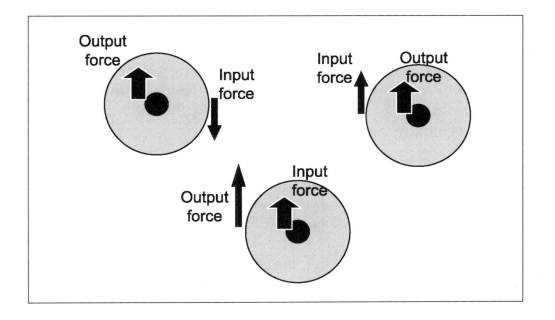

What are **examples** of **wheels and axles**?

The screwdriver is one example The larger the diameter of the handle the greater the torque that can be applied to the screw. Many examples have a rope or chain wrapped around the axle. A sailing ship's steering wheel could be represented by either the left-hand or the center drawing, depending on whether the helmsman pushed down or pulled up on the edge of the wheel. The rope around the axle is connected to the rudder, which is then turned to the right or left.

A similar device would be a device to lift a bucket from a well. The wheel is then replaced by a crank, but the operation of the device is exactly the same. Again, the crank can either be pulled up or pushed down to exert a force on the rope to pull the bucket up.

The rear wheel of a bicycle can be thought of as two wheels on an axle. The large wheel has a rubber tire and exerts a backward force on the road while the smaller wheel is the sprocket that the chain turns. The right-hand drawing above shows that the force of the chain is larger than the wheel's force on the road.

Are **pulleys** really simple machines?

A fixed, or unmovable, single pulley can be considered a wheel and axle where both have the same radius. The mechanical advantage is one, but the pulley changes the direction of the force.

If the pulley is allowed to move and the output force is exerted by the axle of the pulley, then the input force is shared by the two ends of the rope. If you fasten one end and pull up on the other, then you achieve a mechanical advantage of two.

A combination of fixed and movable pulleys is called a block and tackle. Archimedes is said to have pulled a fully loaded ship using a block and tackle.

Energy loss in a block and tackle comes from friction between the axle and its holder as well as the stretching of the rope and rolling friction between the rope and pulley.

Are there **other ways pulleys** can be **used**?

Two or more pulleys, on fixed axles, can be connected together with a belt. If the two pulleys are of different diameters, then the one with the smaller diameter will turn faster, and thus it can exert a larger torque. In your automobile one or more pulley and belt systems are used to deliver torques from the engine to the valve crankshaft, the water pump, the air conditioning compressor, and the alternator.

Continuously variable automobile transmissions are used in a few modern cars to connect the engine to the drive shaft. The torque that can be delivered by an engine depends on the rotational speed. The torque is maximum at an intermediate speed. A transmission is designed to allow the engine to revolve at a speed where it can deliver

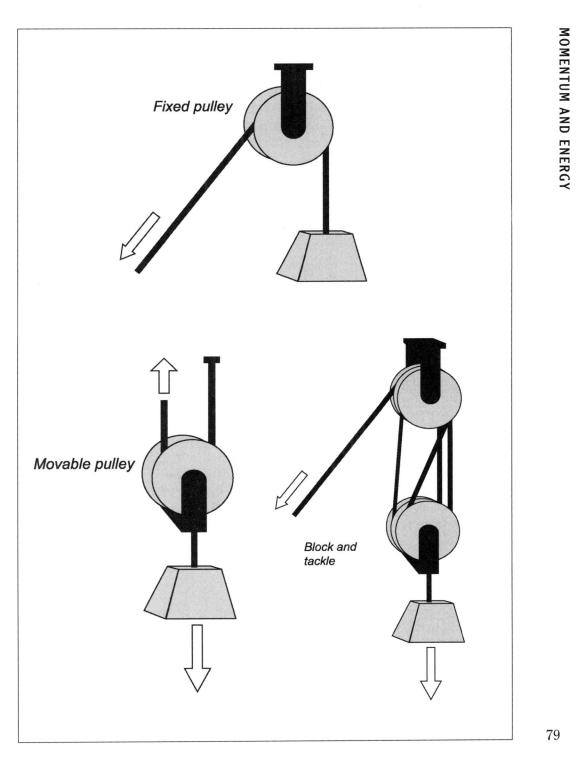

Fixed pulley

Movable pulley

Block and tackle

79

How did gears play a role in an ancient astronomical computer?

In the first century B.C.E. a Roman merchant ship carrying Greek treasures to Rome sank. In 1900 C.E. a storm caused a party of Greek sponge divers to seek shelter on the island of Antikythera. After the storm passed the divers, seeking sponges, found the wreck of the Roman ship. Over the next nine months they recovered much of the treasure, including a badly corroded block the size of a telephone book. A few months later it fell apart, showing remains of bronze gears, plates covered with scales, and inscriptions in Greek. The German scientist Albert Rehm (1871–1949) understood in 1905 that the device was an astronomical calculator. In 1959 the American historian of science Derek de Solla Price (1922–1983) wrote a *Scientific American* article describing some of the details of the device and the calculations it could do. By 1974 he had described 27 gears, including the number of teeth on most. By analyzing the teeth he discovered that the ratio could describe lunar cycles known by the ancient Babylonians.

In 2005 Hewlett Packard and X-Tek Laboratories teamed up to produce astonishing images of the device. HP contributed a camera system that used computer enhanced techniques to show inscriptions that were otherwise invisible. X-Tek brought an eight-ton X-ray machine that could produce extremely high resolution 2-D and 3-D images of the internal mechanisms. The X-ray machine also discovered thousands of Greek letters that described details of the mechanisms. They found thirty gears, but analysis suggests that there must have been at least five more to perform the calculations that move dials on the front and back of the box.

The Antikythera device can predict solar and lunar eclipses, the dates of future Olympic games, and can show the complicated motion of the moon. The precision with which the gears were made was greater than any made in the world for the next thousand years. But who made it and where? It is uncertain, but likely to have been made in one of the colonies of Corinth on the island of Sicily, perhaps by a student of Archimedes decades after he was killed in 212 B.C.E.. (See *Scientific American*, December 2009, and http://www.antikythera-mechanism.gr).

torque to the drive shaft, the axle, and the wheels that are revolving at a variety of speeds. When the auto is accelerating from a stop the wheel rotation speed is slow, and so the transmission needs to match a large-diameter pulley attached to the engine to a smaller one connected to the driveshaft. On the other hand, when the auto is traveling at a high speed, the engine can revolve at the same or even a smaller rate than the driveshaft.

One method of creating a pulley with a variable diameter is to use a v-shaped belt and a pulley with a groove that can be adjusted in width. This kind of transmission is often used with hybrid cars and is being developed as a means of decreasing fuel use.

Who developed **gears**?

As early as 2600 B.C.E. in India and other parts of southern Asia gears were used to open and close doors in temples and to lift water. Around 400 B.C.E. Aristotle (384–322 B.C.E.) described how gears were used. Archimedes described worm gears around 240 B.C.E., and in 40 B.C.E. Vitruvius showed how gears could convert motion from a horizontal axle to a vertical one. In the 1300s C.E. gears were used in clocks in bell towers and on churches.

What are **gears** and why are they **simple machines**?

Gears are toothed wheels which transmit torque between two shafts. The teeth prevent them from slipping, so they are called positive drives. The smaller gear in a pair is called a pinion and the larger one a gear.

Because the teeth mesh, in a given amount of time, the number of teeth engaged on the pinion equals the number on the gear. The number of teeth is proportional to the circumference of the gear that, in turn, is proportional to the radius of the wheel. Therefore if the pinion makes 1 turn, the gear will make (r_{pinion} / r_{gear}) turns. Because energy is conserved, the torque exerted by the gear equals the torque applied to the pinion (r_{gear} / r_{pinion}).

Gears are often used in an automatic transmission, converting the high-speed, small torque output of the engine to the low-speed, large torque output needed to turn the wheels. This kind of gear drive is called a step-down drive. On the other hand, in a windmill, the blades turn at a low speed but provide a large amount of torque. The gear drive in a windmill is a step-up drive. The electric generator, on the other hand, requires high speed, but can work with a smaller amount of torque. Gears also allow a bicyclist to match the speed with which she can rotate the pedal sprocket with the speed needed to drive the wheels.

How are **clocks important** to the development of **gears**?

As clocks have improved, so have the gears used within them. The pendulum

Windmill blades may move slowly, but they can produce a lot of torque. Gears increase the rotational speed but reduce the torque, driving an electric generator. Almost all the energy from the wind is delivered to the generator.

clock uses a type of a gear called an escapement to drive the pendulum, which regulates the time marked out by the clock. Precision gears allow clocks to use less power and have greater accuracy.

How are **gears used today**?

In addition to automobiles and clocks, gears are used in washing machines, electric mixers and can openers, and electric drills, as well as hard drives and CD/DVD drives in computers. Today much of the development of gears is associated with improvements in the materials used. New metal alloys increase the lifetime of gears used in automobiles and industries. Consumer electronics uses plastic gears that require no lubricant and are quiet.

STATICS

CENTER OF GRAVITY

When tossed in the air, why do hammers wobble end over end?

In earlier chapters we considered only "point objects." That is, an object so small that it could be thought of as the tiniest ball. Now we want to expand our considerations to large objects.

First think of a baseball. If a baseball is tossed into the air, the ball follows a smooth parabolic path as described earlier in this book. If, however, a hammer is tossed its path appears much more complex. Why?

All objects are made of atoms. The mass of the object is the sum of the masses of all the atoms. And, the force of gravity on each atom is proportional to the mass of that atom. In a baseball the center may be made of different materials than the surface, but, if you ignore the laces, the ball is made up of the same materials regardless what the direction is. That is, the ball is spherically symmetric. The center of gravity in any object is defined as the average location of its weight. Because the mass is distributed evenly throughout a baseball, the center of gravity is located in the center of the ball. However, for an object such as a hammer, with a metal head and a wooden handle, the center of gravity is not directly in the middle. Since more mass is located in the metal head of the hammer, the center of gravity is closer to that point.

The laws of physics state that the center of gravity follows a parabolic curve when tossed in the air. Indeed, although the ball and the hammer do not appear to have similar motions, their centers of gravity do. If you watch closely both the center of a baseball and the center of gravity of a hammer, you will see that they both follow parabolic paths when thrown.

Because the force of gravity is proportional to the mass, the center of mass is at the same location as the center of gravity.

Where is the **center of gravity** of a **person**?

The center of gravity of a person depends on how that person's weight is distributed. The distribution is typically different in adult males and females. Males have more upper-body mass while females have more mass in the hip region. A male's center of gravity is about 65% of his height while that for a female is about 55%.

Try this. Stand facing a wall with your toes against the wall. Now try to rise to stand on your toes. Can you do it? Standing on your toes moves your point of support in front of your center of gravity, so you will tip backwards. If not against a wall, how can you stand on your toes? You naturally move your arms forward or bend forward to move your center of gravity forward.

If you stand with your back against a wall, can you bend forward so the trunk of your body is horizontal? Most likely if you're male you will tend to fall forward, while if you're female you are more likely to be successful. Again, the question is whether or not the center of gravity of your body is above your point of support—in this case your toes, or in front of it.

Why is it **easy to tip over** some objects?

If you want to tip something over you'll have to rotate it. As you have seen, rotation means you need to apply a torque. An object sitting on a surface has several forces on it. First, is the gravitational force that acts on its center of gravity. Second, is the upward force of the surface. Third, is friction between the object and the surface that exists only if the surface is on a slant.

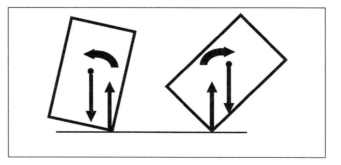

Now suppose you exert a sideways force near the top of a box. That creates a torque and the box begins to rotate. If you now let go will the box continue to roll or will it go backwards? It depends on the relative locations of the center of gravity and the force of the surface as seen in the illustration above.

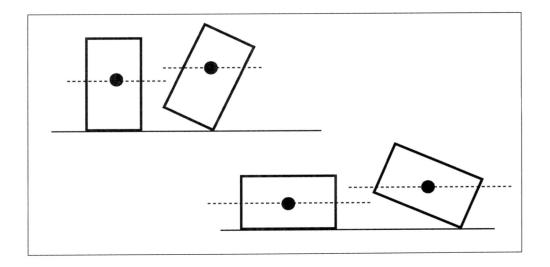

If the center of gravity is above the bottom of the box then the box will return to its upright position. If the center is directly above the corner of the box, then the box won't rotate at all. If the center is outside the bottom of the box, then it will tip over.

Why do **some things tip easily** while others are stable?

As you can see from the example above, when the center of gravity is outside the base of the object it will tip. That suggests the general rule: it will be stable if it is low and wide. That is, keep the center of gravity low and the base wide!

Not only is a larger torque needed to tip an object that is wide and low, but the center of gravity has to be lifted higher for the low and wide object. As a result the work needed to tilt the object is larger.

Left: a "tall and narrow" object can be more easily tipped over.

Right: a "low and wide" object is more difficult to tip.

You can see from the dashed lines the amount the center of gravity must be lifted to tip over the object.

What are applications of this rule? Autos and trucks that are narrower and have high center of gravities are less stable against rollover. Football players and wrestlers are taught to get down low and spread your feet apart. This stance follows the "low and wide" rule making your body more stable and difficult to knock over. Your opponent, in order to knock you over, would have to exert a force to lift up your center of gravity and then push you over. If you simply stood upright, with your feet close together, you would be "tall and narrow" and your opponent would have less difficulty pushing you and your center of gravity.

STATICS

What does it mean to say that an object is **static**?

Static means "not moving." In the fields of engineering and physics to be static means that the object does not move. When static, all the forces acting on a body must sum to zero. That is, the net force on the body is zero so the object does not move.

Why are we **static** when **sitting in a chair**?

As long as you are sitting in a chair and not moving (relative to Earth), you are static. That means that the chair is supporting you with an upward force that is equal to your weight. You would remain static until some external force was exerted on you to start you in motion.

What is the name of the **supporting force** from the chair?

Another term for a supporting force is "normal force." The normal force is always direct-ed perpendicularly out of the surface. The normal force of a chair is straight up if the chair is on a level surface, while the normal force of an incline would be perpendicular to the surface of the incline, and not perfectly vertical. The term "normal" was derived from the geometrical name for a 90° angle and is not the opposite of "abnormal."

What **kind of forces** can you **exert** on an object?

Hold a book in your two hands. What kind of forces can you exert on it? You could pull on the book to try to make it longer or you push it to try to make it shorter. The name for the force on the book that pulls on it is a tension force, while the name for the force that pushes on it is a compressive force.

You could also try to twist it, applying what is called a torsional force. Finally, place one hand on the front cover and the other hand on the back cover. Now push one hand to the right and the other to the left. This places a shear force on the book.

Tension forces are common. If you hang from a chin-up bar, your arms experience a tension force. Pulling on a rope, wire, or cable also exerts tension on them. Materials experiencing tension will stretch, some more and some less. Materials that stretch easily are called pliant; those that do not are called rigid.

Scientists have measured the response of materials to tension and compression. The ratio of the applied pressure (called stress) to the change in length divided by the original length (called strain) is called Young's modulus. Its value varies from 10 GN/m^2 (10 bil-lion newtons per square meter) in wood to 200 GN/m^2 in steel and cast iron. That means that for an equal force on it steel will stretch (or compress) 1/20 as much as wood. If you would hang a 120-kilogram (264 pound) ball from a 6-meter (19.7 feet) long steel cable 2.5 millimeters (0.9 inches) in diameter it would stretch 7 millimeters (0.3 inches).

In the case of the cable above when you remove the weight the cable will return to its original length. It is called an elastic material. If you exert a much larger force the cable won't return to its original length. This is called the plastic region. At some point it will break. The force that breaks it is called the tensile strength. For steel the tensile strength is 1 GN/m^2. Thus for cable in the example above it would take a hanging mass of 500 kilograms (1,100 pounds) to break the cable.

Young's modulus also describes the change in length of an object when a compressive force is exerted on it. When you stand you exert a compressive force on your leg bones and they will shrink somewhat in length. But a 70-kilogram (154 pound) person standing on one leg will compress it by only 0.01% of its length! The Young's modulus for cartilage, the material between the bones in all parts of your body, is one ten-thousands as large as that of bone. So putting the weight of a 70-kilogram (154 pound) mass on a piece of cartilage with same area as the femur in your leg would compress it by 10% of its original thickness.

For compressive forces there is a compressive strength. But, many materials will buckle if too great a compressive force is placed on it. Brittle materials will, on the other hand, break. The compressive and tensile strength of bones is almost exactly the same.

What determines **how much** a **material** will **bend**?

Try to bend a ruler. Hold one end tightly on your desk and push down on the other end. You are exerting a tension force on one surface and a compressive force on the other. The amount the ruler bends depends again on the Young's modulus (Y) of the material as well as its length (L), width (w), and thickness (t). The bending (x) is proportional to the force applied (F) and the length cubed and inversely proportional to its thickness cubed. Or, in the form of an equation, $x = FL^3/(t^3wY)$ That is why the joists supporting a floor are much thicker than they are wide. Typically wood 1–1/2" wide but 10" or 12" thick is used. The larger thicknesses are used if the span between supporting walls is longer.

An "I" beam is often used to support weight. The beam is in the shape of the letter I. The vertical member, tall and narrow, keeps it from bending while the top and bottom members keep the vertical member from twisting. While I beams are most often made of steel, wood beams are now used in houses because they are stronger, lighter, and cheaper than steel.

What materials are used for **static structures**?

The first materials used were made of stone, especially stones like flint that could be chipped to make sharp edges and points. Ancient peoples discovered metals, either in pure form, or in ore, that is mixed with non-metals. Copper, tin, gold, and iron were known before 2000 B.C.E. Copper, tin, and gold are very soft and not very useful as tools or weapons.

A method of hardening metals is to mix two or more different metals forming what is called an alloy. Bronze is made by adding tin to copper and was known by about 4000 B.C.E. in what is now Iran and Iraq. The tin came from southern England. Bronze is a hard metal that can be melted and cast in various shapes, including statues. Today bronze is used for bells and cymbals.

Not much later methods were found to convert iron ore, which is iron mixed with either oxygen or sulfur, to pure iron using a charcoal fire. Iron replaced bronze as a material for tools and weapons because it was cheaper and didn't require long trade routes to obtain the tin. To keep iron from being brittle the carbon must be removed. If the iron is heated to red-heat and then hammered the carbon is forced to the surface where it can be removed. The result is steel that contains less than about 1% carbon.

Steel is a versatile alloy of iron and carbon because additional materials can be added to change its properties. Very hard steel, called tool steel, contains tungsten, molybdenum, and chromium, among other minor additions. Its hardness can be changed by heating it and then cooling it very quickly by quenching it in water or other liquids. Stainless steel used in knives, forks, and spoons has 18% chromium and 10% nickel. Other stainless steels use different amounts of these two metals as well as molybdenum and magnesium. Stainless steel does not rust, but isn't as hard as tool steel.

Pewter is a tin alloy that was developed in England for plates and cups. It adds copper, bismuth, and antimony to tin. Originally lead was used, but because it is poisonous, it is no longer found in pewter.

Brass is typically 80% to 90% copper with zinc added. Actually, brass was made before metallic zinc was isolated! The zinc ore calamine was melted with copper. Brass is used as a decorative metal as well as in bullet casings. If aluminum and tin are added it is resistant to corrosion by sea water.

Gold is another very soft metal. It can be made harder by adding copper and silver in equal quantities. Gold alloys are measured in karets. 24-karat gold is pure gold while 18-karet gold has 18 parts gold and 6 parts other metals. 10-karet gold has 10 parts gold and 14 parts other metals.

Titanium is used in many alloys today because it is lightweight and corrosion resistant. Many alloys of both metals and non-metals are used in electronics. Some LED lamps contain the metals gallium and aluminum and the non-metal arsenic.

What are **composite materials**?

As was mentioned above, brittle materials like concrete have larger compressive strength than tensile strength. For that reason reinforcing bars (rebars) made of steel are embedded in concrete. The steel supports the concrete when it is under tension, reducing its tendency to crack and thus keeping it from failing.

Composite materials have been used since ancient times when straw was added to clay to make bricks.

The most recent composite materials use carbon fibres (less than 0.01 mm in diameter) that are light in weight (they're made of pure carbon) with very high tensile strength but low compressive strength. They are added to plastics to make them more rigid (increase their Young's modulus) without adding weight. Golf club shafts are often made of a plastic filled with carbon fibres, giving the shaft great strength with low weight. By varying the amount of carbon fibres the flexibility of the shaft can be changed.

Reinforcing concrete blocks with rebar (steel bars) helps support the concrete and keep it from cracking. Composite materials such as these blocks are an example of how combining materials that have compressive strength with those that have tensile strength is an effective strategy in construction.

Although it isn't really a composite, glass can be made to be less brittle. The key idea is to keep the surface under compression at all times so it won't crack. Glass sheets are made from the fluid state by cooling. If the surfaces are cooled quickly with strong blasts of air it produces a form of glass called case-hardened that was used for shatterproof lenses in eyeglasses. In the 1970s a chemical method was invented that could be used on cold glass. It involves putting the glass into a bath of potassium salts. The potassium replaces sodium in the surface layers of the glass. The larger potassium atoms expand the surface layers so that the interior portion of the glass compresses the surface, keeping it from developing cracks.

What is **shear**?

Scissors, also known as shears, use each of their blades to move the object it is attempting to cut in opposite directions. Thus they exert a shearing force on the object to be cut. Earthquakes often cause land and roads to experience significant shearing forces. Pictures of torn-up roads after earthquakes show how one side of a street moved one way while the other side of the street moved in the opposite direction, tearing up the pavement as the parts moved past one another.

What **other major force** can be experienced by **structures**?

A torsion force, usually from winds, is responsible for twisting structures. Buildings, bridges, and towers use cross-supports to prevent such forces from damaging the structures. For example, the John Hancock building in Chicago has visible cross-supports.

BRIDGES AND OTHER "STATIC" STRUCTURES

What was the **first type** of **bridge**?

The first type of bridge ever used was a beam bridge. This bridge was probably just a fallen tree that was used to cross a ravine or a small stream; the tree was probably supported by the river bed or by a group of rocks. Beam bridges consist of a horizontal roadbed supported by vertical piers on the shores that are planted in the ground. Beam bridges are limited by the resistance of the roadbed to bending.

How can you **reduce the bending** of a **beam bridge**?

The simplest way to keep the roadbed from bending is to use a king post. In the illustration below, the downward force of the center of the bridge pulls down on the vertical post. This places the diagonal braces under compression. They transmit the force to the piers. The upward force on the post makes the net force on the post zero. It is under tension.

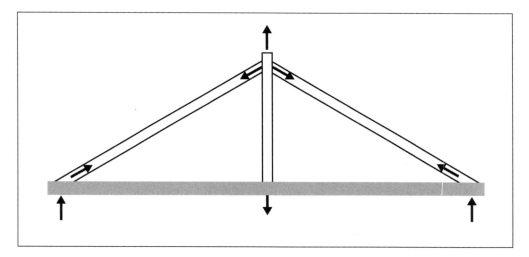

While the king-post bridge can reduce the bending in the center of the bridge, it can do nothing about bending between the pier and the bridge center. One solution is to add a second vertical post and connect the two by a horizontal member, creating a queen-post bridge. But a method that allows much more support on a longer bridge is the truss.

Similar methods of translating downward forces to compression forces exerted on the piers on the ends of the bridge were known to the Romans who worked in stone and concrete. They were famous for the arches used in their massive aqueducts. One such aqueduct, the Pont du Gard, was completed in 18 B.C.E. and was used to carry water a length of 270 meters (886 feet) over the Gardon river valley in southern France.

What is the longest bridge in the world?

The longest suspension bridge in the world is in Kobe, Japan. The Akashi Kaikyo spans a distance of 1,991 meters (6,532 feet). The total length of the bridge is 3,911 meters (12,831 feet). This $3.3 billion bridge took 12 years to build and is designed to withstand 8.5-Richter scale earthquakes and 178 mph winds. It weathered the 7.2-Richter earthquake that killed 5,000 citizens of Kobe in January 1995. The only damage sustained by this incredibly well engineered bridge was that one of the piers and anchorages shifted a little less than 1 meter. This high-tech bridge uses pendulums within the massive vertical towers to counteract dangerous bridge movement produced by seismic activity. These high-tech mechanisms move against the motion of the bridge, stabilizing it and keeping drivers on the bridge relatively safe.

How does the **suspension bridge** support the roadway?

The twenty longest bridges in the world are all suspension bridges. Suspension bridges are able to span huge distances because the long cables suspending or holding up the roadbed are draped over a set of tall vertical towers called pylons. The pylons support the suspension cables from which vertical cables are attached that lift the deck. The ends of the suspension cables must be anchored into the ground at each end of the bridge to exert the tension forces on the cables.

The first known suspension bridge was constructed in the seventh century C.E. by Mayans at their capital Yaxchilan, Mexico. It spanned 100 meters.

Where is the **longest bridge** in the **United States**?

The longest bridge in the United States, a suspension bridge, ranks as the sixth longest bridge in the world. The Verrazano Narrows Bridge is between Staten Island and Brooklyn, New York. This bridge, completed in 1964, spans 1,298 meters (4,260 feet).

The span between the towers of the Mackinac Bridge that links the upper and lower peninsulas of Michigan at 1,158 meters (3,800 feet) is shorter than that of the Verrazano Narrows Bridge, but when measured by the distance between the cable anchorages it is the longest bridge in the western hemisphere The length of the entire bridge, shore to shore, is 5 miles.

What is the **newest hybrid** of **bridges**?

One of the newest, prettiest, and most economical bridges is the cable-stayed suspension bridge. With its sleek lines and thin roadways, it is the perfect bridge for most mid-span designs. The Tatara Bridge, in Onomichi, Japan, completed in 1999 is the longest cable-stayed bridge in the world, with a span of 890 meters (2,919 feet). Cable-stayed bridges suspend the roadbed by attaching multiple cables directly to the deck supporting the roadbed. These cables are then passed through a set of tall vertical towers and attached to abutments on the ground. Such engineering methods reduce the need for heavy, expensive steel and the massive anchorages that are needed to support suspension bridges.

At 4,260 feet, the Verrazano Narrows Bridge in New York is the longest suspension bridge in the United States.

What **challenges** are there in **building skyscrapers**?

The first challenge is to design a foundation that can support the tremendous weight of a large building. The best way is to dig down to the bedrock. This can be as close as about 21 meters (70 feet) in New York City to almost 61 meters (200 feet) in Chicago. If the distance is short holes can be bored and concrete piers can be formed in the holes. More frequently a caisson is required. This is a large hollow waterproof structure that is sunk through the mud, pulling it into and then out of the top of the caisson. A third method is go build a large steel and concrete underground pad that "floats" on the top of a hard clay layer.

The load that the foundation must support includes the weight of the building, its furnishings and equipment, and the changing load of occupants. In addition to the loads, strong winds must also be considered.

The walls of early tall buildings were constructed of masonry that supported the weight of the building. The 16-story 65.5-meter (215-foot) high Monadnock Building in Chicago, built from 1889 to 1891, required 1.8-meter (6-foot) thick walls at the base. It was so heavy that it sank, requiring steps to be constructed between the sidewalk and the ground floor. The second half of the building used a steel frame on which masonry was attached, allowing much wider windows to be used.

The steel frames can be bolted, riveted, or welded together. When the 59-story, 279-meter (915-foot) tall Citigroup building was constructed in New York City from 1974 to 1977 the frame was bolted together, but later computer models showed that if hurricane-strength winds struck the building it would be in danger of collapse. As a hurricane moved up the eastern seaboard in 1978 workers hurriedly welded plates over the bolted joints. Luckily the hurricane moved out to sea, sparing New York.

What makes a building a skyscraper?

The name "skyscraper" is an informal term. The first skyscraper had load-bearing outer walls made of stone and concrete. Today a skyscraper is supported by an internal iron or steel skeleton. Skyscrapers are economical in crowded cities because they take advantage of the more abundant vertical space that is available.

Another effect of winds on tall buildings is to make them sway back and forth. While a variety of braces can reduce the sway, they add weight to the building. Another method is now used. The Citigroup building has a 400-ton concrete damper at the top. The damper moves back and forth, opposing the wind-driven motion of the building and reducing sway. Dampers, both liquid and solid, are used in tall buildings, towers, off-shore oil drilling platforms, bridges, and skywalks. The 210 meter (690 foot) Burj al-Arab hotel in Dubai has 11 mass dampers. The dampers can also mitigate the effect of earthquakes.

Transporting large numbers of people into and out of upper floors is a challenge to those who design the elevator systems. As was demonstrated in the collapse of the World Trade Center buildings on September 11, 2001, stairways can be used in emergency situations, but the simultaneous movement of occupants down and firefighters up the stairways caused severe problems.

Another consideration is the safety of occupants in case of fire. Some buildings have entire floors designed to be especially fire-resistant so that people could gather there and be safer than on other floors.

What is the **tallest building** in the world?

The tallest building, until recently, was the Sears Tower (now the Willis Tower) in Chicago, Illinois, which was built in 1974 and is 443 meters (1,453 feet) high. Three new skyscrapers in Asia now surpass this height. The Petronas Towers in Malaysia, completed in 1996, are 452 meters (1,483 feet) high. The Shanghai World Finance Centre in China, completed in 2008, stands 492 meters (1,614 feet) high—almost half a kilometer into the sky. The Burj Dubai (Dubai Tower), 818 meters (2,684 feet) high, the world's highest, opened in January 2010.

What is the **tallest structure** ever built?

Structures and towers are listed separately from skyscrapers. The Warszawa Radio Tower on the outskirts of Warsaw, Poland, was the tallest structure ever built. Although it needed to be supported by long cables to keep it up, the tower reached 646 meters (2,119 feet) into the sky. Unfortunately, in August 1991, the tower came crashing down during repair work.

The tallest structure ever built that is still in existence is the KTHI-TV tower in North Dakota, which stands—with the help of cables—629 meters (2,063 feet) tall. The tallest self-supporting tower in the world is the Canadian CN Tower, whose tip is 553 meters (1,815 feet) above the city of Toronto.

What is the difference between a **dead load** and a **live load**?

In order to remain static, bridges (and all structures, for that matter) must be able to withstand loads placed on them. A load is simply the engineering term for force. Dead load is the weight of the bridge or structure itself. The live load, on the other hand, is the weight and forces applied to the bridge as a result of the vehicles and people that move across the bridge at any one time. Of course, in order to be safe, engineers account for much higher live loads than would normally occur.

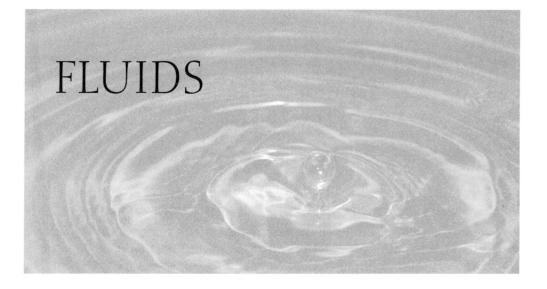

FLUIDS

What is a **fluid**?

Solids retain their shape because strong forces hold the atoms in their places. In a liquid the forces keep the atoms close together, but they are free to move. In a gas the atoms are about ten times further apart than in either solids or liquids and forces between them are very weak. As a result of weaker forces, a liquid or gas can flow freely and assume the shape of their container. They are called fluids.

There are two main areas of fluid study. The field that studies fluids in a state of rest is called static fluids, and the field that analyzes the movement of fluids is called fluid dynamics. We will first explore static fluids.

WATER PRESSURE

What is **water pressure**?

Pressure is the force divided by the area over which it is exerted. In 1647 the French scientist Blaise Pascal (1623–1662) recognized that water exerts the same pressure in all directions. This statement is known as Pascal's Principle.

To understand Pascal's Principle, think of a small cube of water as shown on page 96. The grey arrow in the middle is the force of gravity on the cube. As a result the total downward force of the cube is larger than the upward force.

Now, by Newton's Third Law, the outward force of the water is equal to the inward force on the water. What exerts this force? If the cube is at the edge of the container, then the container exerts the force. You can check that for yourself. Use an empty fluid container—say a juice box or milk carton. Punch a hole in one side. Pour water

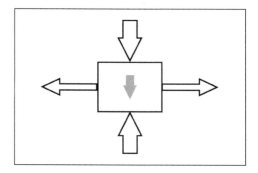

into the container. To keep the water from leaving through the hole you have to exert a force with your thumb. The same would be true if the hole were in the bottom. What exerts the downward force on the top of the cube? If the cube is not at the top surface, then the force is exerted by water above it. If it is on the top, then air pressure, which we will discuss shortly, exerts this force.

Because the downward force exerted by the fluid is greater than the upward force, the force, and hence the pressure, increases as the depth of the fluid increases. So the pressure at the bottom of a container of fluid is greater than at the top. The amount of increase in pressure is given by the product of the density, ρ, the gravitation field strength, g, and the depth, h, or $\rho g h$. That is, the pressure on the bottom, $P_{bottom} = P_{top} + \rho g h$.

What does it mean to say that **water seeks its own level**?

The surface of water placed in a single container (a glass or a bathtub or a lake) will remain at the same level relative to Earth on both sides of the container. Adding water to one side will only make the entire level uniformly rise; there can never be one section of the glass or tub or lake that is at a higher elevation than another section. To understand this fact, consider adding the small cube of water on top of the surface at one location. It would exert a downward force on the water under it. But, because water can flow, water under it would flow outward, raising the level elsewhere in the container until the pressure is equal everywhere.

Water also seeks its own level in other containers. If you fill a hose or tube with water and hold the tube in a "U" shape, the water level will be at the same locations in the two ends. You can use the "U" tube to make a device to show you equal heights at two different locations. You may have a coffeemaker that has a water height indicator on the side. This is a small tube that connects to the water reservoir at the bottom. The water level in the narrow tube and the wide reservoir is the same.

What are the **units** used to **measure pressure**?

Pressure is force divided by area, so in the metric system pressure is measured in newtons per square meter, called a pascal (Pa). One pascal is a very small pressure, so usually the kPa, or 1,000 Pa, is used. In the English system, pounds per square inch (psi) is often used. Another unit used is a measurement of the height of mercury in a glass tube that would create a pressure that balances the pressure of the fluid. That unit that used to be called millimeters of mercury is now called the torr. Here is how these units compare: 760 torr = 14.7 psi = 101.3 kPa.

Why are **water towers** needed on **tall buildings**?

A typical home requires water pressures of 50 to 100 psi. City water systems use pumps to maintain that pressure in the pipes. Vertical pipes are needed to supply the upper floors with water. Each foot of height reduces the pressure by 0.443 psi. Auxiliary pumps at various floors can provide the needed increase in pressure. An alternative is to put a large storage tank on the roof and use pumps to fill it. It then supplies the building with water under pressure due to the height of the tank. It also allows the pumps to be run to fill it at night when electricity rates may be cheaper. In addition, it provides a backup source of water in case of fire.

Water towers such as these in New York City are often placed on top of tall buildings. This way, the force of gravity supplies the water pressure needed to deliver water throughout the building.

Small towns, which often use wells as a water source, use water towers to store water in case there is an interruption in electrical service. It also allows the town to use smaller pumps because the tower can supply the pressure during peak water demands. A typical daily water use is 500 gallons per minute, but this can rise to 2,000 gallons a minute in peak times. A tower typically stores one day's worth of water.

Why are many **water towers placed** on **high towers**?

The height of the water determines the pressure. Since a holding tank is up high, it places a lot of pressure on the water in the rest of the water network. The tank should be as large across as possible so that for the same amount of water the vertical dimension of the tank can be as small as practicable. This design limits the variation in pressure as the tank empties or is filled.

Why do your **ears hurt** when you dive to the **bottom** of a **swimming pool**?

Just as the weight of the air above us creates atmospheric pressure, the weight of water creates liquid pressure. Close to the surface of a pool there is very little water that can push down and increase water pressure. The further a person dives below the surface, however, the greater the water pressure. The eardrums are especially sensitive to the increased pressure, for they do not have the reinforcement that the diver's skin has. In fact, your eardrums can usually feel pressure when diving just 1.5 to 3 meters (5 to 10 feet) below the surface of the water.

97

Why are **dams thicker** at the **bottom** than at the top?

Dams hold back bodies of water, and water pressure increases with the depth of the body of water, so the pressure from the water pushing horizontally on the dam is greater at the bottom than at the top. If holes were bored near the bottom, middle, and top of a dam, the longest horizontal stream of water would fire out through the bottom hole because the water pressure is greatest there.

BLOOD PRESSURE

What does it mean to **measure** your **blood pressure**?

Blood pressure is the pressure your blood exerts on the walls of your arteries. The fluid dynamics of blood play a major role in blood pressure. The heart is the pump that moves the blood throughout the body, with vessels carrying the blood to different sections of the body.

The device used to measure blood pressure is the sphygmomanometer. It is placed around the upper arm, inflated, and then deflated, while a meter measures the pressure passing through that section of the arm and either a person using a stethoscope or an electronic sensor detects the pulse.

Why is your **blood pressure** taken from your **upper arm**?

Liquid pressure is dependent on the depth of the fluid. Since blood pressure can't be measured around the heart, and the depth of the fluid must be the same as the heart, doctors and nurses need to find a location at the same depth as the heart. A convenient location at that level is your upper arm. When lying down, however, your blood pressure can be taken just about anywhere, since most of the blood is at the same vertical level as the heart.

The cuff is inflated until no pulse can be heard. It is then slowly lowered. As the pressure falls below the systolic pressure the pulse can be heard. When it's below the diastolic pressure the pulse gets weaker. The report "120 over 70" means that the systolic pressure is 120 torr, the diastolic pressure 70 torr.

ATMOSPHERIC PRESSURE

How is the **pressure** of a **gas similar** to **liquid pressure**?

The pressure from a gas, acts the same way liquid pressure does. One difference between gaseous and liquid pressure is that gases are about 1/1000 as dense as liquids and therefore apply less pressure. The second difference is that gases can be easily compressed while the compressibility of liquid is very small.

What is the **atmospheric pressure**?

Earth's atmosphere extends approximately 100 kilometers (328,000 feet) above the ground, but it gets less and less dense as the altitude increases. If the density were constant, then the atmosphere would be about 8-kilometers high (26,246 feet; close to the height of Mt. Everest). Some 63% of the atmosphere is below that height. The amount of force on an area of one square meter is about 101,300 newtons. That is, the pressure is 101.3 kPa. Atmospheric pressure varies with temperature and other conditions. Our weather is mostly influenced by high and low pressure regions, which can deviate by about 5% from normal.

The pressure decreases as your altitude increases because the amount of air above you creating the pressure is smaller. At an altitude of 3 kilometers (10,000 feet) above sea level the atmospheric pressure is about 70% of what it is at sea level. At 100 kilometers it is 3 millionths of the sea level pressure!

What are the **bends**?

Nitrogen under normal atmospheric pressure is nearly insoluble in blood. Under pressure, the solubility increases. Thus, as a diver goes deeper the blood holds more and more nitrogen, which dissolves in the blood during gas exchange in the lungs. As the diver ascends, the pressure decreases, and hence the blood is now supersaturated with nitrogen. The supersaturated nitrogen forms bubbles as it comes out of solution in the blood, or cells. These bubbles collect in veins and arteries. They cause pain, and can rupture cell walls and block the flow of blood to cells, causing injury or even death.

A barometer is an instrument used for measuring gas pressures. They are also used to help predict weather, as low pressure systems tend to bring inclement weather while high pressure systems bring fair skies.

The best way to avoid the bends is to rise to the water surface slowly, allowing the liquid pressure from the water to gradually decrease and prevent any physical damage. Most scuba divers use "nitrox" that contains 35% oxygen and 65% nitrogen (rather than normal air that contains 20% oxygen and 80% nitrogen) to reduce, but not eliminate the possibility of getting the bends.

How is **atmospheric pressure measured**?

A device to measure gas pressure is called a barometer. There are two major types of barometers, the mercury barometer and the aneroid barometer. Galileo's secretary, Evangelista Torricelli (1606–1647; the unit of pressure, the torr, was named after him), developed the mercury barometer in 1643. It consists of a thin glass tube about 80 centimeters (31 inches) long, which is closed at the top, filled with liquid mercury and placed upside down in another mercury-filled dish. Depending upon the atmospheric pressure pushing on the mercury in the dish, the level of mercury in the tube will rise or fall because there is no air above it. By measuring the height of the mercury, which would usually be between 737 and 775 millimeters (29 to 30.5 inches) high, atmospheric pressure of the atmosphere can be measured.

The most common household barometer is the aneroid barometer, in which atmospheric pressure bends the elastic top of an extremely low-pressure drum; by measuring the amount the top bends, a measurement of atmospheric pressure can be determined. The aneroid barometer is often used in airplane altimeters to measure altitude. Since atmospheric pressure decreases as altitude increases, the aneroid barometer is an ideal instrument to use. It is much safer than the mercury barometer, because mercury vapor is poisonous and the mercury must be exposed to the atmosphere.

What happens to a **balloon** when it is **submerged in water**?

When an air-filled balloon is placed underwater, the water (which has higher pressure than the air) exerts a force on all sides of the balloon. The pressure from the water causes the air to compress inside the balloon. The deeper the balloon is taken into the water, the greater the pressure, and therefore, the smaller the balloon will become. The pressure from the water will compress the balloon until the air pressure within the balloon can supply an equal amount of force against the water.

If the air pressure is 101,300 newtons per square meter, why don't we get crushed?

Since our bodies have air inside them, the air inside our bodies is at the same atmospheric pressure as the air outside our bodies. Therefore, the pressures are equal and we can move quite freely in our atmospheric environment. The remainder of our body is mostly liquid water and cannot be compressed.

The same cannot be said for divers. As the divers go deeper and deeper, the water pressure increases. At a depth of 10 meters (32 feet) it is about twice atmospheric pressure. A diver can breathe in compressed gas to balance this pressure. If a diver goes to extreme depths and then ascends too rapidly the diver can suffer the "bends."

Why do **closed containers** sometimes **dent** or even **collapse** on **cool days**?

Just as a balloon can become smaller when placed under water, a sealed container can change its shape and even collapse under certain atmospheric conditions. For example, a container that stores gasoline for a lawnmower is usually sealed when it's not being used. If the container were sealed on a warm day, when there was little atmospheric pressure, subsequently on a cool, high-pressure day, the gasoline container could be crushed. A second reason is that gasoline vapor exists in the tank, and at low temperatures more of the vapor condenses to liquid, reducing the pressure of the gases in the tank.

Why do some **athletes** go to **high elevations** to **train**?

Runners have always trekked to the mountains of Colorado to train in higher elevations because of the lower atmospheric pressure that the mountains provide. Since the air is not as dense as it is at lower elevations, the lungs need to work harder to supply the body with a sufficient amount of oxygen. Many athletes feel that training in such conditions gets their bodies used to lower amounts of oxygen. Therefore, when running in a competition at lower elevations, they can compete quite well because their bodies are used to working hard to get a great deal of oxygen.

SINKING AND FLOATING: BUOYANCY

What **distinguishes** objects that **sink** from those that **float**?

An object will sink if the downward forces on it are larger than the upward forces. There are two downward forces: the force of the liquid above the object and its weight

101

(the force of gravity). The upward force is the force of the liquid below it. Let's think of a cubic object of height h and area of the top and bottom A. Its volume, V, is then given by $V = hA$. Density is the mass divided by the volume, or $\rho = m/V$.

Let's start by considering the difference in water pressure between the bottom and top of this cube. Pressure is force divided by area, so, using $P_{bottom} = P_{top} + \rho_{water}gh$, we can write $F_{bottom}/A = F_{top}/A + \rho_{water}gh$ so $F_{bottom} = F_{top} + \rho_{water}ghA$. Now we recall that hA is the volume of the cube and $\rho_{water}hA$ is therefore the mass of the water, m_{water}, whose place is taken by the cube of matter.

The net downward force on the object in the water is then the force on the top plus the object's weight less the force on the bottom. That means

$$F_{net} = F_{top} + m_{object}g - F_{bottom}$$

From the results of the paragraph above, $F_{net} = m_{object}g - m_{water}g$. Therefore, if the object's mass is larger than the mass of the water whose place it takes, it sinks. If the mass is smaller, then it will rise. Weight is the mass times the gravitational field strength, g, so the net force is $F_{net} = W_{object} - W_{water}$. The water whose place the object takes is normally referred to as the "water displaced" by the object.

What is the **buoyant force**?

From the equation obtained in the answer to the last question, you can see that when you place an object in water, its weight is reduced by the weight of the water displaced. The reduction in weight, $W_{object} - W_{water}$, that is the net upward force of the water, is called the buoyant force.

Why does a **stone sink** but **wood float**?

A stone is more dense than water. That is, if you compare the mass of a stone to the mass of the same volume of water, the stone's mass will be greater. Therefore its weight will also be greater, so $W_{stone} - W_{water}$ will be positive, and there will be a downward net force. Therefore the stone will move downward through the water until it rests on the bottom.

Wood, on the other hand, is less dense than water. If the wood is pushed under water $W_{wood} - W_{water}$ will be negative. There will be an upward force on the wood and it will rise. How far will it rise? As some of the wood rises above the level of the water the volume and the mass, and therefore the weight, of the water displaced will be reduced. When the weight of the wood and the weight of the water displaced are equal, the net force on the wood will be zero and it will no longer move.

Why is a **stone easier** to **lift** when it is **in water**?

Even though the stone sinks because it is denser than water, there is still a buoyant force on it given by $W_{stone} - W_{water}$, so the stone's weight is reduced.

Hot air balloons work by expanding the gas within the balloon so that the air within is at a lower pressure than the surrounding atmosphere.

Floating usually refers to a liquid such as water, but **can anything float** in a **gas**?

Remember that there is a difference in pressure of any fluid between the bottom and top of an object in it given by $P_{bottom} = P_{top} + \rho g h$. Even though the density of a gas, ρ, is much smaller than that of a liquid, there is still a pressure difference, and therefore a buoyancy. There is another difference between a liquid and a gas. A gas can be compressed, and so the density of the atmosphere decreases as you rise. When a hot air balloon is launched the air within it is heated and it will rise. As it rises the density of the air decreases, and so the buoyant force decreases. At some height the weight of the balloon and basket equals the weight of the air displaced and the balloon stops rising. The operator of the balloon can go higher by making the gas in the balloon hotter, and thus less dense.

What **major discovery** did **Archimedes make** in the third century B.C.E. and how did he apply it?

Archimedes (c. 287–c. 212 B.C.E.) lived in the city of Syracuse on the island of Sicily, then part of Greece. He was charged by the king of Sicily to find out if his crown was made out of pure gold, or an alloy of gold and silver. Archimedes had to do this without destroying the crown. One day, when bathing in the public baths he noticed that when he stepped into the bath the water rose! He had seen this many times before, but this time he recognized that this common occurrence could help him solve his prob-

A huge steel ship like an aircraft carrier can float instead of sink like a stone because it can displace enough water to compensate for its weight and because the air inside of the ship makes its density actually less than the water around it.

lem. Legend has it that he was so excited that he ran naked through the streets of Syracuse shouting "Eureka!" or "I have found the answer!"

He then did experiments where he hung the crown on one end of a balance and a piece of first gold and then silver from on the other end. When the weight of the crown and that of the metal were equal the balance was horizontal. He then immersed the balance in water. If the crown and the metal had different volumes, the water they displaced would be different and the balance would tip. He found that the balance tipped when both the silver and gold pieces were on the balance. Archimedes had found that the crown was not pure gold, but a mixture of silver and gold. The king had been cheated!

Archimedes had discovered a principle of hydrostatics (liquids at rest) that would one day carry his name: Archimedes' Principle states that an object immersed in a fluid will experience a buoyant force equal to the weight of the displaced fluid.

Why does a **small clump** of **steel sink**, while a **50,000-ton steel ship** can **float**?

In order to remain afloat, a ship needs to displace an amount of fluid equal to its own weight. Therefore, if a clump of steel is placed in water, it will sink because it is much denser than water and its volume isn't large enough to displace a weight of water equal to its own weight. A 50,000-ton steel ship can easily stay afloat because it can

---

How do hippopotamuses sink to the bottom of a riverbed?

Hippopotamuses spend over half their day in the water. In order to eat, the hippo, which can reach almost 3 meters (10 feet) in length and weigh 10,000 pounds, must sink to feed off the vegetation that grows on the bottom of rivers. The hippo, however, has one major problem: his low density forces him to float at the surface, and he is not agile enough to quickly dive down to the bottom and come back up again. In order to reach the bottom, he needs to increase his density, so the buoyant force cannot supply a large enough force to keep the animal afloat. To do this, the hippo exhales, reducing the air in his body to increase his density.

---

displace 50,000 tons of water. That is because its hull isn't all steel, but contains a large amount of air. Therefore its density, its mass divided by its volume, is less than that of water.

What happens to the **buoyancy** of the **ship** when **cargo and passengers** are **added**?

When cargo and passengers are added to a ship its weight increases. As the ship's weight increases is sinks further into the water, displacing a greater weight of water. If it sinks so far that water can spill into the ship, increasing its weight even more, the ship sinks to the bottom of the water.

The amount the ship sinks in the water as a result of cargo and passengers can be critical for navigation and maneuverability. Large cargo and cruise ships have numbers on the bow of the ship that indicate how far the distance is between the water line and the bottom of the ship. This distance is called the ship's draft. If the ship has a 6-kilometer (20-foot) draft and the water is only 5.5 kilometer (18 feet) deep, cargo and passengers must be unloaded to allow the ship to rise.

How much **water** does a **ship need** in order **to float**?

Not a lot of water is needed to keep a ship afloat, just enough so that it can displace enough water to equal its weight. Therefore, if a ship were to enter a canal that was just a bit larger than the size of the ship's hull, it would float as long as there was a small film of water around the entire hull of the ship.

How do **blimps** remain at a chosen **altitude**?

A blimp is a non-rigid airship that floats in the air solely due to the buoyant gas within the giant balloon-like bag. It typically carries over 5,000 cubic meters of helium at a

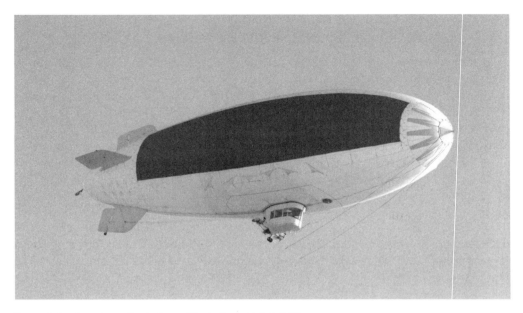

Because helium is a safer gas than hydrogen, it is what is used in today's blimps.

density about seven times less than air. An airship floats in the air in the same manner as a ship floating in water. The weight of the airship must equal the buoyant force of the gas inside the bag. In order to increase the altitude of the blimp, the pilot increases its buoyancy by adding gas from pressurized tanks to the blimp, which expands the flotation bladders, displacing the heavier air and increasing the buoyant force until the reduced weight of the blimp equals the reduced weight of the air. To lower the airship, the buoyancy is decreased by releasing gas from the flotation bladders, which then decrease in size, displacing less of the heavier air.

Why is **helium** used inside **airships** instead of **hydrogen**? Isn't hydrogen more buoyant?

Although hydrogen is twice as buoyant as helium, and would be more effective in lifting an airship off the ground, hydrogen gas is extremely dangerous. The German airship, or zeppelin, the *Hindenburg*, the world's largest airship at the time, was destroyed on May 6, 1937, in Lakehurst, New Jersey, when it exploded into a huge fireball while attempting to land. Thirty-six people died in the explosion.

In 1937, the United States was about the only source of helium in the world, mostly from one gas well in Texas. The Nazis wanted to buy helium for their zeppelins, but the United States refused to sell it to them—as it was considered a strategic resource. Helium is formed from the radioactive decay of uranium and thorium in rocks. It is used today to cool devices and make them superconductive. As a result, the amount of helium available is rapidly decreasing.

What are **airships used** for?

Since the first airship flight in 1852, by Henry Giffard in France, the dirigible, or airship, was used predominantly for military purposes. In the first and second World Wars airships were used for bombing and surveillance on both sides of the Atlantic. Commercial passenger transportation on airships was conducted for only a few years, while today's blimps are used for advertising and for televising sporting events from high elevations.

What happens to the **helium balloon** that a **child releases**?

If the balloon is tied tightly, the balloon will expand as its altitude increases. This expansion is caused by the lower atmospheric pressure at higher altitudes. Eventually, the helium's volume increases so much that the rubber balloon breaks, releasing the helium.

FLUID DYNAMICS: HYDRAULICS AND PNEUMATICS

What is **fluid dynamics**?

Fluid dynamics is the study of fluids in motion.

What is **hydraulics**?

Hydraulics deals with the use of liquid in motion, usually in a device such as a machine. Oil is the most common liquid used in pumps, lifts, and shock absorbers.

How does a **hydraulic lift** work?

Pascal's Principle states that pressure in a liquid is independent of direction. Pressure is force divided by area, so force exerted by a liquid is equal to the pressure times the area. A hydraulic lift has a small piston in a cylinder. If you exert a force on the piston, it will create a pressure in the fluid. The lift also has a second, large area cylinder and piston. The fluid creates a pressure in this cylinder that exerts a much larger force on the piston because of its larger area. Therefore the lift has a mechanical advantage. Of course, energy is still conserved, so the small piston must move much farther than the large piston.

An automobile lift, used in many automotive repair shops, allows the operator to use very little force to lift an automobile off the ground, by pushing liquid from a

This heavy crane boom uses hydraulics to lift the boom.

small-diameter cylinder and piston through a thin tube that expands into a larger-diameter cylinder and piston, which is located beneath the vehicle to be lifted. Since the liquid cannot be compressed like air, the liquid from the small cylinder is pushed into the large cylinder, forcing the large piston to move upward.

What are some other places where hydraulic lifts are used?

Besides their valuable use in auto-repair shops, hydraulic lifts are used in elevating crane and backhoe arms, adjusting flaps on airplanes, and applying brakes in automobiles. It is the non-compressible characteristics of liquids that make hydraulic devices so useful. Oil is used rather than water because it does not freeze.

What is pneumatics?

Whereas hydraulics uses liquids to achieve mechanical advantage, pneumatics uses compressed gas. Since gases can be compressed and stored under pressure, releasing compressed air can provide large forces and torque for machines such as pneumatic drills, hammers, wrenches, and jackhammers.

In what ways do fluids move?

When fluids move slowly they exhibit steady or laminar flow. When the speed increases the flow becomes turbulent. In some cases laminar flow is desired, in others turbulent flow is better.

What is the difference between laminar and turbulent flow?

The film of fluid that touches the container does not move because of friction with the container's surface. But the fluid in the middle of the stream does. In laminar flow the transition from not moving to moving at full speed is continuous. Each thin film of water moves slightly faster than the one closer to the surface. In turbulent flow this transition from not moving to full-speed motion occurs suddenly, and the water moves in tiny circles in this region. Laminar flow has more friction than turbulent. A baseball, for example, has more drag in the laminar flow region.

Why does a river's current run faster when the river is narrower?

When water flows down a river, the current is measured by the volume of water that passes by a cross-section of the river divided by the time taken for the water to pass. For example, if the current of a river is 2,000 m³/min (cubic meters per minute), this means that in every minute 2,000 cubic meters pass by every part of the river. We can write this as $(2,000 \text{ m}^2) \times (1 \text{ m/min})$. If the river narrows, the 2,000 cubic meters of water still must pass in one minute because the water from behind continues to flow downriver. Since the river is narrower, the area (measured in meters squared, m^2) is smaller, so the speed (measured in meters per minute) must increase. This fact is called the principle of continuity.

What makes a **fluid flow**?

As in all of physics, objects move as a result of a net force. In a fluid the net force comes about because there is a difference in pressure between two points. The fluid flows in the direction of decreasing pressure.

AERODYNAMICS

What is **aerodynamics**?

Aerodynamics is an aspect of fluid dynamics dealing specifically with the movement of air and other gases. Engineers studying aerodynamics analyze the flow of gases over and through automobiles, airplanes, golf balls, and other objects that move through air. They also study the effect of moving air on buildings, bridges, and other static objects.

What is **Bernoulli's Principle**?

In 1738, a Swiss physicist and mathematician named Daniel Bernoulli (1700–1782) discovered that when the speed of a moving fluid increases, such as the wind blowing through the corridors of a city, the pressure of that fluid decreases. Bernoulli discovered this while measuring the pressure of water as it flowed through pipes of different diameters. He found that the speed of the water increased as the diameter of the tube decreased (the continuity principle), and that the pressure exerted by the water on the walls of the pipes was less as well. This discovery would prove to be one of the most important discoveries in fluid mechanics.

How does an **airplane wing** create **"lift"**?

The upward force on an airplane wing caused by the air moving past it is called lift. According to Newton's Third Law, if the air exerts an upward force on the wing, then the wing must exert a downward force on the air. There are three causes of this downward force. The first is the tilt, or angle of attack of the wing. The tilt deflects the airflow downward. The second is the Bernoulli Principle. Because of the shape of the wing, the speed of the air on the top surface of the wing is faster than at the lower surface, and so the pressure is lower. Thus the downward force on the wing is reduced. The third reason is the fact that air "sticks" to the surface of the wing. The air coming off the upper surface is moving downward. This exerts an additional downward force on the air. The downward forces produce the lift needed to keep the plane airborne.

What is **drag**?

Drag is a force that opposes the motion of an object through a fluid. An object is often said to be "aerodynamic" when its drag forces are kept to a minimum.

There are two types of drag on an airplane: parasitic and induced. Parasitic drag is the force when an airplane wing, automobile, or any other object moves through a fluid. The amount of drag depends on the density of the fluid, the square of the speed of the object, the cross-sectional area of the object, and its shape. A large fuselage, like that of a 747, has more drag than a small fighter airplane. A tear-drop shaped object has less drag than a rectangular block. A parachute is designed to have a high drag.

Induced drag is a consequence of the lift generated by the wing. It is a function of the angle of attack of the wing—the lower the angle of attack, the smaller the induced drag. It occurs at the outer edge of the wing where the downward motion of the air caused by the wing meets the undisturbed air next to it. Induced drag causes vortices, the spiral motion of air that can be extremely dangerous to planes flying behind or below. Induced drag can be reduced by putting small, tilted surfaces on the wing tips.

An engineer checks the fan in a wind tunnel. Smoke introduced into wind tunnels makes streamlines visible, which in turn help with the analysis of the efficiency of airfoils and automobiles.

What are **streamlines**?

Streamlines are lines that represent the flow of a fluid around an object or through another fluid. Streamlines are often made visible by putting thin films of smoke in the air. They are used in wind tunnel testing of airfoils (wings) and automobiles. A wind tunnel is a closed chamber with vents in the front and rear of the tunnel that allow wind and the streams of smoke that make streamlines visible to pass around an object. If the streamline smoke appears to be flowing in a gentle pattern without breaking up, then the object is considered aerodynamic. If the smoke breaks up upon encountering sections of an object, the flow has become turbulent. If the flow is turbulent the drag may be increased.

How does a **curve ball curve**?

The curve ball experiences the Magnus Force like the golf ball. The pitcher can give the ball a backward spin, creating lift, or a sideways spin, causing the ball to curve sideways. When a pitcher throws a knuckle ball the ball spins very slowly. The laces on the ball create small forces that make the ball move erratically through the air.

Why is it **better** for a **discus** to be **thrown into the wind** instead of with the wind?

In most sports, throwing or traveling with the wind at your back (called tail wind) is a lot easier than working against the wind (called a head wind). In football, teams flip a   111

Why do golf balls have dimples?

Golfing has been played for several centuries, but dimpled golf balls have existed for only a hundred years. The dimples in golf balls, first introduced by the Spalding Company in 1908, can double the distance a golf ball can fly. Without the dimples the flow of air is laminar and the ball drags a thin layer of air completely around the golf ball. The dimples break up this air layer, creating turbulence that reduces drag. Golf balls can also experience lift. When hit with a slight backspin, the air passing over the top section of the ball flows in the direction opposite the motion of the ball. This creates low pressure above the ball. On the bottom of the ball the ball's motion in the same direction as the air and the pressure is higher. According to Bernoulli's Principle, such a pressure difference provides a lifting force, called the Magnus Force on the ball, giving the ball a few more seconds in flight.

coin to determine who will be kicking with the wind. In sailing, it is easier and faster to travel perpendicular to or with the wind and takes more skill to travel against the wind. In track, the world record in the 100-meter dash could more easily be broken if running with a strong tail wind. In most sports, having the wind at your back can be a major advantage.

In the field event of discus throwing, however, the advantage comes when there is a head wind. In fact, it has been documented that a discus can travel up to 8 meters (26 feet) further while experiencing a head wind of only 10 m/s (meters per second). Although the discus still experiences a drag force from the head wind, the lift that the discus gets from pressure differences over and under the disc is substantially more significant than the drag force. Because the discus will remain in the air longer, it will travel farther.

What is the **most aerodynamic shape**?

Some think the narrower and more needle-like an object is, the lower its drag force will be. Although a needle-head cuts easily through the wind, the problem emerges at the tail end, where the wind becomes turbulent and forms small eddy currents that hinder the streamline flow of air. The optimum shape depends on the velocity of the object.

For speeds lower than the speed of sound, the most aerodynamically efficient shape is the teardrop. The teardrop has a rounded nose that tapers as it moves backward, forming a narrow, yet rounded tail, which gradually brings the air around the object back together instead of creating eddy currents.

At high velocities, such as a jet airplane or a bullet may travel, other shapes are better. For turbulent flow, the least drag comes from having a blunt end, which inten-

> ## What happened at Kitty Hawk, North Carolina, on December 17, 1903?
>
> It was on this date that brothers Orville (1871–1948) and Wilbur (1867–1912) Wright warmed up the engines on their Wright 1903 Flyer and took off into the blustery winter air. On its first flight, Orville flew the Flyer for a total of 12 seconds and traveled a distance of 37 meters (120 feet). Later that cold winter day, Wilbur flew for nearly a minute and traveled 260 meters (852 feet). The Wright 1903 Flyer, which weighed only 600 pounds and had a wingspan of 12 meters (40 feet), made only four runs that day. After Wilbur's 260-meter flight, the wind tossed the plane end over end, breaking the wings, engine, and chain guides.

tionally causes turbulence. The rest of the air then flows smoothly over the region of turbulence behind the object.

How are **airplane controls** different from the controls in an automobile?

Automobiles travel on two-dimensional surfaces, and therefore only need two separate controls, the accelerator and brake to control the forward movement, and the steering wheel to control side-to-side movements. Airplanes, on the other hand, travel in three-dimensional space. The forward thrust on an airplane is controlled by the throttle, and the "braking" is achieved by closing the throttle and increasing the drag, usually by deploying the plane's flaps. Yaw, which is responsible for the side-to-side movement of a plane, is controlled by the plane's rudder.

To control the pitch, or up-and-down orientation of the nose, the pilot uses elevators or horizontal control surfaces near the plane's rudder. To roll the plane (the rotation of the plane about an axis that goes from the nose to the tail), the pilot uses control surfaces on the back end of the wings called ailerons. The Wright Brothers recognized that roll control was crucial to successful flight. They invented a method of warping the wings, creating a primitive but useful aileron.

THE SOUND BARRIER

What is a **shock wave**?

Just as a boat moving through the water forms a series of V-shaped waves, airplanes create conical (cone-shaped) waves as they fly through the air. The waves that the airplane produces are waves of compressed air. When an aircraft reaches the speed of

113

sound, Mach 1, the plane's pressure waves that move at the speed of sound overlap each other, creating a shock wave. The shock wave creates one single, loud sonic boom heard by observers on the ground. When the plane travels slower than the speed of sound, the sound waves do not overlap and instead of hearing a sonic boom, observers simply hear the delayed sound of the plane.

You can think of sound waves as being similar to water waves, emanating from a central source and spreading out in a regular pattern unless they are interfered with.

If **Mach 1** is the speed of sound, what is **Mach 2**?

Mach is the ratio of a velocity to the speed of sound, so Mach 2 is two times the speed of sound, Mach 3.5 is three and a half times the speed of sound, etc. Any velocity greater than Mach 1 is referred to as "supersonic."

Who was the **first pilot** to **break the sound barrier**?

On October 14, 1947, Chuck Yeager (1923–) broke the sound barrier in his Bell X-1 test plane, "Glamorous Glennis." In order to reach the sound barrier, the X-1 was carried in the belly of a B-29 bomber to an altitude of 3,658 meters (12,000 feet) where it was dropped. The X-1's rocket engine ignited and Yeager took the plane to an altitude of 13,106 meters (43,000 feet). At this altitude, Yeager was able to break the sound barrier by traveling 660 miles per hour. The X-1 experienced a turbulent set of compression waves just before he broke past the barrier at Mach 1.05. Yeager kept the plane at this supersonic speed for a few moments before he cut off the rocket engine and headed back toward Earth.

Why did **Chuck Yeager** go to such a **high altitude** to break the **sound barrier**?

Sound travels approximately 760 miles per hour in the warm, dense air found close to sea level. The cooler and less-dense air is, however, the lower the speed of sound. Since air is less dense at higher elevations, physicists and engineers felt it would be easier to break the sound barrier at those elevations. Knowing the temperature and density of the air at 12,192 meters (40,000 feet) above sea level, scientists determined that the speed of sound would be reduced to only 660 miles per hour. As an added bonus engineers found that not only was the speed of sound slower at such elevations, but when the air has such low density, the parasitic drag (the drag due to friction), is very low as well. Therefore, to break the sound barrier, Yeager traveled as far up as 13,106 meters (43,000 feet) above sea level to both reduce the sound barrier and decrease the parasitic drag.

What **concerns** did pilots and engineers have about **breaking the sound barrier**?

To reach the sound barrier in an airplane was a major goal for many in the aeronautical field, a goal that carried some uncertainties. Pilots and engineers alike wondered and feared what would happen to a plane's maneuverability when it broke through the shock wave as well as what would happen to the plane itself, structurally.

Near the end of the Second World War there were fighter planes that were very strong and had powerful engines and experienced pilots. A number of pilots died when their planes broke apart in mid-air, often when in dives. There were two problems with these aircraft: first, the wings were not swept back, and second, they were driven by propellers. As the shock wave forms near Mach 1, it bends backward from the nose of the plane, like a bow wave on a boat. If the shock wave encounters the wings (that is, the wing extends through the shock front), there are tremendous forces on the wings. In a supersonic plane the wing is always designed to be fully behind the shock front, because the shock front can tear the wing off the plane. The propeller causes a pulsation in the pressure on the wing: every time one of the blades goes by, it produces a region of slightly higher pressure behind it, followed by a region of low pressure. All of these things came together and helped cause mid-air structural failures of the WWII fighter planes.

SUPERSONIC FLIGHT

Why are the **angles of wings** important for **supersonic flight**?

When a plane breaks the sound barrier, the shock wave in front of the plane has a difficult time moving out of the plane's way. In order to break the sound barrier with less difficulty, aeronautical engineers have designed more aerodynamic fuselages and efficient wing designs. As mentioned above, wings on supersonic planes must remain behind the shock front to prevent structural failure and allow the plane to maneuver safely. The swept-back wing design, as found on both military and commercial airplanes, allows the airplane to accelerate easily and faster before major pressure builds up around the wings. The delta wings, as found on many jet fighters, are large and extremely thin, to keep the wings behind the shock front while increasing lift and reducing drag.

Problems can also occur when using swept-back wings. As a plane moves faster, the center of lift on the wings can move too far backward, causing unbalanced forces on the plane, which can affect the maneuverability and safety of the plane.

Why are there no commercial supersonic planes?

From 1973 to 2000, the Concorde was a symbol of fast and expensive air travel for business people. The plane was a fast but inefficient plane that could carry 78 passengers. The sleek, delta wing design and pinpoint nose, which tilts down during liftoff and landings, could achieve speeds of up to Mach 2.2 at 15,240 meters (50,000 feet) above sea level. But an accident involving a landing wheel, which killed all 109 passengers and crew, grounded the Concorde. Over the next sixteen months the plane was extensively renewed and tested. Unfortunately, the terrorist attacks on September 11, 2001, reduced the demand for fast flights. Both British Airways and Air France suspended commercial flights. The remaining twelve aircraft are in museums around the world.

What is the **fastest aircraft**?

Just like Chuck Yeager's Bell X-1, which was the first plane to break the sound barrier, the X-15A-2, the fastest aircraft ever built, was dropped from the belly of a B-52 bomber. When released, the X-15A-2's rockets ignited, taking it to a maximum speed of 4,534 miles per hour. That speed, which is equivalent to 7,297 kilometers per hour, is Mach 6. The X-15 series flew 199 flights before being retired in 1969. The SR-71 (blackbird) was the fastest known aircraft that can take off under its own power. It was retired in 1998.

THERMAL PHYSICS

What is **thermal physics**?

Thermal physics is the study of objects warm and cold, and how they interact with each other. It is difficult because most of the vocabulary dates from the time before scientists understood what makes an object hot. Terms like "heat," "heat capacity," and "latent heat" suggest that warm objects contain some material that depends on their temperature. As was discussed in the chapter on Energy, it wasn't until the early 1800s that our present understanding was developed. Yet, almost 200 years later, our common usage is based on earlier ideas.

THERMAL ENERGY

Who discovered what **makes** an object **hot**?

Benjamin Thompson, Count Rumford (1753–1814), who was born in the Massachusetts Bay Colony, but did most of his scientific work in the Kingdom of Bavaria, now part of Germany, deserves most of the credit. Before his experiments, most scientists thought that hot objects contained an invisible fluid called caloric. Experiments done before Rumford showed that when you heated an object it didn't gain weight, so caloric must be weightless as well as invisible. This result made many scientists suspicious of the caloric explanation, or theory.

In 1789 Rumford drilled holes in bronze cannons through which a cannon ball would be shot. He found that both the cannon and the metal chips that resulted from the drilling became hot. He determined the amount of water that could be raised to the boiling point by both the cannon bodies and the chips and showed that the caloric theory did not agree with his results. He finally concluded that in hot objects the particles that made up the material moved faster than they moved in cold objects. Using our present terminology, they had more kinetic energy. In their motion they vibrate back and forth; they do not move together like a thrown ball.

What is **thermal energy**?

Thermal energy is the random kinetic energy of the moving atoms and molecules that make up matter. But objects expand when heated, so the bonds holding the atoms together stretch. That means they have more elastic energy. So thermal energy is the sum of the kinetic and elastic energy of the atoms and molecules, and the bonds that hold them together. It is energy that is inside the object, and so it is called a form of internal energy.

TEMPERATURE AND ITS MEASUREMENT

What is **temperature**?

Temperature is a quantitative measure of hotness. In many substances, temperature is proportional to the thermal energy in the object, but the relationship between temperature and energy depends on many factors.

How is **temperature measured**?

Temperature is measured by a device called a thermometer. There are many different kinds of thermometers, but they all have a property that depends on temperature.

The most common type of thermometer contains a thin column of liquid in a glass tube. When the temperature goes up, the liquid expands, and the height of the column increases. Earlier thermometers used mercury, a metal that is a liquid at room temperature. Because mercury is poisonous, all thermometers sold today contain red-colored alcohol. The glass tube has markings on it from which the temperature can be read.

Many thermometers today are electronic. Most contain a tiny bead of a semiconducting material whose resistance varies with temperature. The bead is called a thermistor, or thermal resistor. Others contain a tiny semiconducting diode. The voltage across this diode varies with temperature. To measure very high temperatures, a wire made of the metal platinum is used because its resistance also varies with temperature and platinum is not affected by high temperatures. Another electronic thermometer uses two wires made of different materials that are welded together. Typically the wires are made of copper and a nickel-containing alloy. This kind of thermometer is called a thermocouple and is often used in gas furnaces or water heaters to make sure that the pilot flame is burning when the gas to the main burners is turned on. Voltage and resistance will be discussed later in this book.

A less accurate, but convenient thermometer is a strip of plastic whose color changes with temperature. The plastic contains liquid crystals. The geometric arrangement of molecules in the crystals depends on temperature, and so does their color.

If two different metals are bonded together in a strip when the temperature changes the strip will bend. Because there are two metals, often brass and steel, the device is called a bimetallic strip. This kind of thermometer is often used in household thermostats.

Who **invented** the **first thermometer**?

Although Galileo (1564–1642) is credited with developing the first thermometer in 1592, his thermometer was open to the atmosphere, so it measured a combination of temperature and atmospheric pressure. It was not until 1713 that Daniel Gabriel Fahrenheit (1686–1736) developed the first closed-tube mercury thermometer. Combined with the temperature scale he defined the following year, Fahrenheit made a significant contribution to science.

This thermometer is similar to one developed by Galileo. The objects inside the tube have varying densities that rise or fall, depending on the temperature.

What is the **Fahrenheit temperature scale**?

Temperature scales are artificial in the sense that they are related to temperatures important to humans. The German physicist Daniel Gabriel Fahrenheit developed the first well-known temperature scale in 1714. Equipped with the first mercury thermometer, Fahrenheit defined a scale in which the freezing point of water was 32°F and the boiling point was 212°F.

Why did Gabriel **Fahrenheit define** the **freezing point** of water to be **32°F** instead of 0?

Fahrenheit did not define 32° as the freezing point of water. Instead, he defined 0° as the freezing point of a water and salt mixture. Since salt lowers the freezing point of   119 water, the freezing point for this mixture was lower than it would have been for plain water. Upon defining the degree intervals between the freezing and boiling points of the water and salt mixture, and he found that water itself freezes at 32°F.

Who developed the Celsius scale?

On the Celsius scale the freezing point of water is 0° and the boiling point is 100°. The Celsius scale is named after a person whose life work was dedicated to astronomy. Anders Celsius (1701–1744), a Swedish astronomer, spent most of his life studying the heavens. Before developing the Celsius temperature scale in 1742, he published a book in 1733 documenting the details of hundreds of observations he had made of the aurora borealis, or northern lights. Celsius died in 1744 at the age of forty-three.

What do Celsius temperatures feel like?

If you live in the United States you learn what various temperatures feel like. You recognize that 86°F is typical of a hot summer day and that –4°F is a very cold winter day. But what does 10°C feel like? Here is a handy guide:

Temperature	Celsius	Fahrenheit
Extremely hot day	40	104
Hot summer day	30	86
Room temperature	20	68
Jacket weather	10	50
Water freezes	0	32
Cold winter day	–10	14
Very cold winter day	–20	–4

What is the **Kelvin scale**?

The Kelvin temperature scale, developed by William Thompson, Lord Kelvin (1824–1907), in 1848, is widely used by scientists throughout the world. Absolute zero is the temperature at which thermal energy is at a minimum. Each division in the Kelvin scale, called a kelvin (K) is equal to a degree on the Celsius scale, but the difference is where zero is. In the Celsius scale, 0° is the freezing point of water while in the Kelvin scale, the zero point is at absolute zero. Therefore, 0°K is equal to –273.15°C; 0°C is equal to 273.15 kelvins. The Kelvin scale is used for very low or very high temperatures when water is not involved.

How do the three **temperature scales compare**?

The following table provides some examples for the three commonly used temperature scales:

Temperature	Celsius	Kelvin	Fahrenheit
Absolute zero	–273.15	0	–459
Water freezes	0	273.15	32
Normal human body temperature	37	310.15	98.6
Water boils	100	373.15	212

How do you **convert** from **one scale** to **another**?

Use the conversion chart below to convert between Celsius, Fahrenheit, and Kelvin.

From	To	Formula
Fahrenheit	Celsius	°F = 9/5°C + 32
Celsius	Fahrenheit	°C = 5/9°F − 32
Kelvin	Celsius	K = °C − 273.15
Celsius	Kelvin	°C = K + 273.15

Can a **temperature** be **measured without contacting** the object?

As was described in the chapter on energy, when an object is hot it transfers energy to colder objects. If it is in contact with the other object the transfer is by means of conduction. But all objects at temperatures above absolute zero radiate electromagnetic waves in the infrared part of the spectrum (see the chapter on waves for a description of the spectrum). A hotter object radiates more energy this way, and so there will be a net transfer of energy from the hot object to cold objects around it. An electronic sensor can detect the infrared radiation and convert the amount and wavelength of radiation it receives to a temperature. The sensors can be built into cameras that create a picture that shows the temperature of every location in the picture. Such a picture is known as a thermograph.

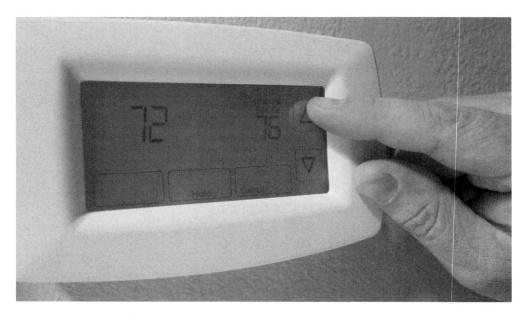

An electronic thermometer within the thermostat in your home or office triggers a switch that turns your furnace on or off, according to the temperature.

How can **temperature** be **controlled**?

A device that maintains a constant temperature is called a thermostat. A traditional home thermostat contains a coiled bimetallic strip. When the temperature drops below a set point, the strip trips a switch that turns on the furnace. More modern home thermostats use an electronic thermometer and electronic circuits that turn the furnace and air conditioning system on and off.

How are **thermographs** used?

Thermographs, which detect the amount of infrared radiation emanating from objects or regions, use colors to display the temperature on an image. Typically, red indicates the warmest temperatures, while blue indicates cooler temperatures. Thermographs are used throughout science, but are well noted for their use in detecting humans in wilderness areas, identifying areas of homes that need more insulation, and in measuring the temperature over regions of Earth.

How do **astronomers determine** the **temperature** of the **sun**?

When iron is hot, you can feel the energy radiating from it. That radiation is in the form of infrared waves leaving the iron. When iron gets extremely hot, it produces a red glow—and when it gets even warmer, it can take on a whitish glow. The temperature of iron and other objects can be measured by the amount of radiation flowing from it as well as by the light it emits.

Scientists measure the temperature of stars and the sun by analyzing the color and brightness of the stars. From such measurements, astronomers have determined that the surface of the sun is approximately 5,500°C (9,900°F).

ABSOLUTE ZERO

What is the **lowest possible temperature**?

The lowest possible temperature is called absolute zero (0 K). It is the temperature at which molecular motion is at a minimum and cannot be further reduced. While absolute zero can never be reached (see the Third Law of Thermodynamics later in this chapter), the present record low temperature is 4.5 nK (4.5 billionths of a kelvin).

Is there a **highest temperature** that can be achieved?

Although there is an absolute zero temperature, there is no highest temperature. The highest temperatures achieved to date have been from nuclear explosions, where the temperature can reach as high as one hundred million kelvins.

What are the average **surface temperatures** of the **planets** in our solar system?

For the planets that have atmospheres (mixtures of gases surrounding the surface of a planet), the average temperature stays relatively constant because the atmosphere acts as a type of insulator. These planets have only small variations in the temperature when a section of the planet faces away from the sun. Mercury, on the other hand, with no atmosphere and an elliptical orbit has very large differences. In the table below the temperatures of Mercury, Venus, Earth, and Mars are taken on the planetary surface, while those of Jupiter, Saturn, Uranus, and Neptune are taken at the tops of the clouds, there being no solid surface on these planets.

Planet	Temperature Range (°C)
Mercury	−184 to 420
Venus	427
Earth	−55 to 55
Mars	−152 to 20
Jupiter	−163 to −123
Saturn	−178
Uranus	−215
Neptune	−217

This plasma lamp—an apparatus you often see at science fairs and novelty shops—emits streams of plasma, electrically charged particles that are found in everything from stars to television displays and fluorescent lights.

STATES OF MATTER

What are the different **states of matter**?

The four phases of matter are the solid, liquid, gaseous, and plasma states. Solid phases are found at lower temperatures; as the amount of internal energy increases, the material changes from the solid to the liquid and then to the gaseous phase and finally, under extreme conditions, to the plasma state. Water, for example, changes from ice, its solid state, to liquid water, and finally to steam, its gaseous state. The temperatures at which the phase changes occur depend on the properties of the material.

In the solid phase the atoms or molecules are held in rigid positions by the chemical bonds between them. They can vibrate, but not change positions.

When the temperature is at the freezing point the solid melts into a liquid. In the liquid phase molecules, or small groups of molecules can move easily past one another. In most liquids the spacings between the molecules is slightly larger than in solids, giving them a lower density. In water, however, the spaces are larger in ice than in liquid, meaning that ice has a lower density than water, so it floats.

When the temperature reaches the boiling point the liquid becomes a gas. In the gaseous phase the atoms or molecules have essentially no forces between them, so

Do all substances go from a solid to a liquid to a gas as the temperature increases?

No. Carbon dioxide (CO_2) goes from the solid state called dry ice directly to its gaseous state. This process is called sublimation.

they are free to move independently. They are about 10 times further apart than in a liquid or solid, meaning that the density of a typical gas is 1/1,000 that of the solid.

To enter the plasma state one or more electrons must be removed from the atom. Plasmas consist of electrically charged particles. Plasmas are found in fluorescent lamps, in some television displays, in the upper atmosphere, in the sun and other stars, and in interstellar space.

What determines the amount of energy required to increase the temperature of a substance?

The amount of energy needed to increase the temperature of a substance with a mass of one kilogram by one degree Celsius is called the specific heat capacity. For example, to raise one kilogram of water (one liter) by a degree Celsius requires 4,186 joules of energy. The following table lists specific heat capacities of common solids, liquids, and gases:

Substance	Specific Heat Capacity (J/kg °C)
Aluminum	897
Copper	387
Iron	445
Lead	129
Gold	129
Silver	235
Mercury	140
Wood	1,700
Glass	837
Water	4,186
Ice	2,090
Steam	2,010
Nitrogen	1,040
Oxygen	912
Carbon Dioxide	833
Ammonia	2,190

How do you **use specific heat capacity information**?

Specific heat capacity is the energy per kilogram per change in temperature. So, to find the amount of energy needed to heat something you need to know the material, its mass, and the change in temperature desired. For example, if you wanted to increase the temperature of 2 liters (2 kilograms) of water from room temperature (20°C) to the boiling point (100°C) you need to multiply the specific heat capacity of water (4,186 J/kg°C) by 2 kg and by 80°C to obtain 669,760 J, or about 670 kJ.

Suppose that the water is placed in an aluminum pan with a mass of 300 g (0.3 kg). How much extra energy is needed to heat the pan? The answer is found by multiplying the specific heat capacity of aluminum (897 J/kg°C) by the mass (0.3 kg) and by the temperature change (80°C) to obtain 144,000 J or 144 kJ. So the total energy needed to heat both the water and the pan is 814 kJ. Note how much more energy is needed to heat the water than the pan. Look through the list of specific heat capacities on page 125 to find other metals from which you could make a pan that would need less energy to heat it.

The high specific heat capacity of water makes it a good material to use, for example, in cooling an automobile engine by circulating the water through the engine where the water is heated and then through the radiator where flowing air can cool the water.

How much energy is needed to change the state of water from ice to water to steam?

The amount of energy needed to change phase is called latent heat. The latent heat involved in the transition from solid to liquid is called the latent heat of fusion while the energy involved in the transition from liquid to gas is called the latent heat of vaporization. For water the latent heat of fusion is 334 kJ/kg. The latent heat of vaporization is 2,265 kJ/kg.

Energy must be added to go from ice to water and water to steam, but if steam condenses to water it produces 2,265 kJ for each kilogram of steam condensed. That's the reason that steam burns are so dangerous. Almost all of that energy is transferred to your skin. If water freezes it releases 334 kJ for each kilogram of water frozen. In a freezer that amount of energy must be removed by the freezing mechanism.

On a warm day, why do water droplets accumulate on the outside of glasses and soda bottles?

The water does not seep through the container, but instead comes from the air surrounding it. Water vapor is the gaseous form of water that is in air below the boiling point of water. As discussed above, it takes a larger amount of energy to vaporize water, so the

What is a calorie?

Heat, like work (processes that transfer energy), is measured in the joule, a unit named after James Prescott Joule. Although the joule is the international standard for measuring energy, heat flow is often measured in calories. A calorie defines the amount of energy needed to increase one gram of water by one degree Celsius. The energy required for one calorie is 4.186 joules, which is a relatively small amount of energy. Nutritionists also use the term "calorie" to describe the amount of energy a particular food can provide a person. A nutritional calorie is really a kilocalorie (1,000 calories), which is also written as the capitalized "Calorie."

Another unit used to measure heat flow is the British Thermal Unit, or Btu. The Btu is the amount of energy needed to increase the temperature of one pound of water by one degree Fahrenheit. This unit, which is used only by countries such as the United States that still employ the English standard method of measurements, is equal to 252 calories.

molecules of water in the air have more thermal energy than do the molecules in the colder glass. So when the water molecules strike the glass they transfer much of their thermal energy to the glass. The colder water molecules join together to form water droplets on the glass. The process is called condensation. Condensation also occurs on window panes when the outside is cold and the interior air is warm and humid.

How do **liquids evaporate**?

A substance does not have to boil to change from liquid to gas state. The boiling point is where the pressure of the water vapor equals the atmospheric pressure. At all temperatures there is a great variation in the kinetic energy of the molecules that make up the substance. When the molecules with high energy reach the surface they have the possibility of escaping into the air. The process of changing from liquid to gas at a temperature below the boiling point is called evaporation.

When the molecules with higher energy leave the liquid the remaining molecules have a lower average energy, and thus a lower temperature. It is cooled by evaporation.

How do **clouds form**?

As warm air rises into the atmosphere through convection currents, the air expands as it experiences less atmospheric pressure. During the expansion, the warm water vapor quickly cools and condenses, forming water droplets in the air. When the droplets begin to accumulate, they attach themselves to dust particles in the air and form clouds.

How can **evaporation** be a **cooling process**?

Because evaporation leaves a cooler liquid, the material on which the liquid rests is cooled. Evaporation is an effective way to cool our bodies. Perspiration leaves a coating of water on our skin that evaporates, cooling the skin. Alcohol evaporates more quickly than water and has an even greater cooling effect. For that reasons parents are advised to rub alcohol on the skin of babies who are running dangerously high temperatures. It can reduce the fever and keep the baby out of danger.

HEAT

What are some ways of **transferring thermal energy**?

Heat is the transfer of energy from a warmer to a cooler object. The transfer can occur by two different processes. If the hot object is in contact with a cooler one, the process is called conduction. The faster molecules in the hotter object strike the slower ones in the cooler one, transferring their thermal energy. The average speed, and thus kinetic energy of the molecules in the warmer object are reduced, while those in the cooler object are increased.

If the warm object is in contact with air or water, it heats the fluid. Hotter fluid has a lower density, and so it rises. Its place is taken by colder fluid, creating moving currents of the fluid, called convection currents. Convection is a very efficient way of transferring energy from hot to cold objects.

The thermal energy can also be transferred even if the warmer object isn't in contact with any other object. The vibrations of the molecules create infrared electromagnetic waves that carry energy. These waves are called radiation. The warmer the object, the more energy in the waves. More energy goes from the warmer to the cooler object than in the reverse direction, so radiation cools the warmer object and warms the cooler one.

What are common **methods** of **heating a home**?

Homes using forced air heat have a furnace that heats the air in it and a fan that blows the hot air into heating ducts that allow the hot air into the rooms. It rises and forces the colder air out of the room through return ducts, the entrance to which are usually near the floor. Hot water heat has pipes carrying hot water that have fins on them. The fins promote convection of air past the hot water pipe. This warmed air then circulates through the room. Electrical baseboard heat works in the same way. Electric resistance wiring in the floor or ceiling can warm the air in contact with these surfaces, again creating convection currents. Convection is the movement of thermal energy through a fluid (such as liquid or gas).

How do **convection currents** create **sea breezes**?

During a day at the shore, the sun warms the ground and the water. The ground has a lower specific heat and so its temperature increases more than the temperature of the water. The ground heats the air above it, which rises in convection currents and cooler air from over the ocean flows toward the shore to "fill in the gap" left by the rising warm air. This flow of cooler air from the ocean toward the shore creates what is known as a sea breeze.

In the evening, when the sun dips below the horizon, the ground cools faster than the water. Therefore the air over the ocean is warmer than the air over the shore, and the reverse takes place. The warmer ocean air rises while a breeze flows from the shore to the water.

How does **heat flow** by **conduction**?

When part of a substance is heated its thermal energy is increased. The fast-moving atoms or molecules strike the slower-moving cooler atoms. They begin to move faster, and so gain thermal energy. Their temperature increases. The ease of conduction depends on the material. Most metals are good conductors—even a small difference in temperature produces heat flow. Other materials are poor conductors. There can be large temperature differences without significant heat flow. In that case one part of the substance is hot, another cold.

Heat conductivity is higher in metals that have freely moving electrons. So copper, silver, gold, and aluminum are good conductors. Stainless steel is a poor conductor.

In non-metals conductivity depends on their ability to transfer vibrations of the atoms. The conductivity of ice, concrete, stone, glass, wood, and rubber is less than 1/100 that of metals. Conductivity depends on the material, its thickness, the area it covers, and temperature difference.

Light gases have better conductivity than heavier gases. For example the heavy gas argon is used to fill the space between dual-pane windows because of its lower heat conductivity.

How does **clothing** keep us **warm**?

Our skin is cooled primarily by convection currents in the air. Clothing, especially wool, traps air in small pockets, which reduces or eliminates convection. There can still be conductive cooling through the cloth, but cloth is a good insulator.

Home insulation, like styrofoam panels, fibreglass or blown-in cellulose act like clothing in reducing heat transmission by convection. Home insulation is rated by its R value of resistance to energy transfer. R is the inverse of conduction: $R = 1/U$.

Air pockets within snow and ice function as excellent insulators. Many small mammals build snow dens to keep themselves warm, thereby taking advantage of the insulating properties of the snow. Natives of the Arctic build igloos that also keep the inhabitants warm by reducing loss by convection.

Many farmers protect their crops during sub-zero temperatures by spraying water on the crops, and when the water freezes, the plants are insulated by the poor conductive properties of the ice.

Why is it often **cooler** to wear **white clothes**, rather than black clothes?

White surfaces reflect all the colors of the rainbow, whereas black absorbs all the colors in the light spectrum. The absorption of this energy heats the black material, increasing its temperature. This, in turn, increases the temperature of the air between the clothing and the person. So white, which absorbs much less radiative energy from the sun, is cooler in hot climates.

THERMODYNAMICS

What is **thermodynamics**?

Thermodynamics is the field of physics that studies changes in thermal energy and the relationship between energy, heat, and work. The field of thermodynamics was developed when people sought to increase the efficiency of early steam engines. There are four laws of thermodynamics, which for reasons of history, are numbered 0 through 3 rather than 1 through 4.

What is the **zeroth law** of thermodynamics?

The zeroth law is so obvious that it wasn't added as a law until after laws one through three were developed. It is based on thermal equilibrium between two bodies. As has been stated, if two objects have different temperatures, heat will flow from the hotter to the colder. If there is no temperature difference, there is no net heat flow. They will be in thermal equilibrium. The zeroth law states that if objects A and B are in equilibrium and B and C are in equilibrium, then A and C are also in equilibrium. Suppose object B is a thermometer. You put it in contact with object A. Heat flows until they are at the same temperature. You then move the temperature to object C. If the thermometer shows no change, then B and C are in equilibrium and we can conclude that objects A and C are at the same temperature.

What is the **first law of thermodynamics**?

The first law of thermodynamics is a restatement of the conservation of energy. It says that the energy loss must equal the energy gain of a system. It relates the heat input and output, the work done on the system and the work the system does, and the change in internal energy, or temperature change.

For example, the cylinder and moving piston of an automobile engine is heated by the burning of gasoline in the cylinder. The piston moves out, doing work, and heat is transferred to the coolant because of the higher temperature of the cylinder and piston. As long as the temperature of the cylinder and piston does not change, the heat input equals the work output plus the heat output. Energy isn't gained or lost, it just changes form.

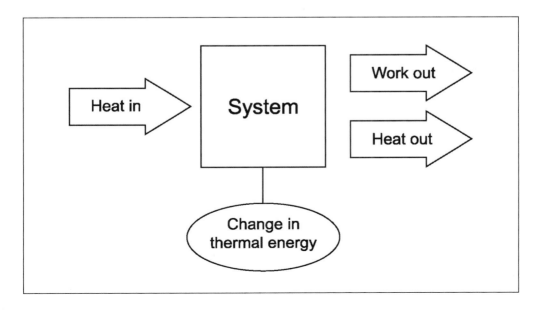

Another way to state the first law is that net heat input equals net work plus change in thermal energy. Note that heat can be either positive (heat input) or negative (heat output). Work can also be positive (work done by the system) or negative (work done on the system). Thermal energy can also go up or down.

What is the **second law of thermodynamics**?

In the early 1800s many scientists and engineers worked to improve the efficiency of steam engines. The French military engineer and physicist Sadi Carnot (1796–1832) tried to answer two questions: Is there any limit to the amount of work available from a heat source, and can you increase the efficiency by replacing steam with another fluid or gas? Carnot wrote a book in 1824 called *Reflections on the Motive Power of Fire* that was aimed at a popular audience, using a minimum of mathematics.

The most important part of the book was the presentation of an idealized engine. This engine could be used to understand the ideas that can be applied to all heat engines. A heat engine is a device that converts heat into work in a cyclical process. That is, the engine periodically returns to its starting point. A steam engine, gasoline or diesel power automobile engine are all heat engines where the pistons return to their starting positions. A rocket is not a cyclic heat engine. A diagram of Carnot's simplified model allowed him to give answers to his two questions. He found that the efficiency, that is, the work output divided by the heat input, depended only on the temperatures at which heat enters and leaves the system, or efficiency = $(T_{hot} - T_{cold})/T_{hot}$. It doesn't depend on the fluid or gas used in the engine. Real engines would have lower efficiencies, but no engine could have a higher efficiency.

A simplified version of the Second Law according to Carnot is that it is impossible to convert heat completely into work in a cyclic heat engine; there is always some heat output. Note that the First Law would allow a heat engine to have no heat output, the heat input would equal the work done. The Second Law can be shown in the diagram below where the input heat is taken from a "reservoir" that maintains a constant high temperature and the output goes to a second reservoir at a constant low temperature.

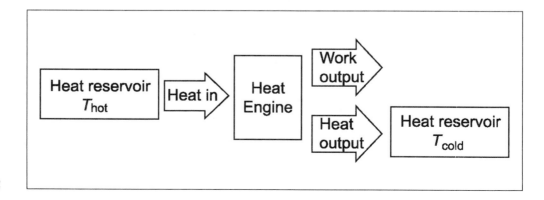

This turbine plant in California uses natural gas to heat water into steam that then produces energy in the turbines that generate electricity. Not all the energy from the natural gas can be converted to electricity, and lost energy escapes as heat.

For example, a steam turbine gets its heat from high temperature steam and puts its output heat into a much colder lake or river.

Carnot's model engine is reversible. That is, it can be run backwards, which is impossible for a real engine. Thus all real engines are less efficient than the ideal one. Friction also lowers the efficiency of real engines.

The Second Law, according to Carnot, means that there is no such thing as a perpetual motion machine. Although such a machine could obey the law of conservation of energy, the heat it puts out means that the machine would eventually stop.

Carnot died in a cholera epidemic at age 36. Because people were concerned with the transmission of this deadly disease, all his papers and books were buried with him after his death. Thus only a few of his works have survived.

Despite the limitations of Carnot's model, his work inspired Rudolf Diesel (1858–1913) to design the engine named after him to achieve higher efficiencies than steam engines. So, Carnot's book was important to the design of practical engines.

A second version of the Second Law can be stated as "Heat doesn't flow from cold to hot without work input." For example, a refrigerator removes heat from the food at a low temperature and outputs heat from the coils in the bottom or back of the refrigerator. It will not move the heat from cold to hot, however, without a motor doing work on the circulating gas.

How do **refrigerators** and **air conditioners** work?

As you know, when a liquid evaporates into a gas, it is cooled. Heat flows into the system. The opposite process, the condensation of a gas into a liquid results in an increase in thermal energy and an output of heat. A refrigerator circulates a refrigerant, a liquid that evaporates at a low temperature, through tubing, The gas is compressed by an electrically-driven compressor. The pressure and temperature of the gas increases. Coils of the tubing outside the refrigerator cool the liquid and heat the air around them. As it cools the refrigerant condenses back to a liquid that goes through a tiny hole, called the expansion valve. The pressure drops, evaporating the liquid, making the gas cold. The tubing containing the cold gas is in the inside of the refrigerator, making it cold, and cooling the food.

An air conditioner works in a similar way. The evaporator is in the unit inside the house and the compressor is outside. Because of the work put into the compressor heat is removed from the air inside the house and transferred to the outside air. The diagram below shows work and heat flows in a refrigerator or air conditioner.

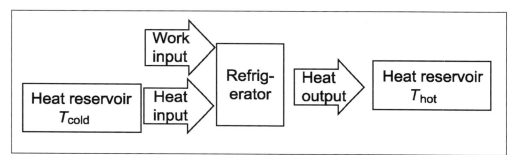

The first home refrigerators used ammonia as a refrigerant, but ammonia is toxic. In the 1930s Freon was first developed by the DuPont Company of Wilmington, Delaware. Freon is a chlorofluorocarbon (CFC). If Freon escapes it carries chlorine atoms to the upper atmosphere. There ultraviolet radiation from the sun separates one of the chlorine atoms from the CFC. That atom converts ozone back to oxygen, contributing to the destruction of the ozone layer, an essential barrier against harmful ultraviolet sunlight.

Freon's destructive nature has been known since the 1970s, but it was not until the early 1990s that legislation was implemented banning the use of Freon in new air conditioners and refrigerators. It has been estimated that in 2002 there was six million tons of Freon in existing products. Unfortunately, when the chlorine destroys an ozone molecule, the chlorine is not destroyed, but instead continues to live for a while destroying more ozone. In fact, more Freon is still headed toward the upper limits of the atmosphere, for it can take several years for Freon to reach such elevations.

DuPont and other corporations have developed replacements for Freon that replace the chlorine atoms with hydrogen atoms. These substances do not harm the ozone layer and are in use in refrigerators, air conditioners, and aerosol cans.

How **efficient** are **electrical generators** and **vehicles**?

In the United States, electrical generators are only 31% efficient, while 75% of energy used for transportation is wasted.

Engineers are working on both improving efficiency and in making use of the "rejected" energy. For example, the warm water that carries away the waste heat in an electrical generating plant can be used to heat homes close to the generator.

What is the **third law of thermodynamics**?

The third law of thermodynamics states that absolute zero, the lowest possible temperature, can never be reached. The entropy of a system is zero at absolute zero. A procedure can remove a portion of the entropy, but not all of it. Thus it would take an infinite number of repetitions of the procedure to reach absolute zero. It's been possible to achieve temperatures as low as a few billionth of a kelvin, but it has never been able to reach absolute zero.

What is **entropy**?

The German physicist Rudolf Clausius (1822–1888) was concerned about Carnot's use of the term waste heat. He developed another version of the Second Law that involves the concept of entropy. Entropy can be defined as the dispersal of energy. The greater the dispersal or spreading the larger the entropy. For example, when salt and pepper are in separate piles the two substances are in distinct locations. If you mix them together they are no longer in separate regions of pure salt and pure pepper, but dispersed throughout the combined pile. Further, it is not possible for the salt and pepper grains to separate themselves. Thus the mixing of salt and pepper increases the entropy of the system. If the salt and pepper could be shaken in such a way that they would separate, then entropy would be decreased, but such an event has never been observed.

Suppose you place ice in water. Ice and water are separate. The water has higher thermal energy, the ice lower, and so the system has low entropy. When the ice melts the two can no longer be separated. The thermal energy is dispersed throughout the system, so the entropy has increased. But wait, you might say, the ice water (the system) has cooled the air around it (the environment), decreasing its entropy because hot gas has greater entropy than cold gas. Calculations, however, show that the increase in the ice/water mixture is greater than the decrease in the air. So a statement of the Second Law is that the entropy of the system and the environment can never decrease.

The increase in entropy suggests a direction of time, sometimes called the "Arrow of Time." The "forward" direction of time is the one in which entropy increases or remains the same.

Clausius had given Carnot's work a firmer foundation. Later work by Ludwig Boltzmann, Josiah Willard Gibbs, and James Clerk Maxwell developed the statistical basis for entropy.

WAVES

What is a **wave**?

A wave is a traveling disturbance that moves energy from one location to another without transferring matter. Oscillations in a medium or material create mechanical waves that propagate away from the location of the oscillation. For example, a pebble dropped into a pool of water creates vertical oscillations in the water, while the wave propagates outward horizontally along the surface of the water.

What are the **types of waves**?

Transverse and longitudinal waves are two major forms of waves. A transverse wave can be created by shaking a string or rope up and down. Although the string moves up and down, the wave itself and its energy moves away from the source, perpendicular to the direction of the oscillations.

The oscillations in longitudinal waves move in the same direction that the wave is moving. The medium in longitudinal waves alternately pushes close together (compression) and separates from each other (rarefaction). The best example of longitudinal waves are sound waves, which are a series of back and forth longitudinal oscillations of atoms or molecules that form alternate regions of high and low pressure in a medium such as air.

Water waves are a combination of transverse and longitudinal waves that move in circles. Just as in the case of transverse and longitudinal waves, energy is transferred but matter is not moved.

What **determines** the **velocity** of a **wave**?

The velocity of a wave depends upon the material or medium in which it is traveling. Typically, the stronger the coupling between the atoms or molecules that make up the medium, and the less massive they are, the faster the wave will travel. All waves of the same type (transverse or longitudinal) travel at the same speed. For example, a sound wave in air at 0°C will travel at 331 meters per second, regardless of the sound's frequency or amplitude. Electromagnetic waves can travel either through empty space or through material. Their velocity depends on the electric and magnetic properties of space or the material but not on frequency or amplitude. The velocity of water waves depend both on the properties of the water and on the frequency of the wave.

What are some of the **terms used** to define the **properties of waves**?

The table below summarizes the different properties of waves.

Type of Wave	Term	Definition
Transverse	Crest	The highest point of the wave.
	Trough	The lowest point of the wave.
Longitudinal	Compression	An area where the material or medium is condensed and at higher pressure.
	Rarefaction	An area that follows a compression where the material or medium is spread out and at lower pressure.
Transverse & Longitudinal	Amplitude	The distance from the midpoint to the point of maximum displacement (crest or compression).
	Frequency	The number of vibrations that occur in one second; the inverse of the period.
	Period	The time it takes for a wave to complete one full vibration; the inverse of the frequency.
	Wavelength	The distance from one point on the wave to the next identical point; the length of the wave.

What is the **relationship** among **frequency, wavelength,** and **velocity**?

Suppose you shake a rope up and down at a constant rate. The rate or frequency is the number of times your hand is at the top of its motion per second. As the waves move along the rope, the distance between the crests of the rope, the wavelength, will remain the same. The wavelength depends both on the frequency of oscillation and on the velocity of the wave along the rope. The relationship is velocity = frequency $\times$ wavelength ($v = f\lambda$) or wavelength = velocity/frequency ($\lambda = v/f$). Therefore, if the frequency of a wave increased, the wavelength decreases while the velocity, being a property of the medium (rope) doesn't change. The frequency and wavelength are inversely proportional to each other.

The following table shows the relationship between frequency and wavelength for a sound wave in 0°C air:

Velocity of Sound	Frequency (Hz)	Wavelength (m)
331	128	2.59
331	256	1.29
331	512	0.65
331	768	0.43

What is the **relationship** between **frequency** and **period**?

Frequency, f, is how many cycles of an oscillation occur per second and is measured in cycles per second or hertz (Hz). The period of a wave, T, is the amount of time it takes a wave to vibrate one full cycle. These two terms are inversely proportional to each other: $f = 1/T$ and $T = 1/f$.

For example, if a wave takes 1 second to oscillate up and down, the period of the wave is 1 second. The frequency is the reciprocal of that, 1 cycle/sec, because only one cycle occurred in a second. If, however, a wave took half a second to oscillate up and down, the period of that wave would be 0.5 seconds, and the frequency would be the reciprocal, or 2 cycles per second. So, you see that a wave with a long period has a low frequency, while a wave with a short period has a high frequency.

On what does the **amplitude** of a **wave depend**?

The amplitude of a rope wave depends on how hard you shake it. For a sound wave it depends on how much compression the loud speaker or musical instrument creates. In other words, it depends on the energy the source put into the wave. It does not depend on frequency, wavelength, or velocity.

Does the **amplitude** of a **wave depend** on **distance** from the source?

The energy carried in a wave depends on the wave's amplitude and its velocity. Waves can be put into two categories: those that spread, like water waves on a pond, sound waves, or electromagnetic waves; and those that are confined to a narrow region, like waves on a rope or electrical oscillations on a wire. A water wave spreads on the surface of a pond, lake, wide river, or ocean. As it spreads its energy is spread over a larger area, so the energy transmitted to a particular location is reduced when the source is farther away. Therefore the amplitude of the water wave is also reduced in proportion to the distance from the source. Sound and electromagnetic waves usually spread in two dimensions. Again, as they spread the energy carried is also spread, so as the distance from the source is increased, the amplitude is decreased, but this time as the square of the distance.

On a rope or wire, the wave doesn't spread, but often a different mechanism reduces the amplitude. In a rope there is friction between the fibers, which changes some of its kinetic energy to thermal energy. If a signal is sent through a wire as an oscillating voltage the resistance of the wire will convert some of the electric energy to thermal energy, reducing the voltage and thus the amplitude of the wave. This loss is reversed by putting amplifiers along the wire, putting more energy into the wave and increasing its amplitude.

WATER WAVES

What type of wave is a **water wave**?

Ocean or water waves look like transverse waves, yet are actually a combination of both transverse and longitudinal waves. The water molecules in a water wave vibrate up and down in tiny circular paths. The circular path of the water wave creates an undulating appearance in the wave.

How **fast** does the **wind** need to blow in order to **produce different types** of **waves**?

Wind rubbing against the water surface is a major cause of waves. Since the water cannot keep up with the wind velocity, the water rises and then falls, creating the familiar wave-like motion. Depending upon the wind velocity and the distance the wind has been able to travel over the water, different size waves are generated.

Type of Wave	Wind Velocity	Effect
Capillary waves	Less than 3 knots	Tiny ripples. The longer they are generated, the larger their amplitude
Chop or regular	3–12 knots	Combined capillary waves that have traveled far and formed larger waves
Whitecaps	11–15 knots	Amplitude of wave must be over 1/7th the wavelength in order to break into a whitecap
Ocean swells	No specific speed	Form over long distances from a combination of different waves

How do **speeds** of **ocean waves** vary?

The speed of an ocean wave depends on the distance between two successive crests, its wavelength. The longer the wavelength, the faster the wave travels. A small surface wave, such as a ripple created by the wind, travels quite slowly because it has such a short wavelength. A swell, the larger, longer wavelength waves created by constant

Waves break as they approach a shoreline because the lower part of the wave moves more slowly than the top part of the wave due to increasing friction with the shallower ocean bottom.

winds, have longer wavelengths and travel at higher velocities. The energy that the wave carries depends on the square of the height of the wave, which explains why high waves can cause so much damage to shorelines.

Why do water **waves break** as they **approach** the **beach**?

Water waves rarely break, or form whitecaps, when they come in contact with a cliff or mountainside shoreline. Waves only break as they approach a gradual decrease in depths, such as a beach. A shoreline with a gradual decrease in depth will produce more spectacular whitecaps than a wave that encounters a steep decrease in depth.

The reason waves break is the result of the way the wave velocity depends on the depth of the water. Consider a water wave with a large amplitude. As the wave moves toward the beach, at first it travels at a constant velocity. As the ocean depth begins to decrease, the bottom of the wave gradually encounters more and more friction with the beach, causing the lower part of the wave to travel slower than the upper part. As the lower part slows down, the crest, moving faster, moves over the trough. When there is not enough water to support the crest, the wave breaks or forms a whitecap.

Where are the **best surfing beaches**?

The best surfing beaches are located along the edges of oceans when wind conditions have produced waves with large wavelengths. Another requirement of a good surfing beach is a gradual decrease in water depth.

141

How can they be similar? Surfing is done on water; skiing on snow. The main similarity is that in both cases the athlete and board travel down a hill. In skiing, the hill is a mountain covered with snow, while in surfing the hill is the rising water of a breaking ocean wave. An ideal surfing wave has a large amplitude as it reaches an extremely gradual decrease in ocean depth. While the surfer moves down the wave, the water on the front edge of a crest continually rises underneath the surfer, allowing the surfer to ride down the wave without actually moving downward.

Some of the best surfing is done on Waikiki beach on Oahu in the summer and the north shores of Oahu and Kaua'i in Hawaii in the winter. In the continental United States the best surfing is in southern California. The Pacific Ocean, famous for its long wavelengths and gradual decreasing depth beaches, has some of the best surfing in the world.

What is a **tidal wave**?

A tidal wave, or tsunami, is not caused by windy conditions or tides, but instead by underwater earthquakes and volcanic eruptions. The seismic disturbances create huge upward forces on the water, the opposite of dropping rocks into water. A tsunami is a series of several waves with a period of more than 30 minutes between each crest. The ocean first recedes from the beach, then water rushes inland at a very high speed. Large tsunamis can be quite destructive upon reaching the shoreline due to their amplitudes.

The most deadly recorded tsunami occurred December 26, 2004, in the Indian Ocean. The earthquake that caused it was off the west coast of Sumatra, Indonesia. That quake released an amount of energy equivalent to 550 million times the energy released in the Hiroshima nuclear bomb. One part of the ocean bottom was lifted by 4 to 5 meters (13 to 16 feet) and moved horizontally 10 meters (33 feet). The tsunami, traveling at a speed of 500 to 1,000 kilometers per hour, had a low amplitude (60 centimeters) in mid-ocean, but when it crashed into the coasts from Thailand to India and as far as South Africa it had an amplitude as high as 24 meters (79 feet). Some 230,000 people were killed and more than a million made homeless.

ELECTROMAGNETIC WAVES

What is an **electromagnetic wave**?

Electromagnetic waves consist of two transverse waves: one an oscillating electric field, the other a corresponding magnetic field perpendicular to it. Light, infrared, ultraviolet, radio, and X rays are all examples of electromagnetic waves.

All electromagnetic waves travel at the speed of light when they are in a vacuum. Electromagnetic waves are characterized by their frequency or wavelength and amplitude. Electromagnetic waves differ from other waves in that they do not need a medium such as air, water, or steel through which to travel.

How is an **electromagnetic wave created** and **detected**?

Electromagnetic waves are created by accelerating electrons that create an oscillating electric field. This field in turn creates an oscillating magnetic field, which creates another oscillating electric field, and so on. The energy carried by the waves radiates into the area around the moving charges. When it strikes a material whose electrons can move freely, it causes these particles to oscillate.

What is the **electromagnetic spectrum**?

The electromagnetic spectrum is the wide range of electromagnetic (EM) waves from low to high frequency. The spectrum ranges from low-frequency radio waves, all the way to gamma rays, which have a very high frequency. In the middle of the spectrum is a small region containing the frequencies of light.

Who **predicted electromagnetic waves**?

In 1861, James Clerk Maxwell (1831–1879) demonstrated the mathematical relationship between oscillating electric and magnetic fields. In his *Treatise on Electricity and Magnetism*, written in 1873, Maxwell described the nature of electric and magnetic fields using four differential equations, known to physicists today as "Maxwell's Equations." Putting the four equations together predicted the existence of the electromagnetic wave.

Maxwell was a professor at Cambridge University in England from 1871 until his death in 1879. He published other works on thermodynamics and the motion of matter as well. He also developed the kinetic theory of gases, and performed research in the field of color vision. Although Maxwell is not widely known to the lay audience, he is revered in the scientific community, and rates in the pantheon of physics greats with Newton and Einstein.

Who **demonstrated** that **electromagnetic waves exist**?

Heinrich Hertz (1857–1894) was a German physicist who was the first person to demonstrate that electromagnetic waves existed. He designed a transmitter and receiver that produced waves with a 4-meter wavelength. He used standing waves to measure their wavelength. He showed that they could be reflected, refracted, polarized, and could produce interference. It was Hertz's breakthroughs in electromagnetic waves that paved the way for the development of radio. In 1930 Hertz was honored by having the unit of frequency, which was cycles per second, replaced by the hertz (Hz).

COMMUNICATING WITH ELECTROMAGNETIC WAVES

How did **radio communications** develop?

An electromagnetic wave with no changes in amplitude or frequency carries no information—it cannot be used for communication. The first method of using these waves to communicate was to switch them on and off in regular patterns. Letters were represented by a combination of long and short pulses using what is called Morse Code after Samuel S.B. Morse, who developed the code to transmit information over wires (the telegraph).

In 1895 Guglielmo Marconi (1874–1937), a twenty-year-old Italian inventor, created a device that transmitted and received electromagnetic waves over a 1-kilometer (3,280 foot) distance. Later improvements to his antenna and the development of a crude amplifier enabled him to receive a British patent for his wireless telegraph. In 1897, he transmitted signals to ships 29 kilometers (18 miles) from shore and in 1901 he was able to send wireless messages across the Atlantic Ocean. As a result of Marconi's work on radio transmitters and receivers, he was the co-winner of the 1909 Nobel Prize in physics. Over the next decade transmitters and receivers were improved enough that they could be installed in ocean-going ships.

Voice communication over the telephone had existed since 1876, but if the distance was to be extended, the voices had to be amplified to be heard. In 1906 Lee DeForest invented a vacuum tube amplifier he called the Audion. It took until 1915 for a radio receiver to be sold using Audions. In 1916 DeForest had developed an Audion-based transmitter that allowed dance music to be transmitted 40 miles. A number of other experimental stations demonstrated music by radio—then called wireless. A large number of radio amateurs made significant advances. When the United States entered World War I in 1917 all stations not owned by the government were shut down and it became illegal for people to listen to any radio transmission.

During the war, radio was used to communicate between ships and between land and the ships. After the war, amateurs were forced to use only one wavelength, 200 meters (1,500 hertz). Wavelengths shorter than that were thought to

Samuel S.B. Morse was famous for inventing the Morse Code that allowed people to first transmit messages over telegraph wires.

be useless for government use. One amateur was able to send signals 3,000 miles. In 1921 transatlantic voice transmissions were made. Companies began to use radio for specialized needs, like to send time information to jewelers to allow them to set their clocks. From 1919 through 1921 radio was mostly used to transmit musical concerts. The first transmission of a football game occurred in November 1919. By 1922 newspapers had developed radio stations transmitting news, weather reports, crop reports, and lectures. Large companies such as General Electric, Westinghouse, AT&T, and RCA. began to be involved in developing commercial broadcasting.

From 1922 to 1923, as the number of stations grew without regulation, chaos reigned. In 1928 the government announced new assignments in the frequency band 550 to 1,600 kHz. Many more assignments were added after World War II, but these regulations are still in use today.

How do **antennas transmit** and **receive signals**?

Antennas for radio and television signals are used to either transmit or receive electromagnetic radio waves. Oscillating voltages produced by the transmitter cause the electrons in a metal wire or rod, the transmitting antenna, to oscillate, creating an oscillating electric field that in turn creates an oscillating magnetic field that creates another oscillating electric field. The combined electric and magnetic wave moves away from the antenna at the speed of light. A receiving antenna is a metal rod, wire, or a loop. When an electromagnetic wave strikes the antenna it causes the electrons in the metal to oscillate at the same frequency as that of the wave. The oscillating electrons produce a voltage in the receiver that eventually results in the sounds and/or pictures produced by a radio or television.

Does the **dimension** of an **antenna** play a significant **role** in the **reception** of an electromagnetic wave?

The length of an antenna determines the frequency that it best receives. The most efficient antennas have a length equal to half the wavelength of the wave it is receiving. This allows the induced electrical current in the receiving antenna to resonate at that particular frequency. If the antenna is a simple rod it is most sensitive when its length is one quarter the wavelength.

A loop or coil antenna are used for the low-frequency, long-wavelength signals in the AM band. A half-wavelength straight wire antenna would be over one hundred of meters long. Shorter wires or rods can be used and are more efficient if coils of wire are used to "load" the antenna.

Home FM radio and television antennas are designed to receive a broad range of frequencies, but with less sensitivity. Antennas for the ultra-high frequencies used in digital high definition televisions are very short and can be easily mounted outside on rooftops or on top of television sets.

What are the different **frequencies** of **radio waves** and **microwaves** that allow for **communication**?

The table below shows the regions of the electromagnetic spectrum and their uses. The unit for frequency is hertz (Hz). Hertz is the name for the number of oscillations or cycles per second of a wave. The letters kHz mean one thousand hertz, Mhz a million hertz, and Ghz a thousand million or a billion hertz or cycles per second.

Frequency Range	Wavelength Range	Name & Abbreviation	Use
Less than 30 kHz	More than 10 km	Extremely low frequency (ELF)	Submarine communication
30 kHz to 300 kHz	10 km to 1 km	Low frequency (LF)	Maritime mobile, navigational
300 kHz to 3 Mhz	1 km to 100 m	Medium frequency (MF)	Radio broadcasts, land and maritime mobile radio
3 Mhz to 300 Mhz	10 m to 1 m	Very high frequency (VHF)	Television broad casts, maritime and aeronautical mobile, amateur radio, meteorological communication
300 Mhz to 3 Ghz	1 m to 10 cm	Ultrahigh frequency (UHF)	Television, radar, cell phones, military, amateur radio
3 Ghz to 30 Ghz	10 cm to 1 cm	Superhigh frequency (SHF)	Radar, space and satellite microwave communication, wireless home telephones, wireless computer networks
30 Ghz to 300 Ghz	1 cm to 1 mm	Extremely high frequency	Radio astronomy, (EHF) radar

PUTTING INFORMATION ON ELECTROMAGNETIC WAVES

What is the difference between **analog** and **digital signals**?

All transmitters create what is called a carrier wave at a specific frequency. For licensed commercial broadcasters, the Federal Communications Commission (FCC) assigns the frequency. The carrier wave transmits no information. For information to be carried, some property of the carrier wave must be changed. The earliest methods of radio broadcasting used analog signals from a source like a microphone. Sound striking a microphone produces a varying voltage output. That is, the output of a microphone has the same shape of the amplitude of the sound wave that strikes it. The smooth graph below represents an analog signal from a microphone.

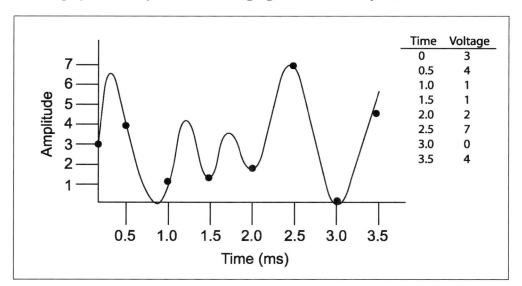

Time	Voltage
0	3
0.5	4
1.0	1
1.5	1
2.0	2
2.5	7
3.0	0
3.5	4

To create a digital signal the analog output of the microphone is converted into a series of numbers. For example, consider the graph above. Suppose the signal is "sampled" every 2,000 times each second. That is, the voltage is recorded each 0.5 millisecond. The dots show the results of the sampling. Voltage can be in only whole numbers. The series of numbers representing the waveform would then be those shown in the table.

Next the voltages are converted into binary digits, a series of 0s and 1s. The table on the right shows the conversion to binary for numbers 0 through 7. The binary digits reading left to right represent the numbers 4, 2, and 1. Thus 5 = 4 + 1 or 101.

Now the binary numbers are converted into a series of voltages to be sent to the transmitter. One method is called *Pulse Width Modulation* or PWM. A narrow pulse represents a 0, a wider pulse represents 1. For example, the first four samples would produce the following wave train:

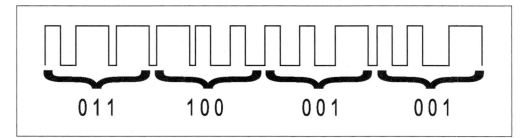

The receiver knows that three bits should be converted into a number. If the receiver is to convert the signal back to analog, then that number determines the amplitude of the analog wave.

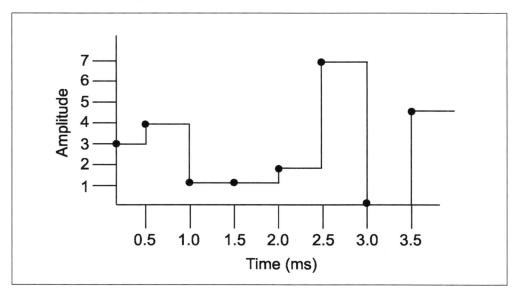

Does this look like the original analog signal? No way! Note that there are two problems. First, the "wiggles" in the wave are missing. Second, with only 8 choices of voltages, the vertical resolution is too small. To make a more accurate conversion you need to sample the wave more often, at least every 0.1 millisecond and to allow more than 8 voltage choices; 32 or more would be better. To obtain the quality of sound in a CD the sampling interval is 22 µs (22 microseconds) and there are at least 4,096 voltage choices.

What do **AM** and **FM** mean?

Traditional radio broadcasting uses analog, not digital, methods. AM, or amplitude modulation, and FM, or frequency modulation are methods of transmitting information on a radio. In each case a property of the "carrier" wave is changed, or modulated. These variations carry the transmitted information.

AM was the first method invented and the simplest to transmit and receive. Edwin Howard Armstrong demonstrated transmission and reception of Frequency Modulation in 1935. Listeners were amazed the way that FM eliminated "static" because noise, like that due to thunderstorms, changes the amplitude, but not the frequency of signals. The FM receiver's output does not depend on the amplitude of the signal. Due to opposition from the radio networks and receiver manufacturers, commercial FM broadcasts were delayed and did not become widely available until after World War II.

Radios once only had AM (Amplitude Modulated) frequencies available; then came FM (Frequency Modulated) in 1939. Today, there is talk among radio professionals that AM stations might become a thing of the past.

Where are **AM and FM broadcasts** found in the **electromagnetic spectrum**?

In the United States commercial AM stations broadcast between 550 kHz and 1600 kHz. Commercial FM stations broadcast between 88 MHz and 108 MHz. In the AM band the frequencies of stations are spaced by only 10 kHz. As a result, the broadcasts are limited to sound frequencies up to only 4kHz, while the ear can detect frequencies as high as 15 kHz. Why are broadcast frequencies so limited? Suppose the analog signal is a 440 Hz tone, the A above middle C on a piano. A typical AM transmitter would have a carrier wave frequency of 1 megahertz (MHz). The result is a signal with three different frequencies: 1 MHz, 1MHz + 440 Hz and 1MHz – 440 Hz. Thus the total signal requires a set of frequencies 880 Hz wide. Sounds with a frequency of 4kHz require a range of frequencies, called the bandwidth, 8 kHz wide, just about the spacing between adjacent stations.

FM stations, on the other hand, were developed to transmit sounds more faithfully, which means that sounds up to 15kHz must be accommodated. That means that the bandwidth of the broadcasts is 30kHz wide. At the very high frequencies used by FM stations there is more bandwidth available and stations are spaced by 200kHz.

The electromagnetic spectrum has many other users. Police and firefighters usually use AM while aircraft, where noise reduction is important, use FM. Television stations used to use AM for the picture and FM for the sound, but as of June 12, 2009, they now all use digital signals. Other users are the military, marine ship-to-shore services, weather broadcasts, commercial mobile phones, the citizen band, and amateur radio operators. They all share the HF, VHF, UHF, and SHF bands.

What **alternative analog methods** are used?

As was described above, modulating a carrier wave produces additional frequencies above and below the frequency of the carrier. These frequencies are called sidebands. There is identical information in the two sidebands, so many radio services filter out one of the two sidebands, resulting in a single-sideband broadcast, or SSB. SSB can work with either AM or FM radios. It has half the bandwidth of a double-sideband broadcast, so more radios can use the same part of the spectrum.

How does an **FM band station transmit stereo** sound?

Stereo sound means that two separate sounds, the left (L) and right (R), are produced from a pair of speakers. Because stereo broadcasts must also be usable by receivers that cannot reproduce stereo the signal that all receivers can detect consists of the sum of the left and right channels ($R + L$). The difference of the two channels ($R - L$) is broadcast 38 kHz above the $R + L$ signals. Mono receivers can't detect these very high frequencies. The stereo receiver, though, can and from the $R + L$ and $R - L$ signals creates separate R and L signals to be amplified and sent to the corresponding speaker.

FM stations have enough bandwidth to accommodate a third signal. This signal can be sold to users to provide background music for stores or elevators. Education stations can deliver lessons to schools, or broadcasts in a second language.

How **far away** can **FM** and **AM** stations be **received**?

All electromagnetic waves travel in a straight line while in a uniform medium such as the lower atmosphere. Therefore, most radio waves only have what is called a line-of-sight range. That means that if a mountain range or the curvature of Earth were in the way of the radio signal, the receiver would be out of range and would not receive the signal. This is why most broadcasting antennae are placed on tall buildings or mountains to help increase the line-of-sight range.

Waves with frequencies below about 30 Mhz are able to reflect off the charged particles in Earth's ionosphere; this is referred to as "skip." Instead of passing through the ionosphere and entering space as higher-frequency electromagnetic waves do, the lower frequencies on the AM band can be reflected back toward Earth to increase their range dramatically. After sunset the ionosphere's altitude permits stations to be heard thousands of kilometers from a transmitting tower. A handful of AM stations are "clear channel" stations. That is, there is only one station in the continental United States broadcasting on that frequency. Those stations can be heard across almost the entire country without interference from other stations. Broadcast FM stations, with frequencies 88–108 megahertz penetrate the ionosphere, and so can only be heard 80–160 kilometers from the transmitter.

How do **cell phones** work?

Cell phones use the UHF part of the electromagnetic spectrum, 800 to 900 MHz, 1,700 to 1,800 MHz, and 2,100 to 2,200 MHz. The service area for a cell phone provider is divided up into hexagonally shaped cells, each one served by base stations with antenna towers at three corners of the cell. The stations can both receive and transmit information to cell phones. The stations are connected to a network that uses fiber optic cables. When a cell phone is turned on it searches for available services according to a list stored in the phone. It selects the correct frequencies to transmit and receive data, then sends its serial and phone numbers to the system, registering itself in that cell. The network makes sure that the phone number is part of its system and that there is money in the account to pay for a call. After registration is complete and a call is made to that phone, the network can direct the call to the correct cell. The cell

The technology behind the cell phones we take for granted today is amazing. Transmitting and receiving messages in the electromagnetic spectrum, cell phones automatically select and correct appropriate frequencies, send serial numbers to servers, and register themselves in the cell in which they are located at the time.

151

phone always searches for the strongest signal from a tower. If the phone moves during the conversation, then the signal strengths will change and the phone will "hand off" the call to a different base station.

Cell phones digitize the voice signals. Special circuits, called digital signal processors, then compress the voice signals and insert codes that can detect errors in transmission. Compression is achieved by sending only the changes in the digital signals, not what stays the same. Cell phone systems also send many different conversations at the same time. One method, called TDMA (time division multiple access), splits up three compressed calls and sends them together. CDMA (code division multiple access) uses TDMA to pack three calls together, and then puts six more calls at two other frequencies. Each of the nine (or more) calls is assigned a unique code so that it can be directed to the correct recipient. Spread CDMA systems use a wide band of frequencies that permit more simultaneous calls.

MICROWAVES

How are **microwaves** used for **communication**?

Microwaves are electromagnetic waves with frequencies above about 3 Ghz. They are used in homes equipped with wireless internet devices, wireless telephones, Bluetooth devices, for satellite radio and television. In addition, industries, governments, and the military use microwaves for communicating from one installation to another.

Microwaves can be transmitted one of two ways. The first is the line-of-sight approach, where the microwave transmitter is pointed at a microwave receiver (these must be no more than 30 kilometers apart). The second method of transmission is to send the signal up to a satellite that receives it and retransmits it back down to a receiving dish.

What are **other uses** for **microwaves** besides communication?

In addition to having a great range of frequencies to transmit information, microwaves are used every day in kitchens around the world. Microwave ovens generate 2.4 GHz waves and scatter them throughout the oven. The microwaves excite water and fat molecules into resonance and cause them to rotate, increasing their thermal energy. Different kinds of molecules absorb the energy at different rates, so some foods are heated more than others. Microwave-safe containers are made of materials that do not absorb microwave energy and so remain cool.

What is the function of the **grating** on the **door** of a **microwave oven**?

People using microwave ovens want to see the food cooking inside the oven, yet not be bombarded by potentially harmful microwaves. In order to prevent the escape of

Can a microwave oven be used to dry things?

Since water molecules are warmed and eventually boil off by microwaves, anything that is wet can be dried in the microwave. However, there is one very important consideration that must be made before placing the object inside a microwave—the object being dried must not contain a great deal of water itself. Microwaves are wonderful at drying wet books, papers, and magazines, but must never be used to dry things like plants or small animals. Living things would be killed by the resonance of water molecules inside their bodies.

microwaves through the plastic or glass door, a grating consisting of small holes is used to reflect the microwaves back into the oven. The microwaves (which have a wavelength of about 12 centimeters [4.7 inches]) are too big to pass through the holes, but visible light, whose wavelength is smaller than the opening, can easily pass though the grating. Although the grating protects people from the microwaves, some microwaves can still leak out through the door seal, however, if it's not cleaned occasionally.

Why **shouldn't metal** objects be placed **inside microwave ovens**?

Manufacturers caution consumers about placing metal containers and aluminum foil inside microwave ovens for two main reasons. The first reason is that metal and aluminum may impede cooking. Microwaves warm food by transmitting energy to water and fat molecules within the food. If food is placed under aluminum foil or in a metal container, the microwaves will be reflected from the metal and won't be able to reach the water molecules and cook the food.

The second reason for not placing metal objects inside a microwave oven is for the safety of the microwave oven itself. Metal acts as a mirror to microwaves. If too much metal is placed in the oven, the microwaves will bounce around the oven in waves that can damage the magnetron that produces the microwaves. If the metal is the correct size it can act as a microwave receiver and the induced voltages can produce sparks that can ignite the food.

THE PRINCIPLE OF SUPERPOSITION

What is the **Principle of Superposition**?

When two waves overlap they don't crash and destroy one another; instead, they pass through each other without interaction. The graphs below show two waves approach-   153 ing, overlapping, and moving away. They continue to move at the same velocity throughout. The arrows show their motions. The dotted drawings show the individual waves while the solid drawings show the resultants.

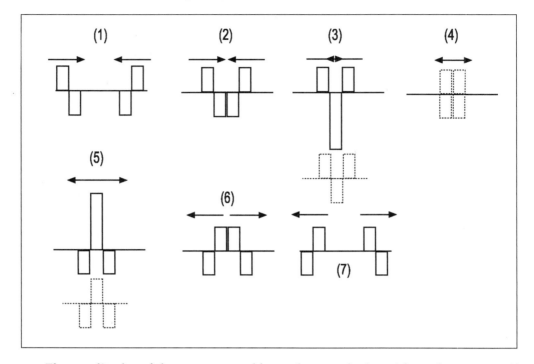

The amplitudes of the two waves add together, producing either a larger wave if they are both positive or both negative. They produce a smaller wave if one is positive and the other negative. In fact, as shown (4), they can produce no amplitude at all. The large amplitudes are called constructive interference. The reduced amplitudes are called destructive interference.

What are **dead spots** in **auditoriums**?

Poorly designed auditoriums can have dead spots. Dead spots are places where destructive interference occurs from the interaction of two or more sound waves. For example, a soloist on stage sends sound waves into the audience. Some of the waves hit the walls of the auditorium, while other waves travel directly to the listeners. In some situations, a direct wave can destructively interfere with a reflected wave so they cancel each other out at that particular location. As a result, the listeners seated in those particular seats would hear nothing from that soloist. Someone sitting a few seats over from the dead spot, however, might not experience the destructive interference and would hear the soloist just fine. (Refer to the chapter on Sound for handy answers dealing with acoustical engineering.)

What is a **standing wave**?

The example used above showed what happened when two single waves going in opposite directions met. A continuous wave is a set of single waves, one after another. You can produce such a wave by shaking one end of a rope up and down at a constant frequency. Now, if the other end of the rope is tied to something that doesn't move, the wave will be reflected back toward you. If you shake the rope at the correct frequencies the two waves will overlap each other and will seem to stand still, producing a standing wave. Two distinct regions on a standing wave can be seen. At certain points the rope won't be moving. That point is called a node. The point where the motion of the rope is largest is called the anti-node.

The nodes are locations of destructive interference where the two waves moving in opposite directions have opposite amplitudes; the crest of one wave and the trough of the other are at the same location. The anti-nodes are at locations of constructive interference where the two waves have both positive or negative amplitudes; that is, both are either crests or troughs.

The frequencies that produce standing waves depend on the length of the rope and the velocity of the wave on the rope. The lowest frequency will have nodes at the two ends and an anti-node in the center. The next higher frequency will have a node in the center and ends and two anti-nodes. At each higher frequency the number of nodes and anti-nodes increases by one.

How are **standing sound waves** generated in **musical instruments**?

Many instruments depend on standing waves to produce their sound. Standing waves are created on the strings of a guitar, piano, or violin, and in the air columns of a trumpet, flute, or organ pipe. The string is caused to oscillate either by plucking it (pulling it aside and then letting go), or by bowing it, where the stickiness of the horsehair on the bow also pulls the string aside and then releases it. In a piano a felt-covered hammer strikes the string, starting it vibrating. In a trumpet or other brass instrument the player's vibrating lips create the traveling sound wave that is reflected when it reaches the open end of the instrument. In a flute or organ pipe that mimics a flute, the player blows air over a hole. The moving air interacting with the hole produces a periodic change in the pressure of the air inside the tube, which creates the travel-

Many musical instruments, such as clarinets and violins, depend on standing waves to produce sound.

155

ing sound wave. In a clarinet or saxophone the player blows through a narrow gap between a flexible piece of bamboo called the reed and the mouthpiece. Oboes and bassoons have two reeds separated by a thin gap. The stream of air causes the reed to vibrate, periodically stopping the air flow and causing the sound wave.

In order to change the pitch produced by an instrument the standing wave inside the instrument must be altered. By changing the length of a wind instrument, or the tension and length of the strings for a string instrument, a different frequency standing wave is produced, which creates a different musical pitch. Pressing a key on a trumpet inserts an additional length of tubing into the instrument. On a flute, clarinet, or saxophone a hole on the instrument is covered or uncovered, changing the effective length of the instrument. On a piano each note uses a string of a particular length. On a violin or guitar the player's finger is used to change the length. Thicker strings have lower pitches than thinner strings of the same length. Increasing the tension on a string increases the frequency of the standing waves, and thus the pitch.

RESONANCE

What is **resonance** and how can it be achieved?

All objects that can vibrate have a natural frequency of oscillation. If you hold one end of a ruler on a desk and push down and then suddenly release the other end you will see it vibrate. The natural frequency depends on the material and its width, thickness, and length.

Resonance occurs when an external oscillating force is exerted on an object that can vibrate. When the frequency of the external force equals the natural frequency then the amplitude of the oscillation reaches a maximum. This condition is called resonance. A very small external force is needed to create a large oscillation amplitude.

You can explore resonance with a mass, like a yo-yo or a heavy metal washer on a string. Hold the top of the string still and pull the object to one side and watch it oscillate at its natural frequency. Then shake the top of the string at the same frequency and watch the amplitude of the oscillation increase. You will have found the resonance frequency. If you raise or lower the shaking frequency you'll find that the amplitude of the oscillation is smaller.

Where can **resonance** be found on the **playground**?

Children discover resonance early in life. When playing on swings, they use their arms and legs to pump themselves back and forth on the swing. They recognize that if they pull back on the chains every time the swing is at its largest backward displacement they will achieve the largest amplitude. If, however, the pull back at other times, or if a parent pushes at the wrong time, the amplitude will be decreased because the external frequency isn't at the natural frequency.

How can **resonance** cause crystal **glasses** to **break**?

Many years ago Ella Fitzgerald performed a physics experiment in an advertisement for Memorex® audio tape cassettes. The company claimed that the famous singer could create a pure tone at just the right frequency to cause a crystal wine glass to break and that the Memorex® tapes recorded and played back sounds so accurately that the glass would break both when Ms. Fitzgerald sang and when the recorded sound was played. "Is it live or is it Memorex®?" was the advertising question.

Although it is hard to think that glass is something that can bend, if you tap the rim of a thin wine glass you can hear it "ping." The shape of the rim of the glass oscillates. When the amplified sound waves pushed on the glass it distorted its

Sound can make glass break when amplified because the sound waves can cause the glass to oscillate and bend. When enough kinetic energy is used, the sound waves can distort the glass to the point where it cracks or shatters.

shape. Some of the kinetic energy in the sound wave was transferred into the kinetic energy of the oscillating glass. When the frequency of the sound wave matched the natural frequency of the glass the amplitude of oscillation was large enough to shatter the glass.

How did **resonance destroy** the **Tacoma Narrows Bridge** in Washington State?

The Tacoma Narrows Bridge, or "Galloping Gertie" as it was often called, was built in 1940 and was known for its unusual, undulating movement. All bridges vibrate to some extent, but to many motorists, the suspension bridge in Tacoma felt more like an amusement park ride than a bridge.

On the morning of November 7, 1940, four months after the bridge opened, the wind was blowing at approximately 42 miles per hour. This moderate wind hit the solid bridge deck and caused the deck to vibrate back and forth as it did almost every day since the bridge had opened. But the bridge began to vibrate more dramatically than ever before. It appeared as though a standing wave had formed between the two

If the energy of a resonant standing wave is large enough, a crystal wine glass can shatter quite easily, but when the amplitude is smaller, the wine glass can produce a sound instead. Take, for example, a person rubbing his finger around the moist lip of a crystal wine glass. The glass seems to sing or hum. The humming is caused by the rubbing of the wet finger on the glass that causes it to have a standing wave pattern. The resonating glass generates enough energy to vibrate the surrounding air and create a steady humming sound. In 1761 Benjamin Franklin (1706–1790) invented the glass harmonica (that he called the armonica) in which wine glasses of various sizes were fastened to a rotating shaft. The musician rubbed his fingers on the appropriate glass to play music. Wolfgang Amadeus Mozart and almost 100 other composers wrote music for the glass harmonica.

towers of the bridge. There was one distinct node in the center of the bridge and an anti-node on each side of the node. After several hours of dramatic vibrations, the bridge deck collapsed into the river below, along with its only casualty, a dog named "Tubby," left in a car by its owner, who narrowly escaped death himself.

There is still an active controversy about the exact cause of the collapse of the bridge. Was it the steady wind, or variable winds? What factor did the solid deck and the solid, high sides to the deck play? Was the deck just not stiff enough? Suffice it to say that more recently built bridges use perforated decks and lower, open sides.

What is a **torsional wave**?

A torsional wave, such as the wave that the Tacoma Narrows Bridge created, is a wave that is not only displaced vertically, but twists in a wave-like fashion as well. The torsional wave on the Tacoma Narrows Bridge achieved resonance in two orientations. The first resonance was seen as the undulating movement that took place over the length of the bridge, while the second resonance, seen as the twisting motion, occurred from side to side on the bridge.

IMPEDANCE

What is **impedance matching**?

Impedance is the opposition to wave motion exerted by a medium. When a wave travels from one medium into another, the impedance changes, causing some of the energy of the wave to be reflected back into the original medium. Therefore not all of the wave's energy travels into the new medium. An impedance matching device between the two media allows for a smooth transition in impedance and reduces reflections.

What are **transformers**?

An impedance matching device is called a transformer. Instead of an abrupt change between the two media, a transformer provides a smoother, gradual transition from the old to the new medium. Depending upon the wave and the medium, different transformers, such as quarter-wavelength and tapered transformers, can be used to help minimize reflection.

An example of a tapered transformer can be found in soundproof rooms or sound studios. Any sound that is produced is supposed to be absorbed by the impedance matching material on the walls. Special foam, tapered in a V-like shape, is used as a transformer to gradually absorb all the sound into the walls. The gradual changeover from the air medium to the wall medium prevents sound from reflecting back into the air.

An example of a quarter-wavelength transformer can be found on many camera lenses and eyeglasses. The quarter-wavelength thick coating on a lens is used to reduce reflections off the lens surface, allowing more light into the lens.

Electrical transformers are also used to match impedances by changing the varying voltages and currents in an electronic circuit. Modern electronic circuits make very minimal use of transformers because of their weight and size.

THE DOPPLER EFFECT

What is the **Doppler Effect**?

The Doppler Effect is the change in frequency of a wave that results from an object's changing position relative to an observer. A well-known example of the Doppler Effect is when an ambulance zooms by you and makes a "wheee-yow" sound. The high-pitched "whee" is caused by sound waves that are bunched together because the ambulance is moving in the same direction as the emitting sound waves. The bunching together of sound waves creates an increase in the frequency and results in a higher-pitch sound. The low-pitched "yow" sound occurs when the vehicle moves away from the propagation of the sound wave. Since the ambulance moves away from the sound wave, the spacing between successive waves becomes greater. This decrease in the frequency of the sound wave results in a lower pitch.

Who was the **Doppler Effect** named after?

Johann Christian Doppler (1803–1853), the Austrian mathematician for whom the Doppler Effect is named, proposed in 1842 that the color of double stars rotating about each other would depend on whether the star was approaching or receding from Earth.

159

The effect was too small to be measured. But in 1845 Christophorus Henricus Diedericus Buys Ballot (1817–1890) set up an experiment using had two sets of trumpeters. One set remained at rest while the other was on an open railway car traveling at the then fantastic speed of 40 miles per hour. Although both sets of trumpeters played the same note, the change in tone was clearly heard. Doppler later extended his theory to the case when both sound source and observer were moving. French physicist Hippolyte Fizeau (1819–1896) later extended Doppler's theory to light.

What is the difference between a **red shift** and a **blue shift**?

The visible color spectrum ranges from the low-frequency red, orange, and yellow, to the higher-frequency green, blue, indigo, and violet. Astronomers observing the planets, stars, and galaxies use the Doppler Effect to measure the velocity at which objects are moving, rotating, or revolving. The faster the object is moving, the more the frequency is shifted. Most galaxies are moving away from us and their light is red-shifted. In general, the further away, the greater the red shift. Recently astronomers have detected more than 400 planets revolving about other stars using the Doppler Effect. The gravitational force of the planet on the star causes the planet and star each to circle around a common point, usually close to, but not in the center of, the star. As a result the star "wobbles" with the same period as the orbital period of the planet.

The effect is truly tiny. Jupiter causes the sun to wobble in a circle with a speed of 12 meters per second. By measuring the Doppler Effect in the light from the star they can find its velocity and how it changes over time. With that information they can determine the period, distance from the star, and mass of the planet. Most discoveries have been of extremely massive planets, but recently a planet with a mass only a few times that of Earth was detected. It is close to a dim reddish star and the temperature of its surface is estimated to be tens of degrees below the freezing point of water. If the planet has greenhouse warming it might be able to sustain life.

What does the fact that most **galaxies** are seen with a **red shift** mean to **astronomers**?

The fact that astronomers observe most of the galaxies in the universe as having a red shift means that overall, galaxies are moving away from our galaxy, the Milky Way. This can only be happening if the universe as a whole is expanding. The expansion of the universe led to the development of the big bang theory of the universe's creation.

How do the **police** use the **Doppler Effect** in **radar guns**?

The police use the Doppler Effect when checking for speeding vehicles. A radar gun sends out radar waves at a particular frequency. As the radar wave hits a vehicle, the wave reflects back toward the radar gun at a different frequency. The frequency of the reflected wave depends upon the direction and speed of the vehicle. The faster the speed, the greater the frequency change. The radar gun determines the speed of the vehicle by measuring the difference between the emitted frequency and the reflected frequency and computing the speed from that measurement.

Radar guns used by police to monitor traffic speeds work by taking advantage of the Doppler Effect.

RADAR

What is **radar**?

Radar is an acronym for "RAdio Detection And Ranging." A radar installation emits electromagnetic waves and detects the waves reflected from an object. It measures the time for the "echo" to return to find the distance of the object. The radar dish is constantly rotating, permitting it to find the direction of the object as well. Radar is used in many different arenas, but was first used in World War II to detect the approach of enemy bombers.

Who developed **radar**?

Radar was developed independently in many countries in the 1930s. But, in 1935, Robert Watson-Watt (1892–1973), a Scottish physicist, was the leader of a group that created the first radar defense system for the British military. Although a large number of nations, from the United States and Canada, to Britain, France, and Germany, to the Soviet Union and Japan, worked to develop radar systems during the 1930s, the British system of ground-based radar stations were the first to use radar effectively in warfare. By the early 1940s radar systems were miniaturized enough to be installed in aircraft so that they could engage other aircraft in fights at night.

Ironically, Watson-Watt became a deserving victim of his own technology nineteen years later. According to Canadian police, Watson-Watt had been speeding on a stretch of Canadian road, and was detected by a police radar gun. Watson-Watt willingly paid the fine and drove away.

What is a **stealth plane**?

Stealth aircraft are planes that are able to avoid radar detection. The materials on the plane's surfaces and their peculiar shapes and angles deflect radar waves away from

161

How are today's fighter jets
using destructive interference to mislead enemy radar?

A French fighter plane developed in the 1990s known as Rafale uses a device to help the jet evade radar. The Rafale uses technology called active cancellation, which receives an incoming wave and sends out the direct opposite pattern of that wave, in this case a radar wave half a wavelength out of phase with the incoming radar. When the two waves interfere with one another, the waves experience destructive interference canceling out the signal. Because there is no return signal, the enemy can't find the location of the plane.

the plane, or in some instances, the plane's outer fuselage can actually absorb the radar waves without reflecting it back to the enemy radar transmitter. (Refer to the Fluids chapter for more information about aerodynamics and aviation.)

How has **radar** been used in **astronomy**?

In radar astronomy electromagnetic waves are aimed at planets. By analyzing the reflected signals, the position, velocity, and shape of objects in our solar system can be determined. In the early 1960s, radar was used to determine the exact distance between Earth and Venus and Earth and Jupiter. Later, radar was installed on the space probe *Magellan* to map the surface of Venus. Radar astronomy has been beneficial in determining distances in our own solar system, but the reflected signals would be much too weak from objects outside our solar system.

NEXRAD DOPPLER RADAR

What is **NEXRAD** Doppler radar?

NEXRAD, or next-generation weather radar, is one of the most recent technological breakthroughs for weather forecasting. NEXRAD relies on the Doppler Effect to calculate the position and the velocity of precipitation. The spherical NEXRAD radar tower emits radar waves 360° around and calculates the frequency shift of the reflected radar waves off rain, sleet, and snow. The NEXRAD computers then translate the information and represent the possible weather problems on a color-coded map for analysis. The maps are readily available in real time over the Web.

The goal and main function of NEXRAD precision radar is to save American money and lives by predicting threatening weather problems and warning the public before tragedy strikes. Meteorologists estimate that this new tool for weather forecasting has saved millions of dollars and many lives through its early warning systems. One of the most impressive advancements has been in pinpointing tornadoes and hurricanes more accurately than what was possible before NEXRAD.

Each NEXRAD station scans a radius of 125 miles with excellent accuracy, and less accurately up to 200 miles. A new system, developed since 1994, is Terminal Doppler Weather Radar, or TDWR. This system, installed at 45 airports, uses radar waves with 5-centimeter wavelength rather than the 10 centimeters (3.9 inches) used in standard weather radar. As a result it can resolve objects with twice as much detail, permitting it to detect wind shear and microbursts. Its range, however, is half that of NEXRAD and it can't see through heavy rain. Radar images are available to the public at www.radar.weather.gov.

RADIO ASTRONOMY

How is **radio astronomy different** from **radar astronomy**?

Radar astronomy measures the reflections of transmitted radio waves to determine an object's size, position, velocity, and surface characteristics. Radio astronomy is like optical astronomy but it uses the VHF, UHF, and microwave portions of the electromagnetic spectrum rather than the infrared and visible portions. Radio waves penetrate the dust that hides the centers of galaxies and obscures regions where stars are forming. They can also detect hydrogen gas that constitutes 85% of the known mass of the universe.

What do **radio astronomers hear**?

Radio astronomers detect what sounds like noise, but is actually signals from atoms, molecules, and ions in stars, galaxies, and particles in interstellar and intergalactic space. In order to detect these signals, radio telescopes are used. These are shaped like large satellite dishes, and are able to detect wavelengths between 1 millimeter and 1 kilometer.

Where is the **largest radio telescope**?

The largest single radio telescope is the 305-meter (1,000-foot) diameter Arecibo Telescope, located in Puerto Rico. The reflecting dish, made of 40,000 perforated aluminum panels, is located in a mountain valley. The actual antennas are in a 900-ton platform, suspended on cables 137 meters (450 feet) above the dish. The antennas, sensitive to frequencies from 50 MHz to 10GHz, detect signals collected by and reflected from the dish, and can be moved to determine the direction from which the signals are coming. Arecibo also performs radar astronomy using a 1 megawatt radar transmitter to send signals to the other planets. The telescope then detects the faint echos.

Highly sensitive telescope arrays like this one are used by radio astronomers to detect signals created by atoms, molecules, and ions from space.

How are **radio telescope arrays** used?

A group of 27 radio telescopes, each 25 meters (82 feet) in diameter, spaced as much as 13 miles apart from each other in Socorro, New Mexico, is known as the Very Large Array (VLA). The signals from this group of radio telescopes are combined by a computer. Constructive and destructive interference between the signals allows the array to make very accurate measurements of the location and size of objects.

Larger still are Very Long Baseline Interferometry radio telescopes. These radio telescopes are located around the world. Cables can't be used to combine the signals, so each telescope needs a very precise clock to determine the arrival time of the signals. Hydrogen masers, whose frequency is stable one part in a million billion, are used. Again, although the signal strength is small, the wide spacing increases the effectiveness of using interference to measure very fine details of the shape, size, and location of stars and galaxies.

SOUND

What is the **source** of every **sound**?

Sound waves are created by some type of mechanical vibration or oscillation that forces the surrounding medium to vibrate. A tuning fork is an excellent example of a vibrating sound source. When struck by a rubber mallet, the tines of the tuning fork vibrate, causing the air molecules around them to move back and forth at the same frequency, creating areas of compressions (where the molecules are close together and air pressure is slightly increased) and rarefactions (where the molecules are spread apart and thus the air pressure is reduced). These pressure changes then move away from the tines creating the longitudinal sound wave.

What **type of wave** is a **sound** wave?

A wave that consists of compressions and rarefactions—such as a sound wave—is called a longitudinal wave. The medium, the material through which the wave is traveling, does not get transferred from sender to receiver; the molecules only vibrate back and forth about a fixed position. The wave does, however, carry energy from its source to the receiver.

SPEED OF SOUND

How **fast** does **sound** travel?

Light travels almost one million times faster than sound—specifically, 880,000 times the speed of sound. Light and all electromagnetic waves travel at a speed of 3×10^8 meters per second, while the speed of sound is only about 340 meters per second or 760 miles per hour on a typical spring day.

The speed of sound compared to the speed of light can be observed at a baseball game. A spectator sitting in the outfield bleachers sees the batter hit a ball before she hears the crack of the bat.

Who determined that sound needs a medium through which to travel?

In the 1660s, English scientist Robert Boyle (1627–1691) proved that sound waves need to travel through a medium in order to transmit sound. Boyle placed a bell inside a vacuum and showed that as the air was evacuated from the chamber, the sound of the bell became softer and softer, until there was no sound.

What did Newton add to the knowledge of sound media?

Although he mainly concentrated on classical mechanics and the principles of geometric optics, Isaac Newton (1642–1727) did make several important discoveries in the field of sound. His major contribution was his work on sound wave propagation. He showed that the velocity of sound through any medium depended upon the characteristics of that particular medium. Specifically, Newton demonstrated that the elasticity and the density of the medium determined how fast a sound wave would travel. Because Newton worked before the field of thermal physics was developed, his theory has errors, but they are not significant to the calculation of sound speed.

How fast does sound travel in different media?

A simple model that explains the main factors affecting the speed of sound is a collection of balls (molecules) connected to each other by springs (bonds between molecules). Vibration from one ball will be transferred by the springs to neighboring balls, and in succession throughout the collection. The stiffer the springs and the lighter the balls, the faster the vibrations will be transferred. The springs are a model of the *bulk elasticity* (how the volume changes when the pressure on it changes) of the material, the balls and their spacing model the density of the material. In general, the speed is slowest in gases, fastest in solids. Even though liquids and solids are about 1,000 times denser than gases, the greater elasticity of liquids and solids more than compensates for the larger density. In gases the speed depends on the kind of molecule and temperature. For air, the speed depends only on temperature. The following table illustrates some examples of the speed of sound in different media.

Medium	Speed (m/s)
Air (0°C)	331
Air (20°C)	343
Air (100°C)	366
Helium (0°C)	965
Mercury	1,452
Water (20°C)	1,482
Lead	1,960
Wood (oak)	3,850
Iron	5,000
Copper	5,010
Glass	5,640
Steel	5,960

What is the **sound barrier**?

The sound barrier is the speed that an object must travel to exceed the speed of sound. The speed of sound is often used as a reference with which to measure the velocity of an aircraft. The speed of sound, 331 meters per second at 0°C, is Mach 1. Twice the speed of sound is Mach 2, three times the speed of sound is Mach 3, and so on.

What is a **sonic boom**?

A sonic boom occurs when an object travels faster than the speed of sound. The boom itself is caused by an object, such as a supersonic airplane, traveling faster than the sound waves themselves can travel. The sound waves pile up on one another, creating a shock wave that travels through the atmosphere, resulting in a "boom" when it strikes a person's ears. A sonic boom is not a momentary event that occurs as the plane breaks the sound barrier; rather, it is a continuous sound caused by a plane as it travels at such a speed, but the shock wave travels with the plane, so we hear it only when the plane is in one location.

All objects that exceed the speed of sound create sonic booms. For example, missiles and bullets, which travel faster than the sound barrier, produce sonic booms as they move through the atmosphere. The shockwave created by an F-15 fighter plane, for instance, is visible.

Supersonic planes, such as this F-15 fighter jet, can go beyond Mach 1. When they do, they generate a "sonic boom" caused by sound waves piling up and creating a shock wave.

Does **sound travel faster** on a **hot** or **cold day**?

Air molecules move faster in hot and humid environments due to their increased thermal energy. Since sound relies on molecules bumping into one another to create compressions and rarefactions, increased speed of molecules makes the sound waves move faster. The speed of sound increases by 0.6 meters per second for every degree Celsius increase in temperature. Water vapor in the air also increases the speed of sound because water molecules are lighter than oxygen and nitrogen molecules. Therefore, on hot and humid days, sound travels faster than on cool, dry days.

The following formula gives the speed of sound in air: Speed of Sound = (331 m/s) $(1 + 0.6\ T)$ where temperature, T, is measured in Celsius.

How **far away** was that **lightning**?

When a lightning bolt goes from cloud-to-cloud or cloud-to-ground it suddenly heats the air through which it passes. This sudden increase in temperature causes thunder that occurs at the same time as the lightning. Although it takes virtually no time to see the lightning, depending upon the observer's distance from the lightning, it can take quite a while to hear the thunder.

The speed of sound at room temperature is about 1,100 feet per second. A mile is 5,280 feet, so it takes about five seconds for sound to travel one mile. So, to determine how far away lightning has struck, when you see the lightning, count the number of seconds it takes before hearing thunder. Divide the number of seconds between the lightning and thunder by five to determine how many miles away the thunder and lightning occurred. For example, if you see a flash of lightning and approximately 10 seconds later you hear the thunder, divide the 10 seconds by the 5 seconds per mile to find that the lightning occurred 2 miles away.

HEARING

How does a person **hear sound**?

The ear is the organ used to detect sound in humans and some animals. The ear consists of three major sections: the outer ear, middle ear, and inner ear. The outer ear, the external section of the ear, consists of a cartilage flap called the pinna. The pinna's size and shape forms a transformer to match the impedance of the sound wave in air to that at the end of the ear canal by gradually funneling the wave's sound energy into the ear. To hear more sound, people can increase the size of the pinna by cupping their hand around the back of the pinna—in effect, increasing its size and funneling capabilities.

Once the sound has entered the ear canal, it moves toward the eardrum, where the longitudinal waves cause the eardrum to move in and out depending upon the frequency and amplitude of the wave. The middle ear includes the eardrum, the hammer, anvil, and stirrup, the three smallest bones in the human body, and the oval window on the inner ear. The eardrum is 17 times larger than the oval window, and this difference in area makes the inner ear act like a hydraulic machine, increasing the changes in pressure on the eardrum to that on the oval window at least seventeen-fold. The three bones link the eardrum to the oval window. They act like a lever system to further increase pressure on the oval window. The mechanical advantage of the lever system varies with frequency, peaking at about 5 in the 1 to 2 kilohertz frequency range. Thus the middle ear is like a complex machine that can amplify the pressure on the eardrum by a factor approaching 100 as it transfers the energy of the sound wave to the inner ear. Muscles connected to the eardrum and the three bones can react to very loud sounds and reduce the sensitivity of the ear, thus protecting it from damage.

The inner ear is a series of tubes and passageways in the bony skull. It consists of the cochlea, a spiral-shaped tube that changes the longitudinal sound wave into an electrical signal on the nerves connecting the ear to the brain. The inner ear also has three semicircular canals that sense the body's motions and give rise to a sense of balance. The oval window is at one end of the cochlea. Sound waves transmitted through the middle ear to the oval window cause a traveling wave in the fluid of the cochlea. This wave in one of the three tubes within the cochlea, the Organ of Corti, causes hair cells, called cilia, to tilt back and forth. The tilting causes chemicals to pass through channels in the nerve, creating electrical impulses that travel along the auditory nerve to the brain for analysis. The further the hair cells are from the oval window, the lower the sound frequency to which they are sensitive. Thus different nerves are excited by different frequencies, allowing the ear to distinguish the sounds' frequencies.

What are the **frequency limits** of the **human ear**?

The ear allows humans to hear frequency ranges between 20 hertz and 20,000 hertz, but it is most sensitive to frequencies between 200 hertz and 2,000 hertz.

169

The lower and upper fringes of this bandwidth can be difficult to hear, but many people—especially younger people—can hear these frequencies quite well. As people age their sensitivity to high frequencies diminishes. Damage to the hair cells caused by exposure to loud sounds also reduces the ear's sensitivity to high frequencies.

What are the **bandwidths** of hearing for **other animals**?

Animal	Lowest Frequency	Highest Frequency
Human	20 Hz	20,000 Hz
Dog	20 Hz	40,000 Hz
Cat	80 Hz	60,000 Hz
Bat	10 Hz	110,000 Hz
Dolphin	110 Hz	130,000 Hz

ULTRASONICS AND INFRASONICS

What are **ultrasonic sounds**?

Ultrasonic sounds are those frequencies above human hearing. Frequencies above 20,000 Hz are not heard by people, but are heard by those animals whose hearing is sensitive to ultrasonic frequencies. For example, dolphins use ultrasonic frequencies to communicate and bats use ultrasonic sounds as a tool for navigation and hunting.

What is **ultrasound**?

Ultrasound is a method of looking inside a person's body to examine tissue-based and liquid-based organs and systems without physically entering the body. Ultrasound sys-

Dolphins can detect frequencies of between 110 and 130,000 Hertz. Their hearing is over six times more sensitive than that of a human being.

tems direct high-frequency sound (usually between 5 and 7 megahertz) into particular regions of the body, and measure the time it takes for the sound wave to reflect back to the machine. By analyzing the pattern of reflections received, a computer can create a visual representation of the interior of the body.

Ultrasound is sometimes used instead of X rays because it does not use ionizing radiation and thus is safer for the person being examined. Obstetricians use ultrasound to examine the progress and/or problems that a fetus might be experiencing. Ultrasound is also used to observe different fluid-like organs and systems in the body such as the nervous, circulatory, urinary, and reproductive systems.

Ultrasound can also be used to pulverize kidney "stones." In this application very intense, tightly focused beams of high-frequency sound are directed at the stone. The stones are shattered and the small pieces can be passed through the urinary track with little or no pain.

What is **sonar**?

Sonar, an acronym for "SOund NAvigation Ranging," is a method of using sound waves to determine the distance an object is from a transmitter of sound. The sonar contains a transducer that converts an electrical impulse to sound when it transmits and a sound wave to an electrical impulse when it receives. Sound waves, usually brief pulses of ultrasound, are emitted from the transducer, reflected off an object, and

171

Tornadoes generate subsonic sounds that cannot be heard by humans but *can* be detected by instruments that can then be used to warn people up to a hundred miles away of approaching danger.

reflected back to the transducer. Electronic circuits measuree the length of time it took for the sound waves' round trip, and use the speed of sound to calculate the distance the object is from the transducer.

Sonar is used predominantly as a navigational tool by humans and animals. Dolphins and bats, among other animals, use sonar for navigation, hunting, and communicating. Machines such as depth-finders on boats, distance meters used in real estate and construction, and motion detectors for security devices all employ sonar.

What is **infrasound**?

Whereas ultrasonic sounds are frequencies above the human bandwidth of hearing, infrasonic (or "subsonic") sounds are those frequencies below the human bandwidth of hearing, or 20 hertz. Infrasounds as low in frequency as 0.001 hertz are produced by a variety of natural sources like earthquakes and volcanoes, as well as man-made structures and nuclear explosions. Elephants are known to make sounds as low as 12 hertz, and can communicate in this way over large distances.

How can **infrasonic sounds** provide **early warning** of **tornadoes**?

Using sound sensors that detect infrasound, scientists discovered quite by accident that the spinning core or vortex of a tornado produces sounds that are a few hertz below the human bandwidth of hearing. The tornado, much like an organ pipe, produces low frequencies when the vortices are large, and higher frequencies when the vortices are small. Since the infrasound waves from tornadoes can be detected for up to 100 miles away, they could help increase the warning time for tornado strikes.

INTENSITY OF SOUND

What is **sound intensity**?

Sound intensity is the power in the sound wave divided by the area it covers. Power is energy per unit time, so the intensity is the wave energy passing through a surface point divided by the time taken. The energy transferred by the wave is proportional to the square of the amplitude of the wave. In the case of sound the amplitude is the amount the peak pressure in compression is greater than the average pressure.

Why does **sound diminish** as you move **farther** from the source?

Sound waves do not travel in a narrow path, but spread out into the surrounding medium as spherical waves. As long as the energy in the sound doesn't change, the power per unit area decreases as the area increases. In some cases, there is some transfer of the sound's energy into thermal energy of the medium. Thus, there can be some loss of sound energy with distance, but this loss is usually small.

How much does a **sound's intensity diminish** as you move away?

When the total energy in the sound wave remains the same as the sound spreads over a sphere the intensity decreases as the area of the sphere increases. The area of a sphere is proportional to the square of its radius, so the intensity is inversely proportional to the square of the distance. That is, sound intensity diminishes according to the inverse square law. For example, if a person stands 1 meter (3.3 feet) away from a source, the sound intensity might be an arbitrary unit of 1. If that same person moved so she was 2 meters (6.6 feet) away from the source, the intensity would be 1 over the square of the distance, or one fourth the intensity. Again, if the listener moved 3 meters (9.8 feet) away, the intensity would be one ninth it was at the 1-meter mark.

Does **sound's intensity** always follow the **inverse-square law**?

Sound doesn't always spread uniformly. You may have experienced this if you shout while you walk through a tunnel. You hear echoes and a person at another part of the tunnel will hear you much more clearly than he or she would in an open area. Whispering chambers are often found in science museums. They are rooms with walls in an elliptical shape. If you stand at one focus of the ellipse and another person stands at the other focus you can hear each other speak even if you whisper while other people in the chamber are talking loudly. Sound can also be transported through the ocean with less spreading using an acoustic waveguide, similar to the way light can travel in an optical fiber (see the chapter on Light). The ocean water can separate into layers at different temperatures. The speed of sound depends on water temperature, and if there is a layer of cold water where sound speed is lower under a layer of warmer water where the speed is higher, then some of the sound moving through the cold layer will be reflected at the boundary back into the cold layer, keeping it from spreading. The name of this channel is SOund Frequency And Ranging (SOFAR) Channel. It is 100 to 200 meters below the surface in the cold waters off Alaska but 750 to 1,000 meters deep in the warmer Hawaiian waters.

What is a **decibel**?

A decibel (dB) is the internationally adopted unit for the relative intensity of sound. The sound intensity of 0 decibel is the threshold of human hearing, 10^{-16} watts/cm^2. This corresponds to a pressure of 2×10^{-5} newtons/m^2 or 2 billionths of atmospheric pressure. The ear is extremely sensitive! The decibel scale is a logarithmic scale, meaning for every 10 decibels, the intensity is increased by a factor of ten. For example, a change from 30 decibels to 40 decibels means the sound will be ten times more intense. A change from 30 decibels to 50 decibels would mean the new sound would be one hundred times more intense.

The following chart shows a typical sound environment, how many times louder those levels are than the threshold of human hearing, and the relative intensity of that sound compared to the threshold of hearing.

Sound	Times More Intencse	Relative Intensity (dB)
Loss of hearing	1×10^{15}	150
Rocket launch	1×10^{14}	140
Jet engine 50 meters away	1×10^{13}	130
Threshold of pain	1×10^{12}	120
Rock concert	1×10^{11}	110
Lawnmower	1×10^{10}	100
Factory	1×10^{9}	90
Motorcycle	1×10^{8}	80
Automobiles driving by	1×10^{7}	70
Vacuum cleaner	1,000,000	60
Normal speech	100,000	50
Library	10,000	40

Why do people hear ringing after leaving a loud rock concert?

After leaving loud rock concerts, many concert-goers often complain of ringing in their ears. The ringing sound is a result of the damage to the cilia by the intense sounds. Usually the ringing is gone the day after the concert, but permanent damage has already been done, because those hair cells will not recover. Although the effects of such hearing loss may take many years and repeated exposure to loud sounds to become apparent, they can nonetheless become very devastating.

Sound	Times More Intense	Relative Intensity (dB)
Close whisper	1,000	30
Leaves rustling in the wind	100	20
Breathing/whisper 5 meters away	10	10
Threshold of hearing	0	0

Are there **federal standards** for using **hearing protection**?

Federal regulations mandating the use of hearing protection in the workplace state that if an employee works for eight hours in an environment in which the average noise level is above 90 decibels, the employer is required to provide free hearing protection to those employees. For example, many high school and college students work for landscapers in the summer. Since the average decibel level for a lawnmower is about 100 decibels, and if the employees are working for eight or more hours per day, free ear plugs or ear muffs must be provided to the employees.

What is the **maximum decibel level** that a person can experience **without pain**?

The threshold of pain for humans depends on the person in question, but typically ranges between 120 decibels and 130 decibels. Such pain can be experienced at extremely loud rock concerts and next to jet engines and jackhammers, for example.

What are the best ways to **protect** one's hearing at **loud rock concerts**?

The first protection against damage to the cilia is to increase the distance from your ears to the speakers. In plain English, the farther away one is from the speakers, the lower the intensity of the sound. By simply doubling the distance, the intensity becomes one fourth of what it was originally. That can work in open areas where the

Loud music from live concerts can cause hearing loss over time by damaging the cilia within the ear. Standing or sitting farther back from speakers can help prevent harm, while musicians themselves often wear earplugs.

sound can spread, but in arenas or halls the sound can reflect off the ceilings and walls and does not decrease with distance.

The second method of protecting one's ears is to dampen the sound waves as they enter the ear. Many musicians, both rock and classical, to avoid gradual hearing loss, now use earplugs to decrease the amplitude of the wave entering the ear. The fluid in the cochlea transfers less energy to the cilia than if the listener were wearing no hearing protection.

What is the **difference** between **loudness** and **sound intensity**?

Sound intensity is a physical property that depends on energy. Loudness describes how a listener responds to sounds and is subjective. The ear doesn't respond equally to all frequencies, even between 20 hertz and 20 kilohertz. So, a sound with an intensity of 60 decibels will sound louder at some frequencies than others. The ear is most sensitive to frequencies between 1 kilohertz and 3 kilohertz and its sensitivity is much less for both low (20-100 hertz) and high (10-20 kilohertz). As people age their ears respond less to all frequencies, but especially frequencies above 5 kilohertz. Loudness also depends on the type of tone, whether a very pure tone, a more complex tone, or noise.

Loudness doubles for each 10-dB increase in sound intensity. It is measured in sones. Normal talking, which is between 40 and 60 dB has a loudness of 1 to 4 sones. Hearing damage from sustained sound, 90 dB, is 32 sones.

ACOUSTICS

What is **acoustics**?

Acoustics is the branch of physics that deals with the science of sound. Although sound has been studied since Galileo (1564–1642) made some predictions in the early 1600s, the ability to study sound has grown tremendously in the past few decades because of the advent of electronic sound generators and measuring devices. In addition to the study of sound itself, the field of acoustics has several applied sub-fields, the most important being speech and hearing, architectural acoustics, and musical acoustics.

What is **architectural acoustics**?

Theaters, concert halls, churches, classrooms, and sound-proof rooms have to be designed and constructed so that their acoustic properties match their uses. In some spaces it is important that the speaker, singer, or musician or group be heard clearly in all parts of the room. A good concert hall has intimacy—music should sound as if it were being played in a small hall. It should have liveness and warmth—fullness of bass tones. The sound should be clear, and the audience should be able to locate the source of the sound. The sound should be uniform over the entire hall, and sounds from the stage should be blended by the time they are heard by the audience. Finally, the hall should be free of noise, from air handling systems as well as outside sounds.

In order to achieve these goals an acoustical engineer must consider not only the size and shape of the hall, but also the acoustic properties of materials used on the floors, ceilings, and walls of the room, as well as chairs and other objects in the room. Even the audience and air humidity affect the acoustic properties of the room! In general hard surfaces such as concrete, plaster, wood, and tile reflect sound while soft materials like carpet, heavy drapes, and plush upholstered chairs absorb sound. Room shapes and sizes that create standing waves should be avoided because sound at frequencies at which the standing waves occur will not be uniform. They will be loud in some places, soft in others.

What is **reverberation time**?

The reverberation time for a sound is the time it takes for the echoing sound to diminish by a factor of 60 decibels, that is, to 1/1,000,000th of its original intensity. The longer the reverberation time, the more echoing is heard because the sound has reflected off walls and other surfaces. The shorter the reverberation time, the less echo is heard.

What **role** does **reverberation time** play in **acoustics**?

Reverberation time plays a major role in the quality of sound heard in a concert hall. Acoustical engineers carefully design concert halls to achieve a typical reverberation time between one and two seconds. Rooms designed for speech should have reverbera-

Sound-absorbing foam panels with bumpy surfaces are commonly used in recording studios to reduce echo.

tion times less than one second, movie theaters a little over one second, and rooms designed for organ music have as much as two seconds. If the reverberation time for middle and high notes is too short, the sound will diminish almost instantaneously and the room will sound "dry." A "full" bass tone requires a longer reverberation time for low notes. If the reverberation time is too long, however, as it is in many gymnasiums, the echoing effects will interfere with the new sounds, making music sound "mushy" and words of a speaker difficult to understand.

What determines the **amount** of **reflection** or **absorption**?

A good absorber of sound matches the impedance of sound waves between air and the new medium, while a poor absorber has a very different impedance and reflects the sound back into the environment. Typical absorbers have an open structure with holes of various sizes into which the sound waves can penetrate and transfer their energy to thermal energy of the material.

What **materials** are **effective absorbers** of **sound**?

Different materials will absorb certain frequencies of sound better than other frequencies, but some of the best absorbers of sound are soft objects. Materials such as felt, carpeting, drapes, foam, and cork are good at matching the impedance of a sound wave and reflecting back very little sound. Materials such as concrete, brick, ceramic tile, and metals, on the other hand, are effective reflecting materials of sound. That is why gymnasiums (with hardwood floors, concrete walls, and metal ceilings) have relatively long reverberation times, while concert halls furnished with upholstered seats, carpeted floors, and long drapes have relatively short reverberation times. People are also effective sound absorbers, so a full concert hall has different acoustic properties than an empty one.

What is the **optimal shape** of a **concert hall**?

Building a concert hall is a mixture of science, engineering, art, and politics. Politics is important because the people who provide the funds have goals for the use of the

What was the first concert hall to be designed by an acoustical physicist?

Boston Symphony Hall, designed by physicist Wallace Clement Sabine (1868–1919), is the first concert hall designed specifically to enhance the sound of an orchestra. Sabine, who designed the hall in the late 1890s, discovered the relationship between sound intensity, absorption, and reverberation time. Sound reflections can either enhance or ruin a sound. Sabine discovered that having strong reflections immediately after a sound was produced would enhance the acoustics, but if sound was reflected midway between the source and the listener it would detract from the acoustics because the time of travel would be significantly different.

Sabine's Boston Symphony Hall, built in 1900, established an excellent reputation for sound quality, mostly due to the choices of sound absorption materials as well as the strategic placement of reflecting material. The goal was to use the sound reflecting materials (high percent reflection ratios) to create strong initial reflections, while using sound absorbing materials (low percent reflection ratios) to absorb most of the energy from sound that would ordinarily reflect off of the high ceiling and the side walls in the rear of the hall.

hall. Consider, for example, the history of Philharmonic Hall in Lincoln Center, New York City. It was originally designed to be similar to the Boston Symphony Hall—long and narrow. But, a campaign led by one of the major newspapers in the city argued that the hall should seat more than 2,400 people, and so the architects made the hall wider. But when it was completed in 1962 critics were very unhappy with the sound. The wider hall did not have enough initial reflections and sounded dry. It was equipped with "clouds," reflectors hung from the ceiling, but reflections from them were too delayed to be effective. Another problem was that performers on the stage couldn't hear each other.

In 1973 Mr. Avery Fischer contributed over $10 million to support a complete reconstruction of the hall, and the hall is now named after him. The changes improved the acoustics, but the large stage reduced the loudness of bass sounds and the initial reflections were still too strong. Curved surfaces on the stage made of extremely tight-grain maple wood have improved bass problem, and reflectors consisting of 30,000 dowel rods were installed on the side walls.

Except for classical music concerts and opera, most performances use electronic amplification. Loud speakers can be located throughout the hall, and their response to different frequencies can be adjusted to compensate for shortcomings in the hall. The signal to speakers far from the stage can be delayed so sound from all speakers arrives at the same time. Further, these changes can be adjusted so that the hall has the best acoustics for any type of use.

MUSICAL ACOUSTICS

What is the **difference** between **pitch** and **frequency**?

Frequency, like sound intensity, is a physical property. Pitch, like loudness, is a description of how the ear and brain interpret the sound. Pitch is primarily dependent on frequency, but depends somewhat on loudness, timbre, and envelope, which will be discussed below. Humans hear pitch in terms of the ratio of two tones. The ear perceives two notes to be equally spaced if the they are related by a multiplicative factor. For example, the frequency of corresponding notes of adjacent octaves differ by a factor of 2. Notes in common chords are related by ratios of 3:2, 4:3, 5:4, etc. In the same way, the perceived difference in pitch between 100 hertz and 150 hertz is the same as between 1,000 hertz and 1,500 hertz.

Does a **musical tone** have a **single frequency**?

It does not have a single frequency, but many. To understand why, consider a stringed instrument like a guitar, piano, or violin. The string can oscillate in response to it being plucked, hit, or bowed. Standing waves will be formed as they would be if you shook a rope back and forth. By shaking it at different frequencies you can make it oscillate in several different modes. The lowest frequency results in nodes only at the ends. Twice this frequency produces nodes at the ends plus one in the middle. Three times the lowest frequency give nodes at the ends plus two nodes at 1/3 and 2/3 its length.

If you pluck, hit, or bow the string of a stringed instrument you cause it to vibrate in many of those nodes at the same time, depending on the location you plucked it. The lowest frequency of oscillation is called the fundamental frequency. Plucking the string 1/4 from one end results in oscillations at 2, 3, 4, 6, and 7 times the fundamental frequency. The higher frequencies are called harmonics. For example, if the fundamental frequency were middle C, 256 hertz, then the second harmonic would have a frequency of 512 hertz, the third 728 hertz, the fourth 1,024 hertz, etc.

The sound made by the vibrating string is very weak. On acoustic stringed instruments the strings pass over the bridge that transmits the oscillations to the top plate of the body of the instrument. Low frequency sounds also excite oscillations in the air and in the bottom plate of the guitar's body. Sounds from the oscillations of the air pass through the sound holes in the top plate into the surrounding air. The amplitude of the fundamental frequency is the largest. The relative amplitudes of the higher harmonics depend not only on the string, but the shape and size of the body of the instrument. Electric guitars will be discussed in the chapter on Magnetism.

The relative intensities of the higher harmonics depend on the instrument. The sound spectrum produced by the instrument is characterized by these relative amplitudes. The spectrum is also called the quality of sound, or the timbre. The sound quality also depends on how the sound starts and stops.

How do **wind instruments** like flutes, saxophones, and trumpets **produce sounds**?

In wind instruments the column of air is the oscillating object. The musician must create the oscillation. Perhaps you have blown over the top of a soda bottle and created a tone. When you blow some of the air goes into the bottle. That increased air pressure is reflected off the bottom of the bottle and returns to the top where it deflects the blown air upward. This process repeats, resulting in a tone whose frequency depends on the length of the bottle. The energy in your breath is converted into the energy of oscillation of the air in the bottle.

A flute works in a similar way, where the player blows over a hole in the side of the flute. The other end of the flute is open. The sound wave is reflected because the impedance in the tube is different from that of the room air. The spectrum of a flute contains all harmonics. If the player blows harder and changes the location of her top lip she can make the flute play one or two octaves higher. In other words, the fundamental frequency of the flute is increased by a factor of two or four.

The frequency of a flute or other woodwind instrument can be changed by opening holes along the side of the instrument. This shortens the length of the oscillating air column, increasing the frequency.

In a saxophone or clarinet the vibrations are caused by a thin piece of wood called the reed. The player blows through a gap between the reed and the instrument's mouthpiece. The pulse of air is reflected off the end of the instrument and returns to the reed, pushing it open to admit another pulse of air. Double-reed instruments like the oboe and bassoon work in the same way. The clarinet is shaped like a cylinder. Its spectrum consists of only the odd harmonics: 1, 3, 5, and 7, etc. A clarinet can be played in a higher register by opening a small hole near the mouthpiece that strongly reduces the amplitude of the fundamental tone. The new pitch is an octave and a fourth higher than the lower register. Saxophones are not shaped like cylinders, but like cones. As a result all harmonics are included in its spectrum, and opening the register key raises the instrument's pitch by one octave.

In a bugle, trumpet, trombone, French horn, or other brass instrument the oscillations are caused by the player's lips. The lips act as a valve, causing pulses of air into the instrument, which causes the oscillations in the air column. In a brass instrument the fundamental tone is absent. By adjusting the tightness of the lips the player can cause the instrument to play at the 2nd, 3rd, 4th harmonic, etc. The valves on a brass instrument add small lengths of tubing, lowering the pitch. In a trombone the length of the tube can be varied continuously, allowing any frequency to be played. The spectrum of a brass instrument depends strongly on its pitch and loudness. The louder it is played, the more energy there is in the higher harmonics.

How does a **synthesizer imitate** any **musical instrument**?

A synthesizer is an electronic device that generates, alters, and combines a variety of waveforms to produce complex sounds. Often a piano-type keyboard allows the musician to select the notes to be constructed. Synthesizers may use electronic circuits to create the tones or use software that controls a circuit that converts a digital number to a voltage. Some synthesizers use a computer rather than a keyboard to select the notes. The computer can then control electronic circuits through the MIDI interface. The most common method of creating the synthesized sounds is to use a frequency modulated synthesis that creates higher harmonics that match

Some synthesizers like this one use keyboards to select the sound to be generated, while others use computers. The ability of a synthesizer to immitate various instruments is accomplished using frequency modulated synthesis to create "voices."

those of the musical instrument being imitated, or create an entirely new musical sound. Instruments also have characteristic attacks, sustain times, and decays. Attack describes how fast the amplitude rises from zero to its full value. Sustain times describe how long the tone amplitude remains the same, and decay describes how the amplitude decreases at the end of the played note. Synthesizers can create hundreds of different sounds, typically called voices.

How is the **human voice similar** to a **wind instrument**?

The vibrating source of the human voice are the vocal cords in the throat. Their vibration, at a relatively low frequency of 125 hertz, creates oscillations in the air that fills the throat and mouth. By varying the size and shape of your mouth and position of the tongue you can change the frequency of the sound as well as the relative amplitude of the harmonics. Try it yourself! Sing a constant pitch, while you vocalize the vowel sounds—"a," "e," "i," "o," and "u." Note how you change the shape of your mouth and position of the tongue when you go from one vowel sound to another.

> ### Why does a singing duet sometimes sound as if there were a third voice contributing to the music?
>
> When two people sing loudly at slightly different pitches, the frequencies can mix, causing a difference tone, or third pitch. You may also experience this phenomenon with fire whistles blaring through a town, or even the beeps or tones emitted by a clock radio. In fact, many difference tones are intentionally created to enhance the sound.

What is the **difference** between an **overtone** and a **harmonic**?

A harmonic is a mode of vibration that is a whole-number multiple of the fundamental mode. The first harmonic is the fundamental frequency. The second harmonic is twice its frequency, etc. Many instruments, especially bells, oscillate in modes that are not whole-number multiples of the fundamental frequency. These higher modes are called overtones. Overtones include harmonics, but harmonics do not include overtones. Another confusing point is that the first overtone is not the fundamental. The second harmonic is the first overtone.

How does the **spectrum of a sound** relate to its **waveform**?

Jean Baptist Fourier (1768–1830), a French mathematician and physicist, made discoveries in a number of fields, including the greenhouse effect. He developed mathematical tools known as Fourier series and transforms that are used in a wide variety of applications. The Fourier Theorem states that any repetitive waveform can be constructed from a series of waves of specific frequencies, a fundamental and higher harmonics. The reverse is also true—if you add together waves of frequencies f, $2f$, $3f$, etc., specifying their amplitudes, you can construct a complex waveform. Using Fourier analysis, then you can record the waveform of a musical instrument and determine the amplitudes of the harmonics of which it is made, that is, its spectrum.

Today Fourier analysis is very easy to do. Most computers either have built in microphones or can use an external microphone. Free software can be downloaded from the web that will display the spectrum.

What are **difference tones**?

Difference tones are frequencies that are produced as a result of two different frequencies mixing with each other. This mixing can only occur if a device is non-linear. That is, if the output is not a multiple of the input. You can create difference tones with two toy slide whistles. Place both in your mouth and blow hard. Adjust their lengths so that the two tones are the same pitch. Then adjust one. As you move its pitch away from the other you will hear a low-pitched sound whose pitch increases as the two whistles' pitches get further apart. For example, if the high-frequency sound was 812 hertz, while the lower frequency was 756 hertz, the difference tone from the interfering sound waves would be 144 hertz. The non-linear device in this case is your ear!

NOISE POLLUTION

Why is **noise pollution dangerous**?

In the past, noise pollution was only thought to create health effects if the intensity was large enough to cause hearing damage. Studies over the past several decades, however, have found that long-term exposure to noise can cause potentially severe health problems—in addition to hearing loss—especially for young children. Constant levels of noise (even at low levels) can be enough to cause stress, which can lead to high blood pressure, insomnia, psychiatric problems, and can even impact memory and thinking skills in children. In a German study, scientists found that children living near the Munich Airport had higher levels of stress, which impaired their ability to learn, while children living further from the airport did not seem to experience the same problem.

What **limits** have been **established** to **reduce exposure** to **noise pollution**?

The World Health Organization has recommended that noise during sleep be limited to a level of 35 decibels, and governments are beginning to place restrictions on noise levels in both residential and business environments. In the Netherlands, for example, regulations specify that new homes may not be built in areas of high noise levels—those that exceed average noise levels of 50 decibels. In the United States, employers must provide hearing protection for those who endure noise levels of 90 decibels for more than eight hours a day.

> ### Does my neighbor's motorcycle have to be that loud?
>
> At times, engineers try to achieve just the right sound or noise for a particular product. The product, whether it's a vacuum cleaner, lawnmower, or motorcycle, needs to be quiet enough so as not to cause stress, yet has to have enough sound to seem powerful. For example, muffler technology has the ability to greatly reduce the noise a motorcycle produces, yet many engineers and manufacturers feel that consumers would not purchase the product if it does not sound "powerful" enough. Hybrid automobiles are so quiet that people with limited sight have difficulties knowing that they are approaching. Engineers are developing noise generating devices for these cars to let sightless people know the location of the cars.

What **methods** are being **used** to **reduce noise pollution**?

Since noise creates stress and can lead to other health problems, industries and governments around the world are working to reduce noise levels, especially around populated regions. One method of reducing noise pollution around airports has been rerouting airline traffic so that it passes over less-populated areas. Sound barriers have been installed along many highways to absorb and/or reflect sound away from houses built alongside the roads. In countries such as Austria and Belgium, roadways are being constructed with a material called whisper concrete that engineers claim reduces noise by 5 decibels. Finally, Swedish engineers have developed a road surface made of pulverized rubber that can reduce the noise level by as much as 10 decibels.

What is **active noise cancellation**?

Active noise cancellation, or ANC, creates a waveform that is the opposite of the noise so that the noise is cancelled. An ANC headset consists of a one or two microphones that detect the noise, electronic circuits that invert the waveform, and a headphone driver. Low-frequency noise is more successfully cancelled than noise at higher frequencies, so passive noise reduction is used to reduce high-frequency noise. As a result, low-frequency noises found in helicopter, jet engine, and muffler noise can be controlled using ANC while the high squealing sounds of jet engine noise are more difficult to cancel out.

If **anti-noise** is produced by the **headphones** can a person hear other people, music, etc.—sounds that aren't "noise"?

The objective of active noise cancellation is to cancel out the noise waveform by producing anti-noise to interfere with the original noise pattern. Since the result is less

185

noise, other sounds are easier to hear. ANC technology is now available for consumers in headsets that reduce noise and allows for easier communication in loud environments. As a result of this new technology, factory workers and helicopter pilots can communicate more easily, and their amount of stress due to noise pollution should be reduced. Airline passengers can listen to music through ANC headphones while noise from the airplane is cancelled or reduced.

What is **psychoacoustics**?

Psychoacoustics, which connects acoustics with psychology, is the study of how the mind reacts to different sounds. This field of study is especially important to consumer product manufacturing, because a consumer associates particular sounds with certain products or sensations. For example, people associate low-frequency rumbling sounds with power and torque, while higher-frequency sounds often represent high speeds and out-of-control occurrences. Psychoacoustics can play a major role in the development and commercial success of many products.

LIGHT

What is **optics**?

Optics is an area of study within physics that deals with the properties of and applications of light. Optics can be used not only with light, but with other parts of the electromagnetic spectrum, including microwaves, infrared, visible, ultraviolet, and X rays.

What were some **early ideas** about **light**?

Does light travel at a finite speed or is it infinite? Is light emitted by the eye or does it travel to the eye? These questions were debated for centuries. In ancient Greece, Aristotle (384–322 B.C.E.) argued that light is not a movement. Hero of Alexandria (10–70 C.E.) said it moves at infinite speed because you can see the stars and sun immediately after you open your eyes. Empedocles (490–430 B.C.E.) said it was something in motion, so it must move at a finite speed. Euclid (325–265 B.C.E.) and Ptolemy (90–168 C.E.) said that if we are to see something light must be emitted by the eye. In 1021 Alhazen (Ibn al-Haitham) did experiments that led him to support the argument that light moves from an object into the eye and thus it must travel at a finite speed. At the same time al-Biruni noted that the speed of light is much faster than the speed of sound. The Turkic astronomer Taqi al-Din (1521–1585) also argued that the speed of light was finite, and that its slower speed in denser objects explained refraction. He also developed a theory of color and correctly explained reflection.

In the 1600s the German astronomer Johannes Kepler (1571–1630) and the French philosopher, mathematician, and physicist René Descartes (1596–1650) argued that if the speed of light was not infinite, then the sun, moon, and Earth wouldn't be in alignment in a lunar eclipse. Despite this misconception, Kepler, in his 1604 book, *The Optical Part of Astronomy*, essentially invented the field of optics. He described the inverse-square law, the workings of a pinhole camera, and reflection by flat and concave mirrors. He also recognized the influence of the atmosphere on both eclipses and the

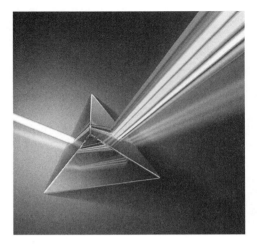

Isaac Newton used a prism to separate white light into a spectrum, leading to the publication of his findings in his *Theory of Colors* (1672).

apparent locations of stars. Willibrord Snellius (1580–1626) discovered the law of refraction (Snell's Law) in 1621. Descartes used Snell's Law to explain the formation of rainbows shortly thereafter. Christiaan Huygens (1629–1695) wrote important books on optics and proposed the idea that light was a wave. Isaac Newton's (1642–1727) celebrated experiments using a prism to separate white light into its colors led to Newton's Theory of Colors published in 1672. He recognized that telescope lenses would cause colored images and invented the reflecting telescope using a concave mirror that would not have this fault. Newton believed that light was made up of very lightweight particles, or corpuscles.

In 1665, a publication by Francesco Grimaldi (1618–1663) described how light could be diffracted when passing through thin holes or slits or around boundaries. In 1803, Thomas Young's (1773–1829) experiments with one and two slits demonstrated the diffraction and interference of light. Augustin-Jean Fresnel (1788–1827) and Simeáon Poisson (1781–1827) did both theoretical and experimental work that firmly established the wave theory of light in 1815 and 1818.

What is the **modern conception** of **light**?

As demonstrated by the work of Young, Poisson, and Fresnel, and later by James Clerk Maxwell (1831–1879), and Heinrich Hertz (1857–1894), light is an electromagnetic wave to which human eyes respond. It is located on the electromagnetic spectrum between infrared and ultraviolet. The limits of human vision define the lower and upper boundaries of light. The lowest frequency is 4×10^{14} Hz, which has a wavelength of 700 nanometers (700×10^{-9} m). Its upper boundary is 7.9×10^{14} Hz, a wavelength of 400 nanometers. Wavelength rather than frequency is commonly used when describing light because until the past three decades only wavelength measurements were possible—light frequencies were too high to measure directly.

Light has all the properties of a transverse wave. That is, it can transfer energy and momentum. It obeys the principle of superposition and can be diffracted and interfere with itself. On the other hand, it also has the properties of a massless particle. While in a medium it moves in a straight line at a constant velocity. It can transfer energy and momentum. A full description of the true nature of light—wave, particle, or both—will be discussed in the "What Is the World Made Of?"

How is **light emitted**?

You undoubtedly have seen the light emitted by hot objects. Whether it is the dull red glow of the heating coil on an electric range or the orange glow of the element in an electric oven or the bright yellow-white of the glowing filament of an incandescent lamp, you have seen light emitted by hot objects. Even the yellow glow of a fire comes from light emitted by hot carbon particles. Energy, usually from stored chemical energy, is converted into thermal energy. That energy is transferred to the surroundings by radiation, including light. Unfortunately, producing light in this way is very inefficient because about 97% of the

A close-up shot of a transparent LED. LED's use the same materials as laser pointers and operate more efficiently than incandescent lights.

energy goes into infrared radiation that warms the environment rather than light that can be seen. Because of their large energy use, many countries will be banning incandescent lamps in the next few years.

Light can also be emitted by gases and solids. Neon signs are one example of a gas that glows because electrical energy is converted into light energy. High-intensity lamps use either sodium or mercury vapor to produce intense light. Fluorescent tubes and compact fluorescent lamps (CFLs) use electrical energy to excite mercury atoms. The ultraviolet emitted by these atoms causes compounds deposited on the inside surfaces of the lamps to glow. The colors can be chosen to emulate incandescent lamps or daylight. CFLs convert up to 15% of the electrical energy into light.

Lasers, mostly used today in CD and DVD players and supermarket bar-code scanners and pointers, usually consist of a small crystal composed of a mixture of elements like gallium, arsenic, and aluminum. The lasers produce single-color, intense light that is emitted as a compact ray. The LED lights that are often used as on/off indicators, traffic lights, car taillights, stop, and turning lights also use electrical energy and the same kind of materials used in laser pointers to produce light that is radiated into many directions. White LEDs that are beginning to be used in home lighting are costly to produce, but are much more efficient and last much longer than incandescent lamps.

How is **light detected**?

Light carries energy, so a light detector must convert light energy to another form of energy. In most cases light is converted into electrical energy. In the eye, which will be discussed more thoroughly later in this chapter, light strikes a molecule called an opsin. The absorption of light changes the shape of the molecule, which results in an electrical signal sent on the optic nerve. In detectors used in digital cameras when light is absorbed in a semiconductor one or more electrons are released. The charge they carry produces a voltage that is then converted into a digital signal. In photographic film molecules consisting of silver and chlorine, bromine, or iodine are used. When light strikes these molecules it transfers its energy to electrons. The molecules are broken apart and a tiny crystal of metallic silver remains.

THE SPEED OF LIGHT

What is the **history of measurements** of the **speed of light**?

In 1638 Galileo (1564–1642) proposed a method of measuring the speed of light. Galileo would have one lamp and an assistant a great distance away would have a second lamp. The assistant was to uncover his lamp immediately when he saw Galileo uncover his own lamp. The speed could then be determined by measuring the time it would take the light to travel from Galileo to the assistant and back again. Galileo claimed to have done the experiment several years before 1638 but there was no record of his results. In 1667 the academy of sciences in Florence, Italy, carried it out between two observers a mile apart. They reported there was no measurable delay, showing that the speed of light must be extremely rapid.

The first measurement of the speed of light in a laboratory was by Hippolyte Armand Fizeau (1819–1896) in 1849. He used a beam of light that passed through the gaps between teeth of a rapidly rotating wheel, was reflected from a mirror 8 kilometers away, and returned to the wheel. The speed of the wheel was increased until the returning light passed through the next gap and could be seen. The speed was calculated to be 315,000 kilometers per second. Leon Foucault (1819–1868) improved on this a year later by using a rotating mirror in place of the wheel and found the speed to be 298,000 kilometers per second. He also used this technique to determine that light travels slower in water than in air.

The American physicist Albert Michelson (1852–1931) greatly improved Foucault's measurement using an eight-sided rotating mirror and a plane mirror located on Mount San Antonio, 35 kilometers (114,800 feet) away from the source on Mount Wilson in California. By measuring the speed of the rotating mirror and the distance between the mirrors, Michelson made the most accurate measurement of the speed of light to that date. In 1907, he was honored by being the first American to win the Nobel Prize in physics. In 1926 he made a new measurement that yielded 299,796 kilometers per second with an uncertainty of 4 kilometers per second.

After Maxwell published his theory of electromagnetism it became possible to calculate the speed of light indirectly from the relationships between an electric charge and its electric field and an electric current and the magnetic field it produces. In 1907 Rosa and Dorsey obtained 299,788 kilometers per second in this way. The uncertainty of 30 kilometers per second made it the most precise measurement to that date.

Research on microwaves used in radar during World War II led to a new method of measuring the speed of light. By 1950 Louis Essen reported a result of 299,792 kilometers per second, slightly more precise than Michelson's result. In the 1970s scientists at the National Institute of Standards and Technology in Boulder Colorado succeeded in directly measuring simultaneously the wavelength and the frequency of an infrared laser. From these two measurements they could calculate the speed of light: 299,792.4562 kilometers per second with an uncertainty of only 1.1 meters per second!

This new technique prompted an investigation of the length standard, the wavelength of light from the gas krypton. The new techniques discovered that the standard was "fuzzy" and had to be replaced. In 1983 the Conférence Générale des Poids et Mesures decided to fix the speed of light in a vacuum at 299,792.458 kilometers per second and define the meter as the distance light travels in 1/299792458 of a second.

Here is a summary of the history of measurements of the speed of light.

Date	Author	Method	Result (km/s)	Uncertainty (km/s)
1676	Ole Rømer	Jupiter's satellites	220,000	
1726	James Bradley	Stellar aberration	301,000	
1849	Hippolyte Fizeau	Toothed wheel	315,000	
1862	Leon Foucault	Rotating mirror	298,000	±500
1879	Albert Michelson	Rotating 8-sided mirror	299,910	±50
1907	E.B. Rosa and N.E. Dorsey	Electromagnetic constants	299,788	±30
1926	Albert Michelson	Rotating 8-sided mirror	299,796	±4
1947	Louis Essen and A.C. Gordon-Smith	Cavity resonator	299,792	±3
1958	K.D. Froome	Radio interferometer	299,792.5	±0.1
1973	Evanson et al.	Lasers	299,792.4562	±0.0011
1983		Adopted value	299,792.458	

Source: K.D. Froome and L. Essen, *The Velocity of Light and Radio Waves,* Academic Press (London, New York) 1969.

What **astronomical methods** have been used to **measure** the **speed of light**?

The Danish astronomer Ole Rømer (1644–1710) measured the orbital period of Jupiter's innermost moon, Io. He found the period was shorter when Earth was approaching Jupiter than when it was moving away from it. He concluded that light travels at a finite speed and estimated that it would take light 22 minutes to travel the diameter of Earth's orbit. Christiaan Huygens combined this estimate with an estimate for the diameter of Earth's orbit. He concluded that the speed of light is 220,000 kilometers per second.

In 1725 the English astronomer James Bradley noted that the location of a star changed with the seasons. He proposed that the shift was due to the addition of the speed of light and the speed of Earth in its orbit. Bradley observed the shift in several stars and determined that light travelled 10,210 times faster than Earth in its orbit. The modern result is 10,066 times faster.

How **long** does it take **light** to **travel** certain distances?

The table below lists some travel times for light.

Distance	Time
1 foot	1 nanosecond
1 mile	5.3 microseconds
From New York City to Los Angeles	0.016 seconds
Around Earth's equator	0.133 seconds
From the Moon to Earth	1.29 seconds
From the Sun to Earth	8 minutes
From Alpha Centauri to Earth	4 years

Some of the destinations above would require light to travel in a circular or curved path, which it does not do.

What is a **light-year**?

A year is a unit of time, but a light-year is a unit of distance. Specifically, it is the distance that light travels in one year. Since light travels at 299,792 km/s and a year consists of 31.557×10^6 s, a light-year is 9.4605×10^{12} km or about 6 trillion miles.

The light-year is used to express distances to stars and galaxies.

Does the **speed of light** depend on the **medium**?

Waves travel at different speeds when traveling in different materials. In the vacuum of space, light travels at 299,792 km/s. When light encounters a denser medium, however, like Earth's atmosphere, it slows down ever so slightly to 298,895 km/s. Upon striking water it slows down rather dramatically, to 225,408 km/s, three quarters of its original speed. Finally, when light passes through the dense medium of glass, it slows to only 194,670 km/s. The ratio of the speed of light in a vacuum to that in a medium is called the *refractive index,* represented by the symbol *n*. The refractive index will be discussed in further detail in the section on lenses.

How is the **brightness** of light **measured**?

There are two separate systems to measure the intensity of light. The first is a physical system that measures energy or power transferred. The second system measures the effect of light on the eye— in other words, how bright we see the light. You are familiar with the watt. It's the rate at which energy is transferred. The equivalent unit for light, the luminous power, is the lumen. If you look at the box a lamp comes in you'll find both the electric power it dissipates in watts (W), and the luminous power in lumens (lm). For example, a 25-W clear bulb emits 200 lm. A 100-W lamp emits 1,720 lm. A 60-W halogen lamp emits 1,080 lm. Compact fluorescent lamps produce more light for the same power: a 25-W lamp is rated at 1,600 lm.

Light travels at slower speeds through mediums denser than a vacuum, such as air or water or glass. Light travels at about 225,400 km/s through water versus 299,792 km/s in a vacuum.

The intensity of light depends on the degree to which the lamp spreads or focuses the luminous power. If the light goes into all directions it won't be as intense as it would be from a reflector spot lamp that reflects light to form a narrow beam. The unit in which light intensity is measured is the candela (cd). If a 100-W, 1,720-lm lamp could spread the light into all directions it would have an intensity of 137 cd. But, if the same lamp were a spotlight, concentrating all the light into a 30° angle, then the intensity would be about 640 cd.

As the distance between the lamp and the surface illuminated increases, the illuminance provided by the lamp decreases according to the inverse square law. Suppose you have a light luminous power of 1,000 lm 1 meter (3.2 feet) away from the surface. If you now moved the lamp to 2 meters (6.5 feet) away, the illuminance would be 1,000 lm divided by the distance squared, or $1/2^2$. That is, 1/4 of the illumination at 1 meter. If you moved the lamp to 3 meters (9.8 feet) away, the illuminance would decrease to one-ninth the original illuminance.

When looking through a pair of polarized sunglasses, why do the rear windows in cars appear to have spots on them?

The spots seen on rear windows when wearing polarized sunglasses are the stress marks of the plastic layer in between the two layers of glass in this safety glass. The spots, created during the manufacturing of the glass, act like polarizing filters and therefore block some of the light, creating small, circular, dark regions in the otherwise transparent glass.

POLARIZATION OF LIGHT

What is **polarized light**?

Light is an electromagnetic wave that consists of an oscillating electric field and an oscillating magnetic field. The two fields are perpendicular to each other. In most cases the direction of the electric field has no preferred direction because different parts of the source produce light with electric fields in different directions. Such light is unpolarized. Some sources, however, like laser pointers, emit light whose electric field is always in one direction. Such light is called polarized.

How does **light become polarized**?

Light can be polarized in two ways. If a shiny surface such as water or an automobile window reflects light, it can be partially polarized. Or, you can use a polarizing filter that passes only the light that has the electric field in one orientation. A Polaroid® filter has long, thin molecules, all oriented in the same direction, embedded in plastic. The molecules absorb light that has an electric field parallel to the long dimension of the molecules. Therefore the light passing through the Polaroid® has an electric field perpendicular to the long dimension of the molecules.

Why are **polarized glasses** useful?

Polarized glasses are useful when driving, sailing, skiing, or in any situation where unwanted glare is present. Glare is caused by light reflecting off a surface, such as water, a road, or snow. Such light is polarized. Navigating your way through some situations could be difficult without polarized sunglasses.

Take, for example, light reflecting off the surface of a lake. The light polarized parallel to the water's surface is reflected with a greater intensity than that polarized perpendicular to the surface. So, sunglasses that pass vertically polarized light will reduce the glare from the water.

How can you **check** to make sure a pair of **sunglasses** is **polarized**?

Polarized lenses are transparent only to light that has its electric field oriented in one direction. Therefore, if a pair of sunglasses is polarized, two pairs of the same sunglasses, when aligned perpendicular to each other, should not allow any light to pass through the lenses.

Light from the sky is also partially polarized due to the scattering of light off gas molecules in the atmosphere, so if you put on a pair of polarized sunglasses and tilt your head so that your ear were near your shoulder, you should see a change in the intensity of the sky on a clear, sunny day. If you see no such change, then the sunglasses are not polarized.

How do **LCD devices** use polarized light?

LCD stands for Liquid Crystal Display. A liquid crystal is composed of long, thin molecules that are free to move like a liquid but organize themselves in a regular array like a crystal. In an LCD display the liquid crystal material is in a thin layer between two glass sheets. The bottom sheet is rubbed in one direction so that the molecules in the liquid crystal touching the surface align themselves with the rubbing. The top sheet is rubbed in the perpendicular direction, aligning the molecules touching it in that direction. As a result, over the thickness of the liquid crystal material the direction of the long axis of the molecules rotate through 90°. Polarizers are placed on the outsides of the glass sheets in the same direction as the rubbing. When light enters the back of the display it is polarized. As it passes through the crystal its polarization direction is rotated through 90° so that it passes through the second polarizer. Thus light passes through the display; it appears bright.

Each glass sheet is coated with a thin, electrically conductive layer. If a voltage is placed across the sheets the molecules align themselves with the electric field. The molecules no longer rotate the direction of polarization of the light, so no light passes through the display. It appears dark. By varying the voltage different degrees of darkness can be produced.

An LCD television screen uses liquid crystals composed of long, thin molecules between two glass sheets. An electrical field rotates the molecules in the prixels so that polarized light either passes through or is blocked. Color filters are added to the pixels to achieve color.

The entire display is composed of tiny pixels, each connected to a source of the control voltage. Thus each pixel can be switched between light and dark. Color filters can be placed over each pixel to produce a full color display. In a 1,080 pixel high-definition television display there are 1,920 pixels in the horizontal direction and 1,080 in the vertical, for a total of 2,073,600 pixels.

OPAQUE, TRANSPARENT, AND TRANSLUCENT MATERIALS

What is an **opaque material**?

An opaque object is something that allows no light through it. Concrete, wood, and metal are some examples of opaque materials. Some materials can be opaque to light, but not to other types of electromagnetic waves. For example, wood does not allow visible light to pass through it, but will allow other types of electromagnetic waves, such as microwaves and radio waves to pass. The physical characteristics of the material determine what type of electromagnetic waves will and will not pass through it.

What is the **difference** between **transparent** and **translucent**?

Transparent media such as air, water, glass, and clear plastic allow light to pass through the material. Rays of light are either not bent or closely spaced rays are bent together. Translucent materials, on the other hand, allow light to pass through, but bend closely-spaced rays into different directions. For example, frosted glass and thin paper are translucent because they let light through, but are not transparent because you cannot see clearly through them.

Is **Earth's atmosphere transparent** to **infrared** and **ultraviolet radiation** as well as to light?

Oxygen and nitrogen, the two principal components of our atmosphere, are transparent to light and to most infrared (IR) and ultraviolet (UV) wavelengths. Infrared from the sun helps warm Earth and long-wavelength ultraviolet is necessary for our bodies to produce vitamin D. Too much UV, however, harms the skin, and may damage our DNA. A high-altitude layer of ozone in our atmosphere protects Earth from all but a small fraction of the UV radiated by the sun. Ozone molecules have three oxygen atoms as opposed to the two in oxygen gas. In recent years large and growing holes in this protective layer have developed due to the emission of chlorofluorocarbon (CFC) molecules from things such as escaped gases from refrigerants and aerosol propellants. Because these gases are no longer being manufactured and existing stocks are gradually being replaced, the the growth of the holes should eventually stop.

Carbon dioxide, methane, and water vapor are transparent to light and short-wavelength infrared radiation, but are opaque to the long-wavelength infrared emitted by warm objects. They are called greenhouse gases. These gases pass the IR rays that warm Earth but reflect the IR rays emitted by the warm Earth back to the ground. Thus they act as an insulating blanket for Earth. Over the past decades the amount of carbon dioxide in the atmosphere has increased dramatically. Much of the increase is due to human activity. This increase is likely to lead to a warmer Earth that could shift weather patterns, disrupting food production, causing shifts in locations of forests and animals, and, by melting polar ice, increasing sea levels. The degree of warming and its impact on Earth and humans is under intensive investigation.

SHADOWS

What are **shadows**?

Shadows are areas of darkness created by an opaque object blocking light. Whether created when someone puts their hand in the light from a movie projector, stands outside in the sunshine, or sees the moon move between Earth and the sun during an eclipse, shadows have always intrigued us.

How is an **eclipse** a **shadow**?

An eclipse is created just as any other shadow, by the presence of an object in the path of light. In a lunar eclipse, Earth blocks the sun's light illuminating the moon. In a solar eclipse the moon keeps the sun's light from reaching Earth. Thus the shadow of the moon on Earth is what we call a solar eclipse.

What is a **lunar eclipse**?

A lunar eclipse occurs when Earth is directly between the sun and the moon. Earth blocks the sun's light rays from hitting the moon, leaving it in complete darkness. For an observer on Earth, a shadow appears on the moon, causing it to become dark. As Earth moves out from between the sun and the moon, the moon is gradually illuminated until the entire full moon is seen again.

What is a **solar eclipse**?

An eclipse of the sun occurs when the moon casts its shadow on Earth. As the moon moves between Earth and the sun, darkness falls upon the small part of Earth where the moon blocks the light.

How dark can it get on Earth during a solar eclipse?

Some observers reported that they saw Venus and some of the brighter stars during a total solar eclipse. It is not entirely dark, however, because light comes from the sun's corona, the extremely faint glow surrounding the sun that is produced by ionized gases emitted by the sun. More light may come from reflection of light by the atmosphere in nearby areas not in the path of totality. The complete eclipse of the sun lasts for an average of about 2.5 minutes, but can last over 7 minutes.

What is the difference between the umbra and penumbra of a shadow?

A shadow produced by a large light source has two distinct regions. The umbra is the area of the shadow where all the light from the source has been blocked, preventing any light from falling on the surface. The penumbra, or partial shadow, is a section where light from only part of the source is blocked, resulting in an area where the light is dimmed, but not totally absent.

During an eclipse, the region of total darkness is the umbra; no direct sunlight can reach this area, resulting in a total eclipse. As Earth rotates, the moon's shadow races across Earth, producing the path of totality. Regions on either side of the path of totality experience the penumbra shadow and see the sun only partially covered by the moon.

Can the size of umbras and penumbras change?

An umbra and penumbra exist wherever there are shadows produced by a large light source. When an opaque object casts a shadow on a surface close to it the shadow will be clear and distinct, because it has a large umbra section and small penumbra. If, however, the object casting the shadow is closer to the light source, there will be

A lunar eclipse occurs when the earth is directly between the moon and the sun.

regions on the surface where light from some of the source is not blocked by the object, producing a smaller umbra and a larger penumbra.

How often do solar and lunar eclipses occur?

Solar eclipses (including partial eclipses) occur more frequently than lunar eclipses; over the entire Earth there are two to five solar eclipses per year, while the average number of lunar eclipses per year is between one and two. A solar eclipse can only be observed by a narrow path on Earth's surface while a lunar eclipse can be seen over a much larger area.

When a solar eclipse occurs, is the entire Earth in the moon's shadow?

The shadow of the moon only covers an area of about 300 kilometers in diameter. This shadow, the umbra, moves along a path of Earth's surface at about 1,000 miles per hour. In the penumbra a partial eclipse is seen, but even this path is quite narrow.

When and where will there be total solar eclipses over the next few decades?

The following map shows the date and location of total solar eclipses that took place from 2001 to 2010, as well as those that will take place into the year 2020:

199

Total and Annular Solar Eclipse Paths: 2001 – 2020

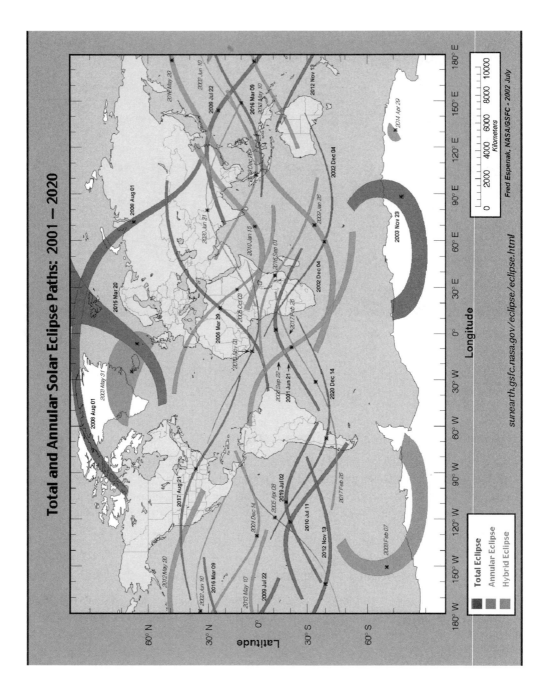

sunearth.gsfc.nasa.gov/eclipse/eclipse.html

Fred Espenak, NASA/GSFC - 2002 July

200

The last total solar eclipse in the United States was seen on February 26, 1979. When will the **United States** see its **next total solar eclipse**?

Although the entire United States will not see the eclipse, it will sweep across the country from Oregon to South Carolina on August 21, 2017. Mark your calendars!

When and **where** will there be **total lunar eclipses** for the next few decades?

The following table shows the date and location of future lunar eclipses.

Date	Where Eclipse Will Be Visible
June 15, 2011	South America, Europe, Africa, Asia, Australia
December 10, 2011	Europe, East Africa, Asia, Australia, Pacific, North America
April 15, 2014	Australia, Pacific, Americas
October 8, 2014	Asia, Australia, Pacific, Americas
April 4, 2015	Asia, Australia, Pacific, Americas
September 28, 2015	Eastern Pacific, Americas, Europe, Africa, Western Asia
January 31, 2018	Europe, Africa, Asia, Australia
July 27, 2018	Asia, Australia, Pacific, Western North America
January 21, 2019	South America, Europe, Africa, Asia, Australia
May 26, 2021	Asia, Australia, Pacific, Americas
May 16, 2022	Americas, Europe, East Asia, Australia, Pacific
November 8, 2022	Americas, Europe, Africa
March 14, 2025	Americas, Europe, Africa
September 7, 2025	Pacific, Americas, West Europe, West Africa
March 3, 2026	Europe, Africa, Asia, Australia
December 31, 2028	Europe, Africa, Asia, Australia
June 26, 2029	Europe, Africa, Asia, Australia, Pacific
December 20, 2029	Americas, Europe, Africa, Middle East

Why is it **dangerous** to **look** at a **solar eclipse**?

Looking directly at the photosphere of the sun (the bright disk of the sun itself), even for just a few seconds, can cause permanent damage to the retina of the eye, because of the intense visible and ultraviolet rays that the photosphere emits. This damage can result in permanent impairment of vision, up to and including blindness. The retina has no sensitivity to pain, and the effects of retinal damage may not appear for hours, so there is no warning that injury is occurring.

Under normal conditions, the sun is so bright that it is difficult to stare at it directly, so there is no tendency to look at it in a way that might damage the eye. During a total eclipse, however, with the sun covered, it is easier and more tempting to stare at it. Unfortunately, looking at the sun during an eclipse is dangerous even if totality occurs only briefly. Viewing the sun's disk through sunglasses or any kind of optical aid (binoculars, a telescope, or even an optical camera viewfinder) is extremely hazardous. The best methods are to use special solar filters, welder's glasses, or a pinhole camera.

REFLECTION

What is the **difference** between **physical** and **geometrical optics**?

Geometrical optics deals specifically with the path that light takes when it encounters mirrors and lenses. Geometrical optics uses the ray model of light in which an arrow represents the direction light travels. The ray model does not consider the wave nature of light. Ray diagrams trace the path that light takes when it reflects and refracts in different media.

Reflection occurs when light bounces off a surface, whether it is a piece of paper or a mirror. Shiny metals and very still water make good reflectors, creating images that resemble objects from which the light is coming.

Physical optics, on the other hand, is the division of optics that depends on the wave nature of light. It involves polarization, diffraction, interference, and the spectral analysis of light waves.

What is **reflection**?

When light strikes an object it can either be absorbed, transmitted, or reflected. Opaque objects absorb or reflect light. Transparent and translucent materials transmit light, but can also reflect it. The energy carried by the light must be conserved. The sum of the energy reflected, transmitted, and absorbed must equal the energy that strikes the material. Light energy that is absorbed increases the thermal energy of the material so the material becomes warmer.

Reflection occurs when light "bounces" off of a surface, such as a mirror or a sheet of paper. A smooth surface, like that of a mirror, reflects light according to the rule that the angle of incidence equals the angle of reflection. Light that strikes the surface at, for example, 30° above the surface is reflected at 30° above the surface. You can test this for yourself with a small flashlight, a mirror, a sheet of paper, and a helper. Hold the piece of paper. Have the helper hold the flashlight, aim it at the mirror, and move it around until the reflected light ray hits your piece of paper. Change the angle with which the light from the flashlight strikes the mirror and see that you have to move the paper closer to the mirror to have the light hit it.

What happens when light hits a sheet of paper? Replace the mirror with a sheet of paper. Darken the room. Aim the flashlight at the paper and use a second sheet to catch the reflection. You'll see that the second sheet is illuminated in many different locations. The light is reflected from the paper, but into many directions. This kind of reflection is called diffuse reflection.

Polished, smooth surfaces that do not absorb light are the best reflectors; examples of reflective materials are shiny metals, whereas non-reflective materials are dull metals, wood, and stone.

MIRRORS

How were the **first mirrors** made?

People have seen their reflections in water for hundreds of centuries, but some of the earliest signs of human-made brass and bronze mirrors have been mentioned in the Bible and in ancient Egyptian, Greek, and Roman literature. The earliest glass mirrors, backed with shiny metal, appeared in Italy during the fourteenth century. The original process for creating a glass mirror was to coat one side of glass with mercury and polished tinfoil.

The German chemist Justus von Liebig (1803–1873) in 1835 developed a method for silvering a mirror. His process consisted of pouring a compound containing ammonia and silver onto the back of the glass. Formaldehyde removed the ammonia, leaving a shiny metallic silver surface that reflected the light. Today, most mirrors are made by evaporating metallic aluminum directly on the glass.

How do **one-way mirrors**, the ones used in interrogation rooms, work?

A one-way mirror seems to be a mirror when seen from one side, but as a window when seen from the opposite side. Thus the window is disguised as a mirror to allow secret surveillance. Physically, there is no such thing as a one-way mirror. That is, the amount of light reflected from one side is the same as that reflected from the other. The light transmitted in one direction is equal that transmitted in the opposite direction. How then does a one-way mirror work? First, the mirror isn't totally reflecting. It transmits half the light and reflects the other half. The second requirement has to do with lighting. It is imperative that the observation room remain dark, because if a lamp were turned on, some of that light would pass through into the interrogation room as well.

What does a **mirror reverse**?

If you look in a mirror you seem to see yourself reversed left-to-right. Your left eye appears to be your right eye in the mirror. Yet the vertical direction is not reversed.

<blockquote>

Why can't you always see yourself in a mirror?

The answer is a matter of angles. The law of reflection states that the angle of light incident on a mirror must be equal to the angle of the reflected light. If you are standing directly in front of a mirror, the angle of incidence (that is, the angle between the direction the incoming light is traveling and the mirror surface) might be 90° and as a result reflects directly back to your eyes at an angle of 90°. If, however, you stand off to one side of the mirror, then the angle of reflectance is much smaller, and so the incident light won't come from your face, but from another part of the room.

</blockquote>

Your chin is still at the bottom of your face. What happens if you look at a mirror while lying down? Now left and right are not reversed—your chin is on the same side of both you and your image. But if your right eye was higher than your left, then in the image the left eye will be higher. What indeed is reversed?

What makes a glove right-handed or left handed? If you examine a vinyl or latex glove you'll notice that it can fit on either hand. How does such a glove differ from an ordinary glove? The thumb is in line with the fingers. In other words, there is no front or back! A mirror reverses not left and right or up and down, but front and back. When you look at yourself you are facing the opposite direction. It's just like facing another person—his or her left and right sides are also reversed.

On the front of many **ambulances**, why is the word **"ambulance" printed backwards**?

The word "ambulance" is printed backwards (left and right reversed) so that when viewed in a mirror—specifically, the rearview mirror of a car—it will appear correctly (that is, forwards), ensuring that the driver can read it and respond as quickly as possible.

How do **day/night rearview mirrors** function in vehicles?

When drivers traveling at night encounter a bright light from the vehicle behind them shining in their eyes, many will flip a tab on the underside of the rearview mirror to deflect the light up toward the ceiling of the car. The silvered surface of the mirror reflects approximately 85–90% of the incident light on the mirror, which is now directed toward the ceiling. The remaining 10–15% of the light is reflected by the front of the glass on the mirror. The mirror is wedge-shaped, thicker at the top than at the bottom. Thus the front surface is angled downward to allow the much smaller amount of light to be reflected into the driver's eyes.

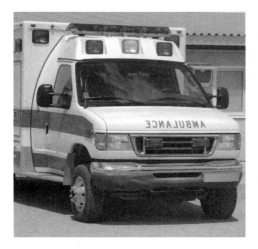

The word "AMBULANCE" is printed backwards on these emergency vehicles so that drivers will see the word correctly in their rearview mirrors.

What is an **image**?

Suppose light falls on your face. Light is reflected diffusely from this surface. That is, it is reflected into a wide variety of directions. Consider a very tiny object, like the end of an eyelash. Light from that point will leave in many directions. That point is called an object. An optical device can cause the light rays from the object to converge again to a single point. That point is called the image.

Does a **mirror** produce an **image**?

The light reflected from an object by a flat, or plane, mirror is not redirected so it will converge again. What you see when you look in a mirror is a virtual image. It is located behind the mirror. Some of the rays from the object are reflected by the mirror and enter your eye. Your eye believes that all these rays came from a single point, and that point is behind the mirror. The image isn't real, it is virtual. Virtual images can also be formed by lenses. They cannot be focused on a screen.

What is a **real image**?

A concave mirror or a convex lens can redirect light from an object so that it does converge to a single point. That point is a real image. It is in front of a concave mirror and on the other side of a convex lens from the object. If you place a screen, piece of paper, or wall at the location of the image you'll see it on that surface.

What is a **concave mirror** and what is it used for?

A concave mirror is curved inward so that the incident rays of light are reflected and can be brought together. If the rays striking the mirror were parallel to each other, as they would be from a distant source, then the reflected rays converge at the focal point. Concave mirrors are typically used to focus waves, whether it is a microwave signal to a receiver, or light to an observer. They can produce real images.

If, however, the object is closer to the mirror than the focal point, as is often the case with a concave bathroom mirror, the image is virtual and upright. If you look at the reflection of a more distant object, like the bathroom wall, you'll notice that it is upside down, or inverted. The wall is farther away from the mirror than the focal point, and its image is real.

**Why do side-view mirrors in vehicles state:
"Objects viewed in mirror may be closer than they appear"?**

This statement, seen on most side-view mirrors, is a very important safety message—the message warns the driver that the mirror is deceiving. Why would an automobile manufacturer put a deceiving mirror on a car? A flat, plane mirror would only show the driver a small, narrow section of the road behind the car; if, however, a convex mirror is used, the driver can not only see behind the car, but to the side as well, reducing his or her blind spot. In the process, however, convex mirrors make objects appear smaller and therefore farther away, so the message is there to serve as a reminder that the image is not exactly as it appears.

What is a **convex mirror** and what is it used for?

A convex mirror is the opposite of a concave mirror in that it is curved outward. The reflected light spreads out rather than converging at a point. Therefore convex mirrors form virtual images. Convex mirrors are used for security purposes in stores because they broaden the reflected field of vision, allowing clerks to see a large section of the store. The images are smaller than the objects, but the mirrors help to see a wide area.

REFRACTION

What is **refraction**?

Refraction is the bending of light as it goes from one medium to another. The most common use of refraction is in lenses. Eyeglass lenses refract light so that the wearer's eyes can focus the light properly. A magnifying glass is used to see enlarged images. Lenses in cameras produce an image on the film or CCD sensor. Refraction also occurs when sunlight strikes Earth's atmosphere and when it goes through water.

How can the **refraction** of light be **determined**?

The extent to which a beam of light bends when it hits a different medium depends on the indices of refraction of the medium as well as the medium from which it came and the angle at which the light strikes the boundary between the two media.

All materials have an index of refraction that depends on the speed of light in the material. The index of refraction is the speed of light in a vacuum divided by the speed of light in the material. A vacuum has a refractive index of 1, water is 1.33, and glass is

Magnifying glasses make images larger by bending light through a glass medium.

around 1.5. The higher the index of refraction, the more slowly light travels through the medium.

Snell's Law of Refraction, named after Dutch physicist Willebrord Snellius (1580–1626), tells us how light behaves at a boundary between two different media. Consider the interface between two media where the refractive index of the top medium is lower than that of the bottom medium. According to Snell's Law, when light hits the boundary between two materials it is bent from its original path to a smaller angle with respect to the line perpendicular to the surface of the interface. As the incoming, or incident angle increases, so does the refracted angle. When light goes from a medium with a higher refractive index into one with a lower index then it is bent away from the line perpendicular to the surface.

What is the **index of refraction** for light traveling through different media?

The following are some sample indices of refraction. The index of refraction, represented by n, is the ratio of the speed of light in a vacuum divided by the speed of light in the material. The larger the index of refraction, the greater the bending that takes place.

Medium	Index of Refraction (n)
Vacuum	1.00
Air (usually rounded to 1.0)	1.003
Water	1.33
Crown glass	1.52
Quartz	1.54
Flint glass	1.61
Diamond	2.42

How does **Earth's atmosphere affect** the apparent locations at which we see **stars**?

The light from stars refracts slightly as it enters Earth's atmosphere. Refraction is largest nearest the horizon. As a result the true position of the stars is a bit off from where we observe them. Refraction of light from the sun results in sunrise being slightly earlier than it would be without the atmospheric refraction, and sunset is

Why does a person standing in a pool of water often appear short and stocky?

The portion of a person's body that is above water does not appear out of proportion because the light entering your eye is not going through a different medium and refracting. The part of the body that is underwater, the person's legs, appears to be short because the light reflecting off their legs is traveling through water and then into the air. Due to this change in medium, refraction occurs. Since the index of refraction for water is larger than the index of refraction for air, the legs of the person appear compressed and stocky.

slightly later. Refraction also distorts our view of the sun and moon when they are very close to the horizon.

What is a **mirage**?

Mirages typically occur on hot summer days when surfaces such as sand, concrete, or asphalt are warm. Mirages look like pools of water on the ground, along with an upside-down image of a building, vehicle, or tree in the distance. As one approaches a mirage, the puddle of water and the reflection seem to disappear.

A mirage occurs because of a temperature difference between the air directly above the surface, which is hot and thus less dense, and the cooler, denser air a few meters above the surface. The denser air has a higher refractive index and that causes the light from an object to bend up toward the observer. As a result, the object is right side up while the refracted image is inverted and underneath the original object. The illusion of water is also a refracted image, the image of the sky. Mirages can only occur on hot surfaces and objects that are at relatively small angles in relation to the observer. Therefore, a person cannot see a mirage from an object that is just a few meters away. A mirage is not a hallucination, but instead a true and well-documented optical phenomenon.

LENSES

When were **lenses first made**?

The word "lens" comes from the name of the lentil bean because of the similarity in shape of the bean and a converging lens. Lenses have been used for over three thousand years. It's possible that ancient Assyrians used them as a burning glass to start fires. A burning glass is mentioned in a play by Aristophanes written in 424 B.C.E.

Roman emperors used corrective lenses and knew that glass globes filled with water were able to produce magnified images. Al-Haitham (Alhazen; 965–1038 C.E.) wrote the first major textbook on optics that was translated into Latin in the twelfth century and influenced European scientists. Shortly thereafter, in the 1280s, eyeglasses were used in Italy. The use of diverging lenses to correct nearsightedness (myopia) was documented in 1451.

Today, lenses used in eyeglasses and cameras are usually made of lightweight plastics that are cheaper and more durable than traditional glass lenses.

What is a **converging lens**?

A converging lens has at least one convex side. Its shape causes the entering light rays to converge, that is, come closer together. A converging lens can create a real, inverted image that may be projected on a screen. When used as a magnifying glass it creates a virtual, upright image.

What is a **diverging lens**?

A diverging lens has at least one concave side. The shape of the lens causes the entering light rays to spread apart when they leave the lens. A diverging lens is often used in combination with converging lenses. Eyeglasses to correct nearsightedness use diverging lenses.

What is the **focal length** of a lens?

The focal length measures the strength of a lens. Consider a converging lens. Rays from a very distant object come together at the focal point. The focal length is the distance from the lens to the focal point. Lenses with very convex surfaces have shorter focal lengths, while flatter lenses have longer focal lengths. For diverging lenses the focal point is virtual. That is, it is the point from which the diverging rays leaving the lens would have come from if a point object were placed there.

How does a **pinhole camera** work?

A pinhole camera is typically made from a box with a small "pinhole" in one side of the box and a screen on the other side. The pinhole is so small that only a very small number of light rays can go through it. The diagram on page 211 shows how a pinhole creates a reproduction of the object on the screen. Note that it is not an image because light rays do not converge on the screen.

Pinhole cameras are easy to make and are often used during solar eclipses because it is very dangerous to look directly at the sun (during an eclipse or otherwise). With the sun at your back, point the hole up toward the sun and view the image of the moon passing in front of the sun on the screen.

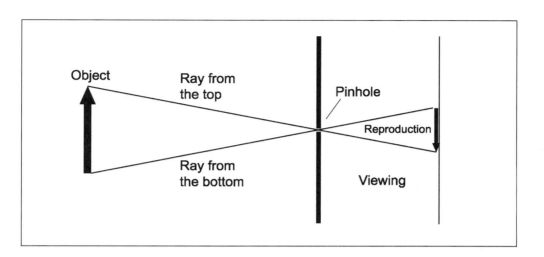

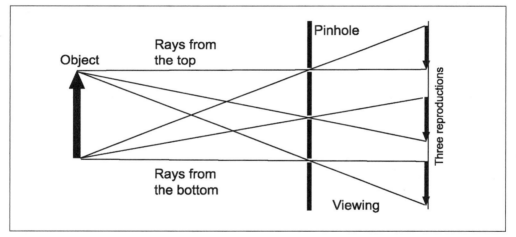

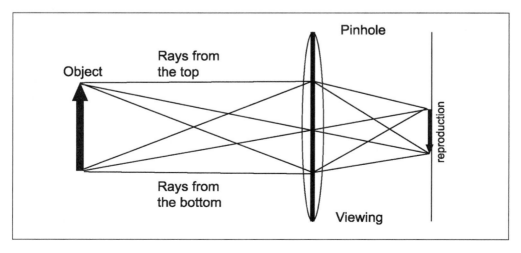

211

Why do diamonds sparkle so much?

Diamond has a very large index of refraction and therefore a small critical angle of only 25°. If the diamond is cut correctly light striking the top of the diamond will be internally reflected and will emerge again from the top rather than the sides of the diamond

How does a **camera lens create** an **image** on the sensor?

The diagram below shows what would happen if there were three pinholes, each creating an inverted reproduction of the object (see page 211).

Now, if a converging lens is placed just behind the pinholes it will bend the rays going through it. If the focal length of the lens and the distance between the lens and screen are chosen correctly, then the three reproductions from the pinholes will all be at the same location. Light rays from the top of the object will converge on the appropriate point on the image. Note that the image is inverted and the same size as the reproductions. What would happen if you had a multitude of pinholes at the location of the lens? The reproductions from all the pinholes would be at the same location, and many more rays from the object would end up at the same place on the image. The image would be much brighter. So, you can model the formation of an image by a lens as a collection of reproductions of pinholes. The larger the lens, the brighter the image.

What are **total internal reflection** and **critical angle**?

Total internal reflection occurs when a ray of light in a medium with a higher index of refraction strikes the interface between that medium and one with a lower index of refraction. When the incident rays are at a small angle with respect to the perpendicular to the interface the rays pass through the interface being refracted to a larger angle. As the angle of incidence increases so does the refracted angle. At the "critical" angle the refracted angle is 90°, that is, the ray's direction is along the interface. If the angle of incidence is increased any more there is only a reflected ray, the refracted ray no longer exists. Because all the light is reflected it is called total internal reflection.

If you open your **eyes underwater**, can you **see out** of the **water**?

Another example of total internal reflection can be seen when you are underwater. If you look straight up out of the water, you will see the sky and any other visible surroundings directly above the water. If, however, you look out of the water at an angle of 48° or more from the vertical, you will not see out of the water, but instead will see a reflection from the bottom of the water. The next time you are in a pool or a lake, try looking up out of the water and you will see a point on the surface where you no longer can see out of the pool, for the light has reached its critical angle.

FIBER OPTICS

How do **optical fibers** use **total internal reflection** to transmit information?

Strands of glass fiber, commonly known as optical fibers, use the principle of total internal reflection to transmit information near the speed of light. The fiber has an inner core of glass with a high refractive index surrounded by a cladding of glass with a lower index. A laser sends light into the end of a strand of fiber. When the light strikes the interface between the core and the cladding, the light is reflected back into the cable, continuing to move down the length of the fiber.

Light or near-infrared radiation can travel kilometers through fibers without significant energy loss. One reason is the total internal reflection. The second reason is that the are made from materials designed to absorb as little as possible of the infrared radiation. A second advantage of optical fibers is that information sent through the fibers is more secure because it doesn't escape the fiber and thus be accessible to those trying to intercept the information.

How did **fiber optics originate**?

The idea that light could travel through glass strands originated as far back as the 1840s, when physicists Daniel Collodon and Jacques Babinet (1794–1872) demonstrated that light could travel through bending water jets in fountains. The first person to display an image through a bundle of optical fibers was a medical student in

Germany by the name of Heinrich Lamm, who, in 1930, used a bundle of fibers to project the image of a light bulb. In his research, Lamm ultimately used optical fibers to observe and probe areas of the human body without making large incisions. From that point on, serious research in optical fibers ensued, expanding later with the development of the laser.

Where are **fiber optics used** today?

The transmission of light information through optical fibers has had a huge

Fiber optics are glass fibers that transmit laser light that can contain digital information more quickly and efficiently than metal cables.

impact on the technological world. The medical field has benefited greatly from the use of flexible fiber optic bundles that enable the viewing of areas of the body that would otherwise be inaccessible.

Communications is probably the field that is benefitting the most from the advent of fiber optic technology. Long-distance telephone, computer, and television signals use fiber optic cables. Some systems even use fiber optics to transmit information directly to the home or business. Fiber optics can transmit large amounts of data at high speeds permitting hundreds of television channels, very high speed internet connections, and telephone conversations to be sent and received at the same time.

DIFFRACTION AND INTERFERENCE

What is the **diffraction** of light?

Reflection and refraction use the ray model of light. But, when light goes through a very small opening the ray model is no longer sufficient. The wave properties of light become important. Suppose you pass light through a round aperture whose diameter you can change. Let the light fall on a screen. As you first begin reducing the size of the hole you'll find the size of the bright spot shrinking. But, when the hole becomes very small a strange thing happens. The spot no longer shrinks, but its outer edge becomes fuzzy. Light begins to bend around the edge of the hole. Diffraction also occurs when light is sent through narrow slits or if there are small objects that cast shadows in a broader beam of light. Diffraction occurs with all types of waves. You can often see it in water waves and it is one reason that sound waves spread when they come through a door or window.

What is the **interference** of light?

We've explored wave interference before in the Waves chapter. The key to the interference of light is that the two (or more) waves interfering must have the same wavelength and phase. While interference can be seen with ordinary light, interference is most easily seen using laser light. If light from a laser passes through two slits that are a small distance apart the diffraction patterns from the two slits will overlap. When they do a pattern of light and dark bands will be seen. The bright bands are where waves from the two slits are constructively interfering. The distance the light has traveled from the two slits will be equal or they will differ by an integer number of wavelengths. The dark bands occur when the waves destructively interfere. In this case the difference in distance light from the two slits will have traveled will be an odd number of half wavelengths. That is, one-half, three-halves, five-halves, etc. Light can also interfere if it is reflected off two closely spaced interfaces, like the two surfaces of a soap bubble or oil film.

Why do **soap bubbles** and **gasoline spills** create different **color reflections**?

Iridescence is the spectrum of colors that are produced when light hits a thin film such as a soap bubble or gasoline layer. The interference of light waves resulting from reflections of light off the two surfaces of the thin film causes iridescence. A soap bubble displays an iridescent pattern because light reflects off the front and back surfaces of the soap bubbles. As the thickness of the layer varies, the interference between the two reflections will vary, causing the color to vary.

Soap bubbles have an iridescent shine when light hits them because light waves reflect off the two surfaces of their thin film.

Gasoline spills are easily seen on wet roads; this is not because people spill more gas when it has rained, but is instead due to the iridescent patterns that result from light reflecting off of the top of the gasoline, and the boundary between the gasoline and water. The resulting pattern appears as the colors of the visible light spectrum in the thin film of gasoline.

COLOR

What is **white light**?

White light is the combination of all the colors in the visible light spectrum. When separated from each other, the different wavelengths have different colors. The longest wavelength light is the color red, and decreasing wavelengths result in orange, yellow, green, blue, indigo, and finally, the shortest wavelength visible color, violet.

Who is **Roy G. Biv**?

The colors of the visible light spectrum, in order from long wavelength to short, can be remembered using the fictitious name "Roy G. Biv," which is an acronym for Red, Orange, Yellow, Green, Blue, Indigo, and Violet. Each individual color has a particular wavelength, and as the wavelength changes, the color changes to the next color on the spectrum. When combined with the appropriate intensities these colors form white light.

What is **indigo**?

Indigo is the color between blue and violet in the spectrum, but almost no one can distinguish that color. Why, then are there seven colors rather than six? It's because

Isaac Newton was drawing an analogy between color and musical tones. There are seven notes in the familiar European scale—A, B, C, D, E, F, and G. So, Newton decided there should be seven colors in the spectrum, and he identified indigo as the seventh. So don't be too disappointed if you can't see it.

How do we see objects?

In order to see an object, light from the object must enter our eyes. We can see stars, lightning, and light bulbs because they are emitting or giving off light. We depend on the light emitted from these sources in order to see objects that don't emit light—we see those objects because they reflect light into our eyes. The paper on which this book is printed, for example, does not emit light. We see it because the paper reflects light into our eyes.

Why do we see specific colors?

When we "see" colors, we are seeing only part of the spectrum of colors that make up white light. The rest of the spectrum is missing. Selecting the color seen can be done in three ways. You can view the object through a transparent material that transmits only some part of the spectrum while absorbing the rest. The object may be colored itself. That is, it reflects part of the spectrum while absorbing the rest. Finally, the spectrum can be physically separated, as it is by interference or by refraction by a prism or rainbow. For example, theatrical "gels" on spotlights produce colored effects for a stage show. Snow reflects all of the spectrum so it appears white, but green paper reflects only the green part of the spectrum. A black cloth absorbs all light, and so it appears black.

Who discovered that white light could be separated into the colors of the rainbow?

By the seventeenth century glass makers had learned to make gem-shaped pieces that were used in chandeliers. Candlelight was refracted in these pieces. When viewing the candlelight different colors were seen, depending on the angles made by the light, the glass, and the viewer's eye. Newton, who was intrigued by the colors that were produced by those chandeliers, decided to examine how a piece of glass shaped as a rectangular prism could create a spectrum of colors. In his own words: "In a darkened Room make a hole in the [windowshade] of a window, whose diameter may conveniently be about [one-third] of an inch, to admit a convenient quantity of the Sun's light: And there place a clear and colourless Prisme, to refract the entering light towards the further part of the Room, which, as I said, will thereby be diffused into an oblong [spectrum]."

To prove that the colors did not come from the prism, Newton expanded the experiment by reversing the procedure and forming white light from the spectrum of colors. He accomplished this by placing a lens in the middle of the spectrum to converge the colors on a second prism in the path of the colors. Sure enough, a beam of white light emerged out of the second prism.

Why does a **prism** separate light into a spectrum of colors?

If all wavelengths of light were refracted by the same amount when entering and leaving a prism then there would be no separation of colors. The refractive index of all materials depends on the wavelength of light. In diamond the refractive index for blue is 1.594, for red 1.571. In flint glass the index varies from 1.528 to 1.514. In crown glass the variation is from 1.528 to 1.514, while in water it is only 1.340 to 1.331. In all cases the refractive index of blue is larger than that of red, so the blue light is refracted through a larger angle than the red. The large difference in diamonds accounts for their "flash."

If white is the combination of the colors of the rainbow, **what is black**?

Black, the exact opposite of white light, is the absence of light or the absorption of all light. It may seem obvious to us, but Newton was the first to recognize this fact. A black piece of paper appears black because all the light is being absorbed in the paper—none is reflected back out to our eyes.

What are the **primary colors** from a light source?

When mixing light (or "additive color mixing") the three primary colors are blue, green, and red. Computer monitors and both cathode ray tube (CRT) and flat-panel television sets use these colors. The combination of these primary colors results in other colors, and when all three colors are combined with equal intensity, white is formed.

What are the **secondary colors**?

When any two of the three primary colors are mixed, secondary colors are formed; they are called secondary because they are by-products of the primary red, green, and blue colors. Red light mixed with green creates yellow light. Red and blue produces magenta. Finally, cyan is formed when blue light and green light are added together.

What are **complementary colors**?

Complementary colors are pairs of one secondary and one primary color that, when mixed, form what is close to white light. For example, yellow and blue light are complementary because when combined, they form white light, as will magenta and green, and cyan and red.

What is **subtractive color mixing**?

As opposed to the mixing of light ("additive color mixing"), subtractive color mixing occurs only when combining dyes, pigments, or other objects that absorb and reflect light. For example, you could shine white light through two colored filters or gels. Or, you can reflect light from colored surfaces. If you shine white light on a blue wall, the wall absorbs red and green but not blue. So if we know what colors are reflected we know the color of the object.

The primary pigments or dyes are magenta, which reflects blue and red light but absorbs green; cyan, which reflects blue and green but absorbs red; and yellow, which reflects red and green but absorbs blue. These are the same colors as the secondary colors obtained when mixing light. Note that if we combine magenta, cyan, and yellow, the red, green, and blue are all removed, leaving nothing; that is, black.

Blue is light in the range of 400 to 500 nanometers. Green is roughly 500 to 560 nanometers. Yellow is 560 to 590 nanometers, orange 590 to 620 nanometers, and red beyond 620 nanometers. It may seem surprising that yellow filter transmits green, yellow, and red. But the eye is not very sensitive to red in comparison to green and yellow. That's the same reason that the transmission of the blue and indigo filters beyond 660 nanometers does not impact what the eye sees. Finally, note that purple is a mixture of blue and red.

What are the **secondary colors** in **subtractive color mixing**?

The secondary colors for dyes and pigments are the same as the primary colors in additive color mixing. Red, green, and blue dyes or pigments reflect their own color while absorbing the other two colors. For example, red would reflect red but absorb the green and blue light.

Color inkjet printers use black, along with yellow, cyan, and magenta, because the colored pigments do not cover the full spectrum. Therefore, they cannot produce a dark black but only a muddy dark brown.

Why do most **color inkjet printers** use **four colors** to print, instead of the three primary colors for subtractive color mixing?

It would seem that a color inkjet printer mixing the three primary pigments of yellow, magenta, and cyan should be able to produce all the other colors, including black. When all three primary colors are combined, however, the mix looks more like a muddy brown color than black. Although these are the primary colors of which other shades can be created, they do not represent all the colors of the spectrum needed to form black. That is, there are gaps in the wavelengths that

What is colorimetry?

Because the perception of color is mostly a neurophysiological function between the eyes and the brain, it can vary slightly from person to person. Further, the subtractive color seen depends on the light source. If you plan to paint a room you should examine the color when illuminated by several different lights sources: sunlight, light from incandescent lamps, and light from fluorescent lamps. The colors may look very different. They will even look different in the sunlight at noon versus that near sunset. Scientists, artists, advertisers, and printers need an objective method of specifying color as it relates to the frequency of light. This technique for measuring the intensity of particular wavelengths of light is known as colorimetry.

these pigments absorb. Therefore, most color inkjet printers have a cartridge with yellow, cyan, and magenta ink, and another separate ink cartridge of just black.

What is the **difference** between **hue** and **saturation**?

Hue is related to the wavelength of a color. Saturation is the extent to which other wavelengths of light are present in a particular color. For example for the hue red, deep red is saturated, but pink is a mixture of red and white.

On **humid summer days**, why does the **sky** take on a **white** or **grayish** appearance?

When high amounts of humidity are in the air, water molecules are more prevalent than on a cool, dry day. Water molecules, which have two hydrogen and one oxygen atom, are larger than oxygen and nitrogen found in the air, and the size of a molecule plays a significant role in what frequencies of light are scattered. When white light encounters a larger molecule or dust particle, larger wavelength light will be scattered, whereas if a smaller molecule is struck by white light, smaller wavelength light will be scattered. Snow, beaten egg whites, and beer foam look white for the same reason.

If the water or smoke is dense enough, then all light waves are scattered many times, and the cloud looks grey.

Why are **sunrises** and **sunsets** often **orange** or **red**?

During the evening and early morning, when the sun is lower in the horizon, the light that the sun emits has to travel through more of the atmosphere to reach us than it does during midday, when the path through the atmosphere is shorter. Since the dis-

If the low-wavelength light is scattered, why do we only see a blue sky and not a blue, indigo, and violet sky?

Lord Rayleigh (John Strutt, 1842–1919) determined that the nitrogen and oxygen molecules in the atmosphere scattered sunlight, allowing us to see the sky. The scattering is strongest at the lowest wavelengths. Why then, don't we see a violet sky? Our eyes are most sensitive to colors in the mid-section of the spectrum, about 550 nanometers. Because blue is closer to this wavelength, our eyes are more sensitive to it than indigo and violet. So, even though all three colors are scattered by the molecules in the air, humans see a predominantly blue sky. Because the shorter wavelength of sunlight are scattered by the atmosphere, the light transmitted has a yellow cast; the sun looks yellow rather than white.

tance through the atmosphere is much larger for sunlight in the morning and evening than during midday, more of the shorter wavelengths of light are scattered out of the direct light from the sun. Thus the sun's color goes from yellow to orange and finally to red. Dust and water vapor in the atmosphere enhance this effect, making sunsets even redder.

Why is the **ocean blue**?

There are two major reasons why the ocean and most bodies of water appear blue. The first can be observed by looking at the water on a cloudy day and then on a sunny day. There is a rather large difference in how blue the water appears to be on the two different days, because the water acts as a mirror for the sky. So on a sunny day with a blue sky, the water will have a richer blue color than on a cloudy day.

The second reason why bodies of water have a blue appearance is that water scatters short wavelength light more than the longer. In fact, water absorbs some orange, red, and the very long-wavelength infrared. As a result it absorbs more energy in the sunlight, increasing its temperature. The much larger amounts of reflected, short-wavelength light results in a crisp blue-colored body of water.

Some bodies of water may take on a more greenish or at times a brownish or black color. Usually this is due to other elements in the water such as algae, silt, and sand. Runoff water from glaciers is very white due to the tiny grains of silt in the water. Still, in the majority of cases, water looks blue.

RAINBOWS

How do **rainbows occur**?

A rainbow is a spectrum of light formed when sunlight interacts with droplets. Upon entering a water droplet, the white light is refracted, and dispersed, that is, spread apart into its individual wavelengths, just as in a prism. The light inside the droplet then reflects against the back of the water droplet before it refracts and disperses as it exits the droplet. The angle between entering and leaving is 40° for blue light, 42° for red.

What **conditions** must be met in order to see a **rainbow**?

There are two main conditions for witnessing a rainbow. The first is that the observer must be between the sun and the water droplets. The water droplets can either be rain, mist from a waterfall, or the spray of a garden hose. The second condition is that the angle between the sun, the water droplets, and the observer's eyes must be between 40° and 42°. Therefore, rainbows are most easily seen when the sun is close to the horizon so the rays striking the droplets are close to horizontal.

Is there such a thing as a **completely circular rainbow**?

All rainbows would be completely round except that the ground gets in the way of completing the circle. However, if viewed from a high altitude, such as an airplane, circular rainbows can been seen when the angle between the sun, the water droplets, and the plane is between the 40° and 42°. In this case, the rainbow is horizontal, meaning that it is parallel to the ground and therefore not blocked by the ground. This is quite a sight!

What is the **order of colors** in a **rainbow**?

The order of colors in a rainbow goes from longest-wavelength red on the outer arc to shortest-wavelength blue on the inside of the arc. The full order from outer to inner is: red, orange, yellow, green, blue, indigo, and violet.

Who was the **first person** to **explain** how **rainbows** are formed?

Newton was not the first person to understand the optical characteristics of a rainbow. In fact, it was a German monk in the early fourteenth century who first discovered that light refracted and reflected inside a water droplet. To demonstrate his hypothesis, the monk filled a sphere with water, sent a ray of sunlight through the sphere, and observed the separation of the white light into colors along with the reflection on the back of the water droplet.

What is a **secondary rainbow**?

The secondary rainbow has its color spectrum reversed, is outside of the original rainbow, and is significantly dimmer than the primary rainbow. A secondary rainbow occurs because an additional reflection of the light takes place inside the water droplets. Instead of reflecting once in the water droplet, the light reflects twice inside the water, reversing the order of the colors. The secondary rainbow appears between the angles 50° and 54°.

EYESIGHT

How does the human **eye see**?

The eye is really an extension of the brain. It consists of a lens to focus the image, an iris to regulate the amount of light entering the eye, and a screen called the retina. The cells of the retina do some preliminary processing of the information they receive then send signals along the optic nerve to the brain.

The cornea is a transparent membrane on the outer surface of the eye. Between the cornea and lens is a fluid. Light refracts when going through the convex surface of the cornea into the fluid. In fact most of the focusing of light in the eye occurs at the cornea. Light passes through the iris that opens and closes in response to the amount of light entering the eye. The iris can only change the amount of light to a going through it by a factor of twenty, while our eye can respond to differences in light level of ten trillion! The major task of the iris then can't be to control light intensity. In addition, when the opening in the iris shrinks the eye can keep objects in focus from a wider range of distances. After passing through the iris the light goes through the lens.

The lens consists of layers of transparent fibers covered by a clear membrane. In order to focus in on objects that we want to see, our eye changes the shape, and thus the focal length of the lens and cornea by contracting or relaxing the ciliary muscle around the eye. Light then passes through a liquid called the vitreous humor that fills the major volume of the eye and falls on the retina. The cornea and lens have created an inverted image on the surface of the retina.

The retina is composed of a layer of light-sensitive cells, a matrix of nerve cells, and a dark backing. There are two kinds of light-sensitive cells: cones and rods. The 7 million cones are sensitive to high light levels and are concentrated around the fovea, the part of the retina directly behind the lens. Surrounding it are some 120 million rods that are sensitive to low light levels. The entire retina covers about five square centimeters. There aren't 127 million nerves in the optic nerve that goes to the brain, so a system of nerves in the retina do some preliminary processing of the electrical signals produced by the rods and cones before sending the results to the brain.

What is the difference between **cones** and **rods**?

Cones are cone-shaped nerve cells on the retina that can distinguish fine details in images. They are located predominantly around the center of the retina called the fovea. The cones are also responsible for color vision. Some cones respond to blue light, being most sensitive to 440-nanometer wavelengths. A second kind has peak sensitivity in the green: 530-nanometer light. The third is sensitive to a wide band of wavelengths from cyan through red. Its sensitivity peaks in the yellow, 560 nanometers.

As the distance grows from the fovea, rod-shaped nerve cells replace the cones. The rods are responsible for a general image over a large area, but not fine details. This explains why we look at objects straight on when examining something carefully. The image will be focused around the fovea, where the majority of cones pick up the fine details of the image. The rods, being much more sensitive in low light levels, are used a lot for night vision.

What **wavelengths of light** are our **eyes most sensitive** to?

Our eyes are most sensitive to the wavelengths corresponding to the yellow and green colors of the spectrum. Flashy signs and some fire engines are painted in a yellowish-green color to attract our attention. Even simple objects such as highlighters, used to emphasize words or phrases while taking notes, are typically bright yellow and green. When we glance over something or see an object out of the corner of our eyes, we are more likely to notice bright yellowish-green objects than red or blue objects, because the eye is less sensitive to these wavelengths.

What is the **shape** of a **lens** when **focusing on objects** far away and objects up close?

The ciliary muscle, responsible for changing the shape of the lens, adjusts its tension to focus on different distances. When focusing on objects far away, the lens needs a large focal length, so the muscle is relaxed in order to make the lens relatively flat. When an object is closer to the eye, however, a shorter focal length is needed. The ciliary muscle contracts, reducing the focal length of the lens by making it more spherical. The process of adjusting the shape of the lens to focus in on objects is called "accommodation."

What is color blindness?

Some people are unable to see some colors due to an inherited condition known as color blindness. John Dalton (1766–1844), a British chemist and physicist, described color blindness in 1794. He was color blind himself, and could not distinguish red from green. Many color-blind people do not realize that they cannot distinguish colors. This is potentially dangerous, particularly if they cannot distinguish between the colors of traffic lights or other safety signals. Those people who perceive red as green and green as red are known, appropriately, as "red-green color blind." Other color-blind people are only able to see black, gray, and white. It is estimated that 7% of men and 1% of women are born color blind.

When **swimming underwater**, why is **vision blurred** when you open your eyes, but clear when wearing swimming goggles?

Although the eye's lens changes shape to focus images on the retina, most of the refraction of light takes place during light's transition from air to the cornea. When water is substituted for air, the angles through which light is refracted is reduced, producing a blurred image on the retina.

How **close** can an **object** be before it **appears blurry**?

There is a limit as to how close an object can be to the eye before the lens can no longer adjust its focus. Up to about thirty years of age, the closest an object can be focused is approximately 10 to 20 centimeters (4 to 8 inches). As one grows older, the lens tends to stiffen and it becomes more difficult for the person to focus on close objects. In fact, by the time a person reaches the age of seventy, their eyes cannot focus on objects within several meters of their eyes. As a result, most aging adults need reading glasses to focus on close objects.

What is **nearsightedness** and what can be done to correct it?

Nearsighted vision means that a person can only clearly see objects that are relatively near the eye. Images from distant objects are focused in front of the retina. Nearsightedness, or myopia, is most often caused by a cornea that bulges out too much. The lens cannot be flattened enough to compensate, and so distant objects appear fuzzy.

To correct for the short focal length of the lens, a concave lens is used to make the light rays diverge just enough so that the image will fall on the retina. So contact lenses to correct for myopia are thicker at the edges than at the center.

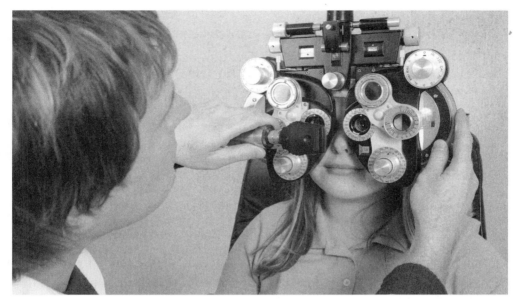

Farsightedness and nearsightedness are common eyesight problems. The former occurs when images entering the eye are focused behind the retina, and in the latter images focus in front of the retina.

What is **farsightedness** and what can be done to correct it?

Farsightedness, or hyperopia, occurs when the lens of the eye can see objects far away, but cannot focus in on objects at closer range. The cornea and lens of the person with farsightedness cause the image to focus behind the retina, resulting in the images from objects close to the eye to be blurred. In order to correct for farsightedness, a convex lens is used to converge the light rays closer together, permitting the image to fall on the retina. The rigidity in the eyelens that affects older people, making them unable to focus on close objects, is called presbyopia.

What allows **nocturnal animals** to **see better** in the **dark** than humans can?

There are three main reasons why some animals can see better than humans can at night. The first reason is that their eyes, relative to body size, are larger and can gather more light than human eyes can. More light results in a brighter image.

The next reason has to do with the rods and cones in the nocturnal animal's eyes. Cones are used for detail and work best in bright light. A nocturnal animal has little need for the color vision provided by the cones and therefore has more room for the rods that detect general information such as motion and shapes.

The third reason why nocturnal eyes excel in the absence of light is due to the *tapetum lucidum*, a membrane on the back of the retina that reflects light back to

225

How do 3-D movies and television work?

When a three-dimensional movie is filmed, two cameras film the movie from slightly different positions. When the film is projected on the screen, each projector uses a separate polarizing filter. The left projector might use a horizontally polarized filter, while the right projector uses a vertically polarized filter. The viewer wears polarized glasses. The glasses allow the left eye to see only the image produced from the horizontally polarized image of the left projector, while the right eye sees the image produced by the vertically polarized right projector. This arrangement simulates the different perspectives that each eye sees when looking at a real-life 3-D scene, allowing the brain to interpret the difference as depth (the third dimension).

The newest methods of producing three-dimensional views use digital methods rather than film. In the method best suited to 3-D television the images from the left and right camera lenses alternate at a rapid rate. The viewer wears glasses in which each lens can be switched from transparent to opaque on command from an infrared signal sent from the television set. Thus each eye sees only the frames captured by the appropriate camera lens. Another method that is more suited to movies, again uses digital images that alternate between those captured by the right camera lens with those captured by the left. A device placed in front of the projector lens switches the polarization of the light coming from the projector so that left images are polarized one way, right images are polarized the other way. The movie viewer wears polarized glasses so that each eye sees only the appropriate frames.

the retina to double the retina's exposure to light. The reflective tapetum can be seen in the light reflected back out of animals' eyes at night when you shine a flashlight on them.

What is **three-dimensional** (3-D) **vision**?

Seeing in three dimensions, which is how a person with normal eyes sees, means that in addition to perceiving the dimensions of height and width (such as seen on a piece of paper, a poster, or a TV or movie screen), one can see the third dimension of depth. We see real objects in 3-D because we have two eyes that see slightly different perspectives of the same view. The combination of these views, when interpreted by our brain, gives us the ability to perceive depth, the third dimension.

If you close one eye, your ability to perceive depth is eliminated. With only one eye, the world won't look very different to you, but you'll experience difficulty in judging distances.

CAMERAS

How is a **camera similar** to, and different from, the **eye**?

A camera performs many of the functions of the eye. It has a lens to form an image on a photosensitive surface. The lens must be able to form sharp images of objects both close and far away. The amount of light reaching the photo detector must be controlled to make the exposure correct. In older cameras the photosensitive surface was film. Today cameras use a digital sensor. These sensors are small—most are between 3/8" and 1" in size—but contain as many as 10 million separate light detectors called pixels. Each pixel is covered by a red, green, or blue color filter so the camera can produce full-color images.

A camera's lens isn't flexible like the one in the eye, but the distance between the lens and the sensor can be varied. Bringing a distant object into focus requires that the lens be closer to the sensor. A close object requires that the lens be moved further away. The amount of light is controlled two ways. One is to have an aperture that can be opened to admit more light or closed down to reduce the amount of light. The second is a shutter that controls the amount of time light is allowed to reach the sensor. While leaving the sensor exposed for a longer amount of time is needed when the light is dim, it also will cause a blurred image if the object is moving. Thus it is important to select the correct combination of aperture and shutter speed to take good pictures.

What causes **red-eye** in **photographs**?

Red-eye occurs when a flash is used because there is not enough light for a good exposure. Under normal conditions, in order for enough light to enter the eye, the pupil dilates. But when a flash is fired, the pupils are not expecting the bright light and do not have a chance to constrict. As a result, a large amount of light enters the eye and reflects off the blood vessels that supply the retina in the back of the eye. The redness on the pupil is actually the reflection off these blood vessels captured by the camera.

What is used to **reduce red-eye**?

Red-eye reduction is a feature found on many modern cameras. It simply attempts to constrict a person's pupil so that not as much light can be reflected back from the

Camera lenses are similar to human eyes in that both have lenses that focus images onto a surface, but camera lenses are not flexible like an eye lens.

retina. There are several methods of accomplishing this. One method is to have a smaller light that illuminates before the real flash; another method is to have a quick burst of five or six mini-flashes that cause the pupil to contract before the picture is taken.

TELESCOPES

Who invented the first telescope?

There are a number of conflicting claims for the first person to combine two lenses to "see things far away as if they were nearby." The Dutch eyeglass makers Hans Lippershey, Sacharias Jansen, and Jacob Metius were some of the first. Lippershey described the design and applied for a patent on October 8, 1608, but was turned down. Copies of Lippershey's device, which was constructed from a convex and a concave lens and had a magnifying power of 3, were common in the Netherlands that year. Galileo (1564–1642) heard about the invention in June 1609, in Venice. By the next day he had figured out how it worked and as soon as he returned home to Padua he constructed one. A few days later he demonstrated it to the leaders in Venice, who, in return awarded him a lifetime position at the University in Padua. Over the next year Galileo improved his instruments, and in 1610, using a telescope with a magnifying power of 33, he discovered the moons of Jupiter, the rotation of the sun, phases of Venus, spots on the sun, and mountains on the moon. The Galilean telescope with convex and concave lenses produces an upright image.

Based on the ideas of astronomers Johannes Kepler (1571–1630) and Christoph Scheiner, the telescope was improved by using two convex lenses separated by a distance equal to the sum of their focal lengths. Such a telescope inverts the image. To achieve high magnifications one of the focal lengths had to be very large. Refracting telescopes proved to be cumbersome and difficult to use. The prism-like shape of the lenses introduced colors into the images that weren't there. This defect, called chromatic aberration, was eliminated 120 years later using a lens made of a combination of two glasses. But this invention did not stop the weight of the large lens from causing it to sag, creating distorted images.

Reflecting telescopes that use mirrors to focus light, were invented by Isaac Newton (1642–1727) in 1668. Today, telescopes, both refractors and reflectors, are relatively cheap; the average person can set one up in his or her backyard and gaze up at the heavens with much better equipment than Galileo or Newton ever dreamed possible.

What is a refractor telescope?

The refractor telescope was the first telescope ever created. It employs one lens to gather, refract, and focus light toward an eyepiece. The eyepiece contains one or more

What was wrong with the Hubble Space Telescope when it was first put into orbit?

An error (1/50th the thickness of a human hair) in the curvature of the main mirror caused major focusing problems for the Hubble Space Telescope. The 2.4 meter (94.5 inch) diameter mirror, was not able to focus all the light it collected to the correct point in the telescope. NASA suffered great embarrassment for this multi-million dollar mistake.

lenses that create an image that they eye can see. The larger the diameter of the lens, the more light the telescope can gather. The weight of the lenses then limits the practical size of refractor telescopes.

What is a **reflecting telescope**?

A reflecting telescope uses a mirror to gather light and focus the light toward the eyepiece. It usually consists of two mirrors: one large curved mirror at the end of the telescope to gather light and a smaller mirror used to direct the light to the eyepiece

What are some of the **largest reflecting telescopes**?

The larger a reflecting telescope, the more light it can gather. The following is a list of some of the largest reflecting telescopes in the world.

Name	Effective diameter	Type	Location	Date Completed
Large Binocular Telescope (LBT)	11.8 m (464.6 ft.)	Multiple mirror	Arizona	2004
Gran Telescopio Canarias (GTC)	10.4 m (410 ft.)	Segmented	Canary Islands	2006–9
Keck 1	10 m (400 ft.)	Segmented	Hawaii	1993
Southern African Large Telescope (SALT)	9.2 m (362 ft.)	Segmented	South African Astronomical Observatory	2005
Hobby-Eberly Telescope (HET)	9.2 m (362 ft.)	Segmented	Texas	1997
Subaru (JNLT)	8.2 m (323 ft.)	Single	Hawaii	1999

What is a **segmented mirror telescope**?

When mirrors exceed 8 meters (26 feet) in diameter, they are no longer rigid enough to maintain the same shape when they are tilted. The Keck I telescope was the first to be built of 36 hexagonal mirrors. The tilt of each mirror is controlled electronically so that they all focus their light rays at a single point. The electronics can position each corner of the mirror to an accuracy 4 nanometers to create the final image. Constructing the telescope of multiple smaller mirrors greatly reduced the cost of the telescope. The Large Binocular Telescope consists of the Keck I and Keck II 10-meter (33-foot) telescopes. When they are used together they can detect the interference of light from a star and thus determine its size and location much more precisely.

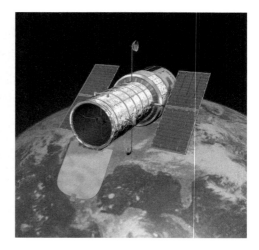

The Hubble Space Telescope can see deeper into space than earthbound telescopes because there is no atmospheric distortion to interfere with how it sees images.

What are the advantages of the **Hubble Space Telescope**?

The Hubble doesn't have a mirror as large as the new Earth-based telescopes, but being in space it is not limited by the distortions caused by variations in the refractive index of air above the telescope. In addition, a space telescope can detect the infrared and ultraviolet rays blocked by Earth's atmosphere. Other space-based telescopes are designed to detect X rays and gamma rays from extremely energetic stars and galaxies.

What was done to **correct** the **Hubble's vision problems**?

Three years after the Hubble was placed in orbit around Earth, a team of astronauts from the space shuttle Endeavor installed two tiny mirrors that would correct the focusing problems that the Hubble was experiencing. The telescope has since had several servicing calls, the most recent in 2009. Upgraded detectors and spectrometers have been installed, faulty positioning gyroscopes replaced, and its batteries replaced. The batteries are used when Earth's shadow blocks the sunlight on the solar panels.

Since its repairs, the Hubble Space Telescope has aided research into the age of the universe and the rate at which it is expanding, and has enabled observation of other stars and galaxies that previously were never seen by earthbound telescopes. More than 8,000 scientific papers have been published using Hubble data. Equally important, the beautiful photos taken by Hubble have fascinated the public and broadened its understanding of the incredible range of objects in our universe.

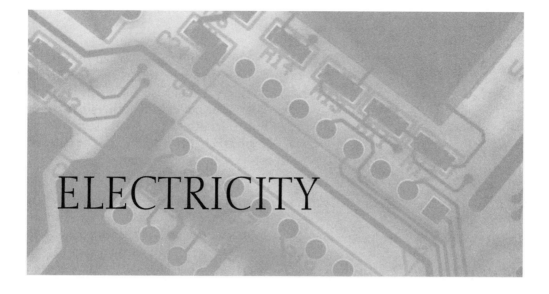

ELECTRICITY

What is **electrostatics**?

Electrostatics is the study of the causes of the attractive and repulsive forces that result when objects made of two different materials are rubbed together. Electrostatics is the study of what is often called static electricity.

What can you **discover** about **static electricity**?

How about exploring the basic ideas of electrostatics? All you'll need is a roll of cellophane tape. Any brand will do—the cheaper the better. Pull off a strip about 5 inches long, then fold over about 1/4 inch at one end to serve as a handle. Press the tape on your desk or a table. Mark the strip with the letter "B." Make a second identical tape and press it down next to the first.

Holding the two tapes by their handles, quickly pull them off your desk. They'll probably be attracted to your hands, so shake them until they hang free. Then bring them closer together. What do you see? You should see them bending, evidence that there is a force between them. If they don't bend, stick them on the desk again and again pull them off. Do they provide evidence that there is an attractive or repulsive force between them?

We'll say that pulling them off the table caused them to be "charged," although we have no evidence with what they are "charged." They were obviously charged in the same way, so we can conclude that objects with like charges repel each other. By the way, we'll work toward an explanation why they're attracted to your hands.

Press the two strips back on your desk. Now make two more strips the same length and press them on top of the first two strips. Mark these strips "T" to identify them as the top tapes, as opposed to the "B" or bottom tapes.

Electrostatics—or static electricity—involves the attractive and repulsive forces that result when objects made of two different materials are rubbed together.

Slowly pull the T + B pair of tapes off the desk together. If they are attracted to your hand then use the other hand to gently pat both sides of them over their entire length. That should remove any residual charge from the pair of tapes. If not, pat them down again.

You have a pair of objects with no charge. Holding the two handles of the pair rapidly pull them apart. Again, if they are attracted to your hands, shake them until they hang freely. Bring them closer together. Is there evidence of a force between them? Is it attractive or repulsive?

You started with a pair of objects with no charge. Pulling them apart caused them to be charged, but not in the same way, because they didn't repel each other, but attracted. Thus you can conclude that they must be charged differently, and objects with different charges attract.

To keep your charged tapes you can hang them from the edge of your desk or a desk light. Make a second T+B pair and see if the two T (top) tapes are charged alike or differently. If the tapes stop interacting you can repeat the charging procedure as often as you like.

Hang a T and a B tape so you can bring objects near them to see if there are forces between them. Make a list of the objects you tried and whether they attracted or repelled the T tape and the B tape. Try your finger. Then try rubbing a plastic pen on a piece of wool. Try plastic rubbed by silk or polyester. Try glass and metal.

Do some objects attract both tapes? Repel both tapes? Attract one and repel the other? If they do the latter, you can characterize them as being charged like the T tape or like the B tape.

We'll come back to understand why some objects can attract both kinds of charge, but no objects can repel both.

What is the **history** of **electricity**?

Pre-historic people valued and traded amber, a gem-like material that is petrified tree sap. Surely more than once a person would have rubbed amber on his or her fur clothing and noticed that fur was attracted to the stone. Perhaps she rubbed it hard enough to produce sparks. The Greek philosopher Thales of Miletus wrote about these effects around 600 B.C.E.

But it wasn't until 1600 c.e., some 2,200 years later, that William Gilbert (1544–1603), an English physician, named this effect "electricity" after the Greek name for amber: "elektron." Gilbert showed that sulfur, wax, glass, and other materials behaved the same way as amber. He invented the first instrument to detect what we now call the electrical charge on objects called a versorium, a pointer that was attracted to charged object. Gilbert also discovered that a heated body lost its charge and that moisture prevented the charging of all bodies.

In 1729, the English scientist Stephen Gray (1666–1736) determined that charge, or what he called the "electric virtue," could be transmitted over long distances by metals, objects that couldn't be charged.

How was **electricity used** as a form of **entertainment?**

In the mid-1700s demonstrations of electrostatics were extremely popular, especially in Parisian salons, where wealthy men and women gathered to discuss events of the day. Benjamin Franklin (1706–1790) was a popular guest. In Stephen Gray's most famous demonstration, called the Flying Boy experiment, a boy was suspended horizontally using two silk threads hung from hooks placed on the ceiling. When a charged tube was held near his foot, pieces of metal foil were attracted to his face and to his outstretched hands.

Louis-Guilliaume le Monnier discharged a Leyden jar through a chain of 140 courtiers in the presence of the King of France. Jean-Antoine Nollet (1700–1770) attempted to measure the speed of electricity by having a line of monks 1 kilometer (3,280 feet) long hold hands. The monks at the ends of the line touched a machine that produced charge. They all jumped simultaneously when they felt the painful shock, so he concluded that electricity moved instantaneously.

How do **fluids** model **electric charges?**

How could these results be explained? Charles-François Dufay (1698–1739) concluded that there were two types of electricity. He named them "vitreous" (meaning glass, precious stones) and "resinous" (amber, sealing wax, silk). Friction separates the two types. When they are combined they neutralize each other. Jean-Antoine Nollet modeled these types as two fluids, each composed of particles that repelled each other. Charging amber gave it an excess of resinous fluid. Charging glass with silk gave it an excess of vitreous fluid. When the two were touched together the fluids combined with each other leaving the objects uncharged.

Benjamin Franklin believed there was only one fluid. When glass was rubbed the fluid filled the glass. When amber was rubbed the fluid left the amber. He called an object with an excess of fluid "positive" and one with too little fluid "negative." When they were touched the fluid flowed from the glass to the amber, leaving each with its

233

proper amount of fluid. The flow was likened to water in a river. The "electrical tension" (difference in potential) and "electrical current" were analogous to the difference of water levels between two points and of the amount of water transferred.

What makes an object positively charged, negatively charged, or neutral?

The massive nucleus of an atom consists of positively charged protons and uncharged neutrons. It is surrounded by a cloud of negatively charged electrons. Normally atoms are neutral: the number of electrons equals the number of protons in the nucleus. A negatively charged object is an object that has an excess of electrons. A positively charged object has fewer electrons than protons in the nucleus.

What combination of charges causes attractive and repulsive forces?

As you observed, unlike gravitational forces, which only attract masses to each other, electrostatic forces can either attract or repel charges. Like charges (positive–positive or negative–negative) repel each other. Unlike charges (positive–negative) attract each other. A common phrase describing many human social relationships, "opposites attract," holds true for electrostatic forces.

What are the two ways to charge an object?

When a rubber rod is rubbed with fur, the fur transfers electrons to the rubber rod. The rod and fur, originally neutral, are now charged. If an object touches the rod some of the excess electrons on the rod can move to the object, charging it. The rod, which is now negatively charged because it has excess electrons, can attract positive charges. This method is called charging by contact.

But, as you observed with the cellophane tapes, your hand and other neutral objects attract both positively and negatively charged objects. How does this happen? The rod attracts positive charges and repels negative charges. Neutral objects contain equal numbers of positive and negative charges. In a conductor the charges are free to move and so the electrons can be pushed to the far end of the object making it negatively charged and leaving the close end positively charged. An object that is neutral but has separated charges is polarized. Is there a net force on a polarized object? And can it exert a net force on the charged object, like the cellophane tape? Yes, because the electrostatic force is stronger at closer distances. Thus the attractive force between the unlike charges is stronger than the repulsive force between the like charges, and there is a net attractive force.

In non-conducting materials the charges cannot be widely separated, but they can move within the atoms or molecules. So insulators, like pieces of paper, dust, or hair, can also be attracted, even though they are neutral.

Why is it important to beware of excess electrostatic buildup when working with computer equipment?

If you have ever installed a circuit board or card into a computer, the product probably was shipped in a "static-free" bag. This bag is designed to keep all excess static charge outside the bag. Many electronic circuits are sensitive to the electrostatic buildup, and can be damaged if such a charge accumulates on sections of the circuit. Therefore, when installing the circuit board, the instructions usually encourage you to neutralize yourself by touching a grounded piece of metal to discharge your body and tools or to wear a grounding strap on your wrist to keep you at ground potential.

Did you ever see a piece of paper attracted to a charged rod, touch it, and then jump away? How would that happen? If it touched the rod, it became charged with a charge like that of the rod, and so it would now be repelled. A conductor can also be charged after being polarized, but without touching the charged object. If you bring a large metal object, like a pie plate, near a charged rod, the positive charges will move to the far end of the plate. If you now touch this end briefly with your finger the positive charges will be pushed even further away into your finger. When you remove your finger the pie plate is negatively charged. This process is called charging by induction.

Rubbing a glass rod with silk will achieve the same effect. The glass rod is positively charged, while the silk receives the excess negative electrons. The glass rod can still pick up small objects, but attracts the negative charges in those objects instead of the positive charges. When the pie plate is charged by induction it will be positively charged.

Why does a rubber **balloon** that has been **rubbed** in your hair **stick** to a **wall**?

The attraction between a charged balloon and a wall is the result of electrostatic forces. When rubber is rubbed on human hair or a wool sweater, electrons transfer easily to the rubber balloon. The balloon is charged by rubbing. The hair or sweater fuzz may stand up as a result of the excess positive charges repelling each other. When the balloon is brought near the wall, it polarizes the wall, moving the positive sources toward it and repelling the negative charges away. The negatively charged balloon is attracted to the many positive charges in the wall. As long as the electrostatic force and frictional force between the balloon and the wall are stronger than the gravitational force pulling the balloon down, the balloon will remain on the wall.

Why do you sometimes get a **shock** when **touching a doorknob**?

This annoyance happens usually on dry days after walking on carpeted floors. The friction between the carpet and your shoes or socks causes charges to be moved between

235

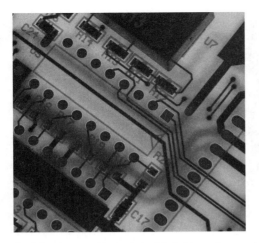

Circuit boards are sensitive to electrostatic buildup, and can be damaged if such a charge accumulates on sections of the circuit.

your body and the carpet. Usually your body becomes negatively charged. When your hand approaches a doorknob the negative charges in your hand are attracted to the positive charges in the doorknob (created by polarization), causing an electrical spark when the two charges meet.

What are some **good conductors** of electricity?

In order to be an effective conductor, a material must allow the electrons to move easily throughout it. The atoms in good conductors, such as most metals, have one or two electrons that can be easily freed from the nucleus to move through the material. Water is a fair conductor, but when salt is added it becomes a better electrical conductor.

What is a **good insulator** of electrical charge?

In an insulator the electrons are strongly bound to their nuclei and thus cannot move through the material. Good insulators are non-metals, such as plastic, wood, stone, and glass. Your skin is a good insulator, unless it is wet.

How is the strength of an **electrical force measured**?

British philosopher, theologian, and scientist Joseph Priestley (1733–1804) suggested that the force caused by static electricity might depend on distance the same way gravity does. Using Priestley's idea, the French physicist Charles Coulomb (1736–1806) made quantitative measurements of the force of attraction and repulsion between charged objects using an apparatus shown in the accompanying illustration. He found that the force depended on the charge of the two objects and the distance between them. The relationship he found is called Coulomb's Law and the unit of measurement of charge as the coulomb (C).

What is **Coulomb's Law**?

Coulomb's Law describes the strength of the electrical force between two charged objects. The formula is $F = k \ (q_1 \ q_2/r^2)$, where k is a constant equal to 9.0×10^9 Nm²/C² (newton-meters squared per coulombs squared). The charges q_1 and q_2, measured in coulombs, represent the charges on the objects that cause the force F, measured in newtons. Finally, r is the distance between the centers of the two charged objects. A negative force is an attractive force, while a positive force is repulsive.

What is a **coulomb** of charge?

A coulomb of charge is equal to the charge of 6.24×10^{18} electrons (negative) or protons (positive). A coulomb is a very large charge. Objects that are charged by rubbing or induction have typically a microcoulomb (10^{-6} C) of charge.

What is an **electroscope**?

An electroscope is a device used to measured the charge on an object. It consists of two metal leaves (either thin aluminum foil or gold leaf) attached to a metal rod. If you touch a charged object to the metal rod the two leaves will be charged with like charges, and so they will repel each other. The larger the charge, the greater the angle will be between the leaves.

What is an **electric field**?

As discussed before, a gravitational field surrounds Earth or any object with mass. Another object with mass that is placed in this field will experience a gravitational force on it. In the same manner, an electric field surrounds a charged object. Another charged object placed in that field will experience a force. If a positive force creates the field then the force caused by the field on a negative force will be toward the source of the field. A positive charge will experience a force away from the source. The English physicist Michael Faraday (1791–1867) was the first to use the concept of a field to describe the electrostatic force.

LEYDEN JARS AND CAPACITORS

What is the **Leyden jar**?

Water can be stored in a jar. In what can charge be stored? In November 1745 Ewald Jurgen von Kleist (1700–1748), dean of a cathedral in Pomerania, put a nail into a small medicine bottle and charged it with an electrical machine. When he touched the nail he received a strong shock. In March 1746 Pieter van Musschenbroek (1692–1761), a professor at the University of Leyden in Holland, performed a similar experiment with the device, now called the Leyden jar.

How does a **Leyden jar work**?

A Leyden jar is an insulating container with conductors on the inner and outer surfaces. When charging the Leyden jar the source of charge is connected to a rod touching the

The modern capacitor is an updated version of the Leyden jar, consisting of two conductors and an insulator.

inner conductor while the outer conductor is connected to ground. The inner and outer conductors become oppositely charged. It takes energy to move additional charges to the jar as the charges overcome the repulsive forces of the charges already on the conductors. The jar stores this electrical energy. If the inner and outer conductors are connected by a wire the charges flow and make the two conductors neutral again.

What were **uses** of the **Leyden jar**?

In the late eighteenth and nineteenth centuries, people attempted to use the Leyden jar in a variety of ways. Some felt that it could cure medical ailments, and many doctors used the jar as primitive electroshock therapy. Others used it as a demonstration device and for entertainment purposes. Still more people felt that it could be used in cooking. Try cooking a turkey with an electrical spark!

What is the **modern-day version** of a **Leyden jar**?

The capacitor is the modern version of the Leyden jar. Like the jar, it consists of two conductors separated by an insulator. The insulators used can be air, a thin plastic film, or a coating of oxide on the metallic surface. One use of a capacitor is to store the energy needed to fire a flash lamp on a camera. A battery-powered circuit slowly charges the capacitor. When the flash lamp is triggered the capacitor's energy is quickly transferred to the lamp, creating a brief, intense flash of light. Capacitors are also used in electronic devices from telephones to televisions to store energy and reduce changes in voltage.

What did **Benjamin Franklin's** famous **kite experiment** prove?

Benjamin Franklin is probably most famous for flying kites in thunderstorms. In the mid-1700s there were three different phenomenon that had similar effects. You could draw sparks with frictional or static electricity. Lightning appeared to be a giant spark, and electric eels could cause shocks like static electricity. But no one knew if these three had the same or different causes. Franklin touched a Leyden jar to a key tied to the string of his kite. When sparks jumped from the cloud to the kite, the charges went down the string and charged the Leyden jar. Thus Franklin showed that lightning and frictional electricity were the same.

Does Benjamin **Franklin's definition** of **positive** and **negative** agree with today's understanding of **charge**?

Franklin decided that sparks given off by an object charged by a glass rod (vitreous electricity) looked more like fluid leaking out than did the sparks from an object charged by a rubber rod (resinous electricity). Thus he decided that glass had an excess of electrical fluid.

Today we know that electric charge is mostly carried by electrons. Electrons are charged the same way that rubber or plastic is (negatively). Thus we say that they have a negative charge. Because they are transferred much more easily than are the more massive positively charged nuclei, when there is an excess of electrons the object is negatively charged; when there is a lack of electrons it is positively charged. So even though Franklin made the wrong choice, we still follow his convention.

Benjamin Franklin did not discover electricity, but he did show that lightning and frictional electricity were the same thing with his famous kite experiment involving a key and a Leyden jar.

How can you **construct** your **own Leyden Jar**?

You can use either a glass or plastic container that has a tight-fitting cap. Use a small nail to make a hole in the center of the cap. Straighten a paper clip and push it through the hole. Make sure the end of the clip reaches the bottom of the jar. Cover the outside of the jar with aluminum foil and fill the jar about 2/3 full of water. Make sure that the jar cap is dry. Now rub a plastic pen with wool and touch the pen to the paper clip. Repeat the rubbing and touching several times. Then touch the clip with your finger. You should feel a very tiny shock. The jar has stored the charge that you gave the pen when you rubbed it.

VAN DE GRAAFF GENERATORS

What is the **Van de Graaff generator**?

Named after its American creator, Robert Jemison Van de Graaff (1901–1967), the Van de Graaff generator has been the highlight of many electric demonstrations in both physics classrooms and museums around the world. The device, created in 1931, con-

239

sists of a hollow metal sphere that stands on an insulated plastic tube. Inside the tube is a rubber belt that moves vertically from the base of the generator to the metal sphere. A metal comb attached to the base almost touches the belt. The rubber belt carries negative charges from the comb up the tube and into the metal sphere. There, a second metal comb captures the charges. They repel each other and spread over the exterior surface of the metal sphere. As more and more charge is carried upward, it takes more and more energy from the motor to move them up because of the repulsive force of the charges already there. The energy of the charges can reach up to a million joules per coulomb of charge. That is, up to one million volts.

What happens if you **touch** the **Van de Graaff generator**?

If you place your hands on the upper sphere while the generator is charging it the electric charges accumulating on the sphere move onto your body as they are repelled by the other charges. When your body has enough charge your hair may stand up on end because the electric charges on the hair repel each other. You won't be hurt because the current through your body is very small. Just don't touch anything or anyone else!

What happens if you get **close** to a **charged Van de Graaff**?

The sphere on the Van de Graaff is a conductor surrounded by an insulator (the air). While there are strong forces on the negative charges on the sphere, they're not strong enough to break down the insulating properties of the air. If, however, you bring another object with less negative charge close to the generator the forces on the air molecules can become strong enough to rip them apart, separating their negative electrons from the positive nuclei. A spark will jump. If that object is your finger, you'll feel a shock when the charges carried through the spark move through your body. While the shock can be painful, one produced by the kind of Van de Graaff in physics classrooms is not harmful.

An old, damaged photograph of British physicist Michael Faraday, who was the first to describe an electric force in terms of a field. He also invented the Faraday Cage, which permits electric charges on the outer shell but not within the cage itself.

How much **charge** is **inside** the **sphere** of a **Van de Graaff generator**?

Zero. When negative charges leave the rubber belt, they move immediately to the outer surface of the sphere. Negative charges like to be as far away from each other as possible, so they move to the outer surface of the Van de Graaff generator's sphere.

What is a **Faraday Cage**?

A Faraday Cage, named after British physicist Michael Faraday, is a cage, metal grating, or metallic box that can shield electrical charge. Charges gather on the outer shell of the cage because they are repelled by one another and can be further from each other if they are on the outside of the cage. This results in no charge within the Faraday Cage. The metal sphere of a Van de Graaff generator is a Faraday Cage. Cars and airplanes can be Faraday Cages as well, and may provide some protection from lightning during an electrical storm.

LIGHTNING

What is lightning and how is it created?

Lightning is an electrical discharge in the atmosphere, like a giant spark. There is still debate about the cause of the separation of charges needed to create the discharge. Atmospheric scientists believe that strong updrafts in the clouds sweep droplets of water upward, cooling them far below the freezing point. When the droplets collide with ice crystals the droplets become a soft mixture of water and ice. As a result of these collisions the ice crystals become slightly positively charged and the water/ice mixture becomes negatively charged. The updrafts push the ice crystals up higher, creating a positively-charged cloud top. The heavier water/ice mixture falls, making the lower part of the clouds negatively charged.

The ground under the cloud is charged by induction. The build-up of negative charges on the underside of a thundercloud attracts the positive charges in the ground. The negative charges are repelled further into the ground, leaving a positively charged surface.

How do the **charged regions** of **clouds** and the **ground** act as a **giant capacitor**?

A capacitor consists of two conducting plates with opposite charge separated by an insulator. When a wire is connected between the two plates, a large electric current flows the charges rapidly from one plate to the other, neutralizing the capacitor.

The charged regions of the clouds act as conducting plates while the air between them acts as the insulator. The same thing occurs between the lower section of the cloud and the ground. The air between these sections acts as the insulator, but when the forces exerted by the charges on the air molecules are large enough, they can rip the electrons from the molecules. The result is a positively charged molecule, called an ion, and a free electron. The air is changed from an insulator to a conductor. The mobile electrons gain more energy, creating more and more ions and additional free

241

In atoms and molecules the negatively-charged electrons are attracted to the positively charged nucleus. It takes a considerable amount of energy to remove an electron from an atom or molecule. The electric fields produced by thunderclouds have enough energy, and so they can pull an electron from an atom, creating a positively charged atom, or ion, and an electron that is free to move.

electrons. When the electrons and ions combine again light is emitted. The tremendous amount of energy released rapidly heats the surrounding air, producing thunder.

Does **lightning always strike** the **ground**?

Although most people think of lightning when it goes between Earth and clouds, the most common type of lightning occurs inside and between thunderclouds. It is usually easier for lightning to jump between the clouds than it is for it to jump from the clouds to Earth. As a result, only one quarter of all lightning strikes actually strike the ground.

How does the **air become** a **conductor**?

When the charges have enough energy to begin to ionize the air a the free electrons will form a negatively charged "stepped leader" that will go from the cloud and make its zigzagged and often branched trip toward the ground. This process is slow, taking a few tenths of a second. The leaders are also weak and usually invisible. The atoms in the air near the ground, feeling the attractive force from the electrons in the stepped leader, separate into ions and free electrons. The positively-charged air ions from tall objects, such as trees, buildings, and towers leave in streamers. When a stepped leader and streamer meet, a channel of ionized air is created, allowing large amounts of charge to move between the cloud and the ground. The return stroke of charge back to the cloud is the brightest part of the process.

How much **energy** is **contained** in a **lightning flash**?

An average lightning bolt transfers about five coulombs of charge and about half a billion joules of energy. The transfer takes about 30 millionth of a second and the electrical power in a bolt can be as large as 1,000 billion watts.

Where in the world does **lightning** occur **most frequently**?

Satellite lightning detectors show that over the entire Earth lightning strikes about 45 times each second, or 1.5 billion strikes each year. In the eastern region of the Democ-

---

How many people are killed or injured by lightning?

Of the 40 million lightning strikes per year in the United States, 400 of those strikes hit people. Half die as a result of the strike, while many of the others sustain serious injuries.

---

ratic Republic of the Congo in Africa every year each square mile, on average, has some 200 lightning strikes. A section of Florida known as "lightning alley" is a 60-mile wide hot-spot of lightning activity in the United States. On average there are 50 lightning strikes in each square mile per year.

SAFETY PRECAUTIONS

Is it true that **lightning never strikes** the same place **twice**?

This is absolutely false. The Empire State Building in New York City is just one example of where lightning has struck more than once. In some thunderstorms, the tower on the Empire State Building has been hit several dozen times.

Why is a **car** often the **best place to be** when lightning strikes?

It is not because of the rubber tires! Many people think the rubber tires of a car provide insulation from the lightning striking the ground. If this were the case, wouldn't riding a bicycle do the same thing? The real reason why a car is a safe place to be when struck by lightning is because most cars have metal bodies, which act as Faraday Cages, keeping all the electrical charge on the outside of the car. Since the charge is kept on the outside of the vehicle, the person sitting inside the car is kept perfectly neutral and safe. It is the shielding of the metal car body, and not the rubber tires, that protects people in automobiles.

What happens to an **airplane** when it is **struck by lightning**?

Airline pilots tend to avoid thunderstorms, but when a plane is struck by lightning, the passengers inside the plane are kept perfectly safe, for they are inside a Faraday Cage, which shields them from the massive electrical charge. The lightning can, however, disturb and even destroy some of the sensitive electronics used to fly the plane.

Studies were performed by NASA in the 1980s in which they flew fighter planes into thunderstorms to see how the planes would react to lightning. The scientists quickly found that the planes actually encouraged lighting, because the planes caused

243

Studies in the 1980s showed that airplanes actually attract lightning. While lightning storms can still be dangerous, passengers inside airplanes are safe from electric shock because they are actually inside a Faraday Cage.

increases in the electric field of the cloud, which in turn caused the lightning to hit the plane's metal body.

What are some **things you should do** if caught in a **lightning storm**?

The safest place to be during an electrical storm is inside a building (where you should stay away from electrical appliances such as the phone and television, as well as all plumbing and radiators) or car, but if you are unable to shield yourself in this way, the following precautions should be taken:

- Crouch down on the lowest section of the ground, but do not let your hands touch the ground. If lightning strikes the ground, the charges spread out sideways and can still reach you. If only your feet are on the ground (especially if you're wearing rubber-soled shoes), this might limit the amount of charge that passes through your body. If you must lie down because of an injury, try to roll up into a tight ball.
- Take off and move away from all metal objects unless they act as Faraday Cages (refer to the question about Faraday Cages).
- Move away from isolated and tall trees.
- Avoid the tops of hills or mountains and open areas such as water and fields.
- If out on a lake or on the ocean, get back to shore as quickly as possible. If that is not practical, get down low in the boat and move away from any tall metal masts or antennas.

Why shouldn't you stand under a large tree during a thunderstorm?

During thunderstorms, many people stand under trees in an effort to stay dry. However, this can have dire consequences. In the spring of 1991, a lacrosse game at a Washington, D.C., high school was postponed after lightning was observed in the sky. Over a dozen spectators ran for shelter under a tall tree to protect themselves from the rain. A few seconds later, lightning struck the tree, injuring twenty-two people and killing a fifteen-year-old student.

Trees are tall points where positive streamers can originate and attract the stepped leader, starting a lightning bolt. By standing under a tree, holding an umbrella, swinging a golf club, or batting with an aluminum bat, people are making themselves part of a lightning rod.

Why are **lightning rods effective** in keeping tall trees and homes safe from lightning?

Lightning rods are pointed metal rods that are installed above a tree or rooftop to protect the object. The rod, connected to the ground by a metal wire, both encourages and discourages a lightning strike. The rod discourages the lightning strike by "leaking" positive charges out of its pointed top to satisfy the need for positive charge in the clouds. If the rod cannot leak out enough charge to satisfy demand the stepped leader from the cloud is instead attracted to the rod, and a flash of lightning occurs. Therefore, the rod attempts to discourage lightning, but if it cannot satisfy the negative charge, it attracts the lightning to the rod instead of the tree or house.

If the lightning rod doesn't have a good connection to the ground through the wire it can increase the danger to the building. Often these heavy grounding wires come loose from the lightning rods, and if the rod is then hit by lightning, the charges will flow along the surface of the building to the ground and could cause a fire. The rods can become disconnected from lack of routine maintenance; it is wise to check these connections on a regular basis.

Who invented the lightning rod?

Although a Russian tower built in 1725 had what would now be called a lightning rod, credit is usually given to American inventor Benjamin Franklin (1706–1790). He invented the lightning rod in 1749 to protect houses and tall trees from being destroyed by lighting bolts.

CURRENT ELECTRICITY

When was it thought that there were **different kinds** of **electricity?**

As you have seen, Benjamin Franklin's kite experiment showed that lightning and static electricity were the same. Since ancient times humans knew that certain fish, such as the electric eel, could shock a person. Was this "animal electricity" the same as static electricity? According to legend the Italian physician Luigi Galvani (1737–1798) was making frog-leg soup for his sick wife. Whenever a nearby static electricity machine created a spark the legs jerked. After completing several experiments, in a 1791 paper Galvani reported that when one metal touched the muscle of a frog's leg while another metal touched the nerve, the muscle contracted. Thus Galvani helped to show that there was a connection between static electricity and electric effects in animals.

How did **Luigi Galvani's experiments** lead to the **development** of **current electricity?**

Galvani believed that the flow of charge from the nerve into the muscle caused the contractions. His fellow scientist at the University of Bologna, Alessandro Volta

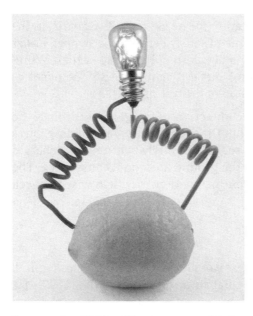

You can use the acidic juice within a lemon—or a potato, if you're short on lemons—to create a simple battery. Insert two different metals (zinc and copper) and electrons will flow from the zinc to the copper to create a current weak current. Connecting multiple lemons gives you a stronger current.

(1745–1807), recognized that Galvani's frog leg was both a conductor and a detector of electricity. In 1791 he replaced the leg with paper soaked in salt water, a conductor, and used another means of detecting the electricity. He found that charges flowed only if the two metals touching the paper were different. The combination of two different metals separated by a conducting solution is called a galvanic cell after Galvani.

Volta went further. He found that the two metals that produced the greatest electrical effect were zinc and silver. In 1800 he stacked alternating disks of zinc and silver, separated by a card wetted with salt water. He found that this device, called the voltaic pile, was a continuous source of charge flow. Sir Humphrey Davy showed that the charge flow was due to a chemical reaction between the metals and the conductive solution in the cards.

How can water be used to model voltage and current?

Think of a river flowing downhill from one lake into another. The water flows from a higher to lower altitude. Electrical potential difference is like the change in height of the water on the two ends of the river. Electric current, the flow of charge, is like the water current, the flow of water over the falls. If there is no difference in altitude of the two lakes, then there will be no flow of water. If there is no difference in voltage, there will be no flow of charges.

What **causes** the **flow** of charges in a **voltaic pile**?

Volta invented the term "electromotive force" (*emf*) to describe what causes the separation of charges. The more disks there were in the voltaic pile, the greater the *emf*. Unfortunately, the word "force" is an incorrect use of that term, because there is no mechanical push, measured in newtons, on the charge. The correct term is potential difference or voltage, the energy change per unit charge separated. Both the term *voltage* and the unit in which it is measured, the *volt* (V), are named after Volta.

What is **potential difference** or **voltage**?

In a voltaic pile, today more commonly called a battery, chemical energy is converted into increased energy of electric charges. Positive charges at the positive terminal of the battery have a greater energy than those at the negative terminal. The quantity that is important is not the total energy difference of charges at the two terminals of the battery, but the energy difference divided by the charge. This quantity is called the electric potential difference, or more commonly, the voltage.

What is **current**?

Current is the flow of charge. It is measured in amperes (or "amps"), named after the French mathematician and physicist André-Marie Ampère (1775–1836). One ampere is equal to one coulomb of charge passing through a wire divided by one second. The greater the voltage difference across the wire, the larger the current.

RESISTANCE

What is **resistance** to current flow?

All objects encounter friction when moving. Electrons are no different, but we refer to the friction that electrons encounter as resistance. The electrons collide with the atoms in a wire and are deflected from their paths. For the same voltage difference,

247

the greater the resistance the smaller the current. Resistance causes the electric charges to lose energy. The energy goes into the thermal energy of the wire or other conductor. That is, they get hot! The thermal energy can produce heat in a toaster and heat and light in an incandescent lamp.

What **factors** determine the **amount of resistance** of a conductor?

The resistance of a wire or other conductor depends upon the following:

- The length of the conductor (the longer the wire, the more resistance).
- The cross-sectional area of the conductor (the thinner, the more resistance).
- The properties of the material (for metals, the fewer the number of free electrons, the more resistance).
- The temperature of the conductor (for metals, the warmer, the more resistance; for carbon the cooler, the more resistance).

What are **resistors**?

Resistors are devices used in electrical circuits to put a definite resistance in a circuit. Normally they are made of graphite or a thin carbon film coated on glass. Larger resistors are cylindrical and have four color bands that encode the value of the resistance. On computer boards they are tiny, rectangular devices barely a millimeter on a side with conductive ends that are soldered to the board. If they are designed to dissipate a large amount of power, they are made of high-resistance wire.

SUPERCONDUCTORS

What is a **superconductor**?

Superconductors allow electrical current to travel without resistance, and therefore no voltage drop across them or energy loss within them. Superconductors must be cooled below their critical temperatures to have no electrical resistance. Some elements, compounds, and alloys that are superconductors are lead and niobium nitride, and a niobium-titanium alloy. All these require liquid helium to cool them to their critical temperatures. In the 1980s some ceramics were found to have much higher transition temperatures that could be reached using much cheaper liquid nitrogen. The first found was yttrium barium copper oxide. As of 2008 a family of materials including iron, such as lanthanum oxygen fluorine iron arsenide, was developed.

Who discovered **superconductivity**?

The creation of materials without resistance was thought to be impossible, but a Dutch physicist by the name of Heike Kamerlingh Onnes (1853–1926) proved it was

Who won the Nobel Prize for their work in superconductivity?

Three American physicists, John Bardeen, Leon N. Cooper, and John R. Schrieffer, explained why superconductivity occurs in metals and alloys. Their development of the BCS theory for superconductivity was cited when they won the Nobel Prize in 1972.

Fifteen years later, two other physicists won the Nobel Prize for discovering superconductive materials that achieved zero resistance at temperatures thought to be too high for superconductors. Physicists Georg Bednorz and Alex Müller of IBM found that a ceramic substance called lanthanum barium copper oxide became a superconductor at 35 kelvins. This was a much higher temperature than anyone thought possible at the time.

possible in 1911. Onnes lowered the temperature of different metals, including mercury, close to absolute zero. He then measured the electrical resistance of the materials at such low temperatures and found that mercury, at only 4.2 kelvin (–277.2°C), had zero resistance to electrical current.

What **technologies** have developed as a result of **superconductivity**?

Superconductors are most commonly used in large electromagnets. With no resistance, once the current is started, it will continue forever without change. Therefore the magnets dissipate no power and do not heat up. These magnets are most often used in magnetic resonance imaging (MRI) machines. An MRI allows a doctor to view the inside of the human body without using harmful radiation. They are also used in particle accelerators that reveal the fundamental structure of matter by smashing the nuclei of atoms together. The most powerful accelerator is the Large Hadron Collider (LHC) in Switzerland. Another application of superconductivity is the SQUID (Superconducting QUantum Interference Device) that is an extremely sensitive detector of magnetic fields used in geological sensors for locating underground oil.

OHM'S LAW

What is **Ohm's Law**?

In the early 1800s, Georg Simon Ohm (1789–1854), a German physicist, developed the law that bears his name: the resistance of an object is independent of current through it. Many materials do not obey Ohm's Law. For example, when the current through

249

the tungsten filament wire in a lamp is increased the temperature of the wire increases and so does its resistance.

What are the **units** and **symbols** used for current, voltage, and resistance?

Quantity	Unit	Symbol
Current (I)	Ampere (amp)	A
Voltage (V)	Volt	V
Resistance (R)	Ohm	Ω

What is the **relationship** between **current, voltage**, and **resistance**?

Voltage = Current $\times$ Resistance, or $V = IR$. That is, the voltage difference across a resistor is equal to the current through it multiplied by its resistance.

What levels of **current** are **dangerous**?

Approximately 1 mA (0.001 A) is enough to produce a tingling sensation. 10 mA is painful. 12-20 mA is enough to paralyze muscles, making it impossible to let go. 60-100 mA causes ventricular fibrillation of the heart. That is, the heart is beating in such a way that it cannot pump blood through the circulatory system. Greater than 200 mA causes the heart to clamp down and stop beating.

How much **resistance** do **our bodies** have to **electrical current**?

On average, the human body has an electrical resistance between 50,000 and 150,000 ohms. Most of this resistance is across the skin. If the skin is wet the resistance drops to about 1,000 ohms. If the skin is broken, then resistance across organs in the body is on the order of a few hundred ohms. In this condition 10 volts is sufficient to cause serious, if not fatal damage.

Do **electric eels** really use electric fields to capture their prey?

Electric eels do indeed set off electrical pulses to stun and even kill their prey. These eels have special nerve endings bundled together in their tails that can produce 30 volts in small electric eels to 600 volts in larger eels. Besides using the electrical shocks for hunting, the eels produce a constant electrical field for use in navigation and self-defense. Most people do not have to worry about encountering electric eels, however. This variety of eel is native only to the rivers of South America.

Birds can perch safely on electric wires, as long as they don't come into contact with objects that have two different voltages.

Why don't **birds** or **squirrels** on **power lines** get electrocuted?

In order to get electrocuted on a bare wire, a bird would have to be in contact with objects that had two different voltages. The difference in voltage along the wire over the distance between the animal's feet is very small. The animal would be in danger only if it made contact with both a high-voltage wire and the ground or a wire connected to ground (low voltage). Then there could be a large current through its body.

Why do **electricians** work with **"one hand behind their back"**?

When working on high-voltage circuitry, many electricians like to place one hand behind their back because this way there is little chance for each hand to touch objects of different electrical potentials and cause a shock.

Does **voltage shock** you?

Signs around power plants and breaker boxes often state, "CAUTION: High Voltage Area." It is not voltage that can hurt you; it is the electrical current that flows through your body that can produce serious and sometimes fatal consequences. The Van de Graaff generator creates hundreds of thousands of volts, but produces such a low amount of current that the sparks it emits only cause muscles to tingle.

ELECTRIC POWER AND ITS USES

What is a **watt**?

A watt is the unit of power, or the rate of energy transfer. Light bulbs, toasters, hair dryers, televisions, and other electrical appliances are rated by the power they use. In the context of electricity, the watt is the unit for electrical power and is often found on light bulbs and other devices used in electrical circuits. The formula to find the power used is $P = I \times V$ (power equals current times voltage).

Who **invented** the electric **light bulb**?

The first person to generate light from a wire filament was the British Sir Humphry Davy. Davy was known for his work on electric arc lamps in the early 1800s, and his breakthrough discovery was the first electric light. His lamp used a very thin piece of platinum wire that had high resistance and emitted a soft glow. The lamp didn't last long, and so it was not practical, but it did pave the way for others.

What contribution did **Thomas Edison** make to the electric light?

The American inventor Thomas Alva Edison (1847–1931) tried literally hundreds of different materials as filaments for the lamp. He found that heating a cotton thread in the air left a thin length of almost pure carbon. The carbon filament was connected to wires sealed in a glass bulb from which the air was removed. In 1878 this first practical electric lamp lasted for hours more than any other lamp.

What is the difference between a **kilowatt** and a **kilowatt-hour**?

A kilowatt, 1,000 watts, is the unit used to describe the power, the rate at which the energy is being converted. Energy is the power multiplied by the time it is used, in this case hours. Therefore a kilowatt-hour, the product of power and time, is a unit of energy. The utility company charges you for the number of kilowatt-hours of electricity you use in a month.

For example, a 100-watt light bulb uses 100 watts (or 0.1 kilowatt) of power. If that light bulb were left on for an entire month, the energy that the bulb consumed would be 0.1 kilowatts × 24 hour/days × 30 days/month, which equals 72 kilowatt-hours/month. If the energy cost is $0.12 per kilowatt-hour, the bill for that one light bulb would be $8.64 per month. Replacing the 100-watt incandescent lamp with a 23-watt compact fluorescent lamp that is equally bright would cost only $1.98 per month. An LED lamp equally bright has a power rating of only 13 watts, and therefore would cost $1.12 to light.

Why is a 100-watt bulb brighter than a 25-watt bulb?

A light bulb transforms electric energy first into the thermal energy of the heated filament and then into light and heat. The rate at which these energy changes occur is determined by the way the bulb is constructed. The 100-watt bulb has a lower resistance filament—the thin wire in the bulb that gets hot. Assuming that both bulbs are connected to a 120-volt outlet, there will be more current through the 100-watt bulb than the 25-watt bulb. Lower resistance is created by making the filament out of thicker wire. The lower resistance means higher current, which in turn means higher power and more light and heat output.

CIRCUITS

What is needed to **create a circuit**?

A circuit is a circular path through which charge can flow. So, the first requirement is a complete, unbroken conducting path. Second, there must be a source of potential difference—most often a battery. The battery provides the voltage that will produce the current in the circuit. With only a battery and wire connecting the two ends of the battery the resistance in the circuit will be almost zero, and the amount of current will be very high. This situation is called a short circuit. The wire will become hot enough to burn you. So, for a useful circuit there must be a third element—a device with resistance. This may be a resistor, lamp, motor, etc.

In terms of energy, the energy input to the circuit is the chemical energy stored in the battery. When the circuit is complete, the chemical energy becomes electrical energy in the wires. That energy is then converted to thermal energy in the resistor or lamp, or kinetic energy in the motor. The hot resistor or lamp then radiates heat and light into the environment.

In our households the battery is replaced by the electric generating station operated by the utility company. It may use the chemical energy in fossil fuels such as coal, oil, or natural gas to boil water (i.e. produce thermal energy). The steam from the boiling water then turns the generator, converting the energy to rotational kinetic energy. The generator then converts this energy of rotation into the electrical energy that is transmitted to the home. A nuclear power plant uses the nuclear energy in the nucleus of the uranium atoms to heat the water and produce steam. From that point on the nuclear and fossil fuel power plants are essentially the same.

What is an **open circuit**?

A circuit is a closed loop through which electric charges can flow. If the loop is opened, then it is called an open circuit and charges no longer flow. A switch is com-

The Indian Point Energy Center on the Hudson River in New York state uses nuclear fission to heat water and produce steam. Except for the fuel source, nuclear power plants operate just like any other type of power plant.

monly used to open and close a circuit. If a wire breaks, the connection between the wire and another part of the circuit fails, or if a lamp burns out (its filament breaks) then the circuit is opened and there is no current.

What is the danger of **short circuits** in a home?

As described above, if there is only a tiny amount of resistance in a circuit the current is very large and the wires get extremely hot. If, for example, the insulation on the wires in an appliance fails and the wires touch each other, the resistance drops and current rises. The wires, including those in the walls, can get hot enough to cause a fire. Household circuits are protected by fuses or circuit breakers. They are designed to open when current exceeds a predetermined limit. With the circuit now open, current stops and the wires will cool.

AC/DC

What is a **DC circuit**?

In a DC, or direct-current, circuit charges travel only in one direction. The voltage source, a battery or direct current power supply, has one positive and one negative terminal, so there is current in only one direction.

Did Thomas Edison use AC or DC when he used his incandescent lamps to light a neighborhood?

Edison invented the first practical incandescent lamp in 1878. In 1882 he connected 59 customers in the neighborhood around his New York City laboratory to a DC generator that supplied 100 volts. The customers used the electrical power for lamps and motors. The relatively low voltage matched the resistance of the lamps and was not believed to be very dangerous. Unfortunately, in order to carry much power, at low voltage the current must be high, and this heated the wires. Customers had to be within two miles of the generating system to avoid serious loss of energy in the wires.

What is an **AC circuit**?

In an AC, or alternating-current circuit, the polarity of the voltage source changes back and forth at a regular rate. In the United States one terminal of the source changes from positive to negative and back to positive 60 times each second. Therefore the flow of charge also alternates in direction 60 times a second as the electrons in the circuit vibrate back and forth. An alternating current is usually found in wall outlets in buildings. Most of our electrical appliances run on alternating current.

What important contributions did **Nikola Tesla** and **George Westinghouse** make in the late nineteenth century?

Nikola Tesla (1856–1943) worked for Edison's company in Europe. Edison offered Tesla a large sum of money if he would invent an improved generator. When Tesla did, Edison refused to pay, saying he had been joking. Tesla quit.

In 1884 George Westinghouse (1846–1914), a prolific American inventor and businessman, formed the Westinghouse Electric Company in 1884. Four years later he persuaded Tesla to join his company and purchased his patents from him. Westinghouse had improved a transformer invented in France, but Tesla's AC induction motor, invented in 1883, and three-phase generator were essential for Westinghouse's dream of transmitting electric energy efficiently over long distances.

Why is **AC preferred** over DC?

In the late 1880s Edison and Westinghouse battled over the relative merits of DC and AC power distribution systems. After the electric chair had been developed Edison attempted to name electrocution "Westinghousing" because it used AC, but he failed. The AC system was victorious because transformers could raise the voltage to thou-

> ### Are holiday lights in series or parallel circuits?
>
> **M**any years ago holiday lights used large bulbs designed to work on 120 volts. Those strings were wired in parallel. Today holiday lights are wired in series so that the tiny bulbs have only low voltages across them. If a lamp burns out because the filament fails, there will be no current through the string. But, there will be 120 volts across the failed bulb. The bulb has a wire touching the two thick wires that deliver current to the filament. The wire is covered with a thin insulating film. The film remains insulating when the voltage difference across the bulb is small, but when it becomes 120 volts, sparks break the insulating film and weld the wire to the thick wires. This short-circuits the bulb, completing the circuit through the remaining lamps.

sands of volts for transmission, then lower it for use in homes and businesses. At the high voltages the current needed to transmit large amounts of power is reduced, and so is the energy lost to thermal energy because the heating of transmission wires depends on current and resistance.

SERIES/PARALLEL CIRCUITS

What is a **series circuit**?

A series circuit consists of electrical devices such as resistors, batteries, and switches arranged in a single line. There is only one path for the charges to flow through, and if there is a break anywhere in the circuit, the current will drop to zero.

What is a **parallel circuit**?

A parallel circuit allows the charges to flow through different branches. For example, the wire from the battery would be connected to one terminal of each of three bulbs. The other terminals are connected together and to the negative terminal of the battery. The charges now have three separate paths through which they can flow. If one bulb burns out or is removed from the socket, that bulb would no longer light, but the current through the other two lamps would not change. They would continue to glow.

What happens to a **series circuit** if more **bulbs** are **added**?

If more light bulbs or other resistors are placed in a series circuit, there is more resistance in the circuit, and so the current, and the brightness of the lamps would be reduced.

What happens in a **parallel circuit** if **more bulbs** are added?

In a parallel circuit the current goes through separate branches. If another branch is added with another bulb, the current has an additional path to take. But, the battery (or generator) produces a constant voltage, so the current through the original bulbs does not change, and neither does their brightness.

Are **series** or **parallel circuits** used in our **homes**?

Each circuit, by itself, contains a series connection of a switch and lamp or appliance. The circuits themselves are connected in parallel so that they can be used independently.

ELECTRICAL OUTLETS

Many outlets have **three holes**—what is the **purpose** of each hole?

Outlets have two slots, one longer than the other, and a "D"-shaped hole. The contacts in the shorter slot are connected to a black wire. This is the "hot" connection that carries the 120 volts. The contacts in the longer slot are connected to a white wire, called the neutral wire. The white wire is connected to ground in the electric distribution box. Thus there is a potential difference of 120 volts across the two contacts. The third hole is attached to a green wire that is at ground potential. Why do you need two connections at the ground potential? Because when the appliance plugged in draws current, there is current through both the black and white wires. Each wire has resistance, so there will be a voltage drop across the white wire, and it will be above ground potential at the outlet. While this voltage will be small, it could be dangerous. The green wire, which carries no current, will remain at ground. It can be connected to the metal case of the appliance, assuring that the case will remain at ground potential.

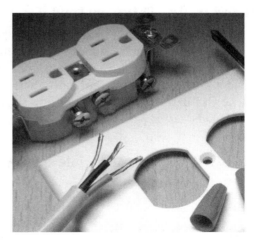

What if the tool or appliance has a **three-prong plug** but you have only **two-slot outlets**?

Do not use the appliance if you do not have the proper outlet for the device. Cutting off the grounding prong will defeat the safety feature of the separate ground wire.

When installing an electrical outlet, note that the green wire (okay, you can't tell in this black and white photo, but it is there!) is the grounding wire, which must be attached to the grounding contact on the outlet.

257

What is the little **green wire** or plate on the three-to-two adapter?

The green wire or metal tab attached to adapters is the grounding wire. Since the adapter is circumventing the ground prong, an alternate means of grounding is needed. If the screw on the outlet plate is grounded, the green wire on the adapter should be attached to it. This way, if there is an electrical short, the current can still flow through the grounding wire. If the screw is not grounded, then the adapter should not be used. An outlet tester that is available at most hardware stores can be used to make sure the screw is grounded.

What is a GFI, or **Ground Fault Interrupter**?

A ground fault interrupt outlet is now required by building codes for outlets within six feet of a sink or in any other environment where water could be close to the outlet. Normally the currents in the black and white wires will be equal, but if the water provides an alternative current path, then the two currents will no longer be the same. The GFI detects this difference and shuts off the circuit within milliseconds. GFIs should be tested periodically to make sure the electronic circuit is still working.

Why is it dangerous to operate **electrical devices** in **bathtubs, showers,** and over full **sinks**?

Although water reduces the resistance of the human body and thus makes it more susceptible to electrical shock, it is the plumbing that is the main hazard. Take, for example, a person who likes to watch a plugged-in TV while sitting in the bathtub. If the TV is not connected to a GFI-protected circuit and fell into the tub, the water would come in contact with the 120-volt wires in the TV. With the metal plumbing of the tub connected to ground, the grounding path would cause a current through the water.

Unfortunately, this translates into a bad day for the bather; an mp-3 player or battery-powered smart phone is much safer.

MAGNETISM

When was **magnetism discovered**?

The discovery of rocks that attracted certain metals is lost to history. As was the case with electrostatics, Aristotle (384–322 B.C.E.) credited Thales of Miletus (625–545 B.C.E.) with the first scientific discussion of the attractive power of the rock later called lodestone. The word "magnet" comes from the region of Greece where lodestone is found. But the power of lodestone was found by other people around the same time. At the time of Thales' life an Indian surgeon, Sushrata, used magnets to aid surgery. In the fourth century B.C.E. the Chinese *Book of the Devil Valley Master* says "Lodestone makes iron come."

In the eleventh century C.E. the Chinese scientist Shen Kuo wrote about the use of a magnetized needle as a compass in navigation. By the next century the Chinese were known to use a lodestone as a shipboard compass. One hundred years later the British theologian, Alexander Neckham, described the compass and how it could be used to aid navigation. Some people thought that the Pole Star attracted the compass, while others thought that the source was a magnetic island near the north pole. In 1269 the Frenchman Petrus Peregrinus wrote a detailed paper on the properties of magnets. But the most comprehensive and famous work was written by William Gilbert in 1600. Gilbert concluded that Earth was a giant magnet.

What are the **properties** of **magnets**?

You've probably played with magnets since you were a child. It is likely that you found that magnets attract some materials but not others. You may have found that you can use a magnet to magnetize items like paper clips, nails, and screws. If you played with two magnets you found that they could either attract or repel each other.

Whether you played with metal bar-shaped magnets, rectangular or circular ceramic magnets, you found that the magnet exerted stronger forces at the ends or faces of the magnets. Those regions are called "poles." If you hang the magnet from a string so it can rotate freely you'll find the magnet orienting itself north-to-south. The end facing north is called the north pole, the other the south. Like poles repel each other while unlike poles attract, but either end can attract other materials.

Magnetic poles always come in north-south pairs called dipoles (two poles). Some theories predict the existence of isolated north or south poles, called monopoles. But, there have been extensive searches for monopoles over the past decades and none has ever been found.

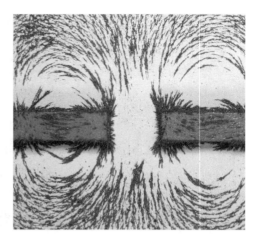

Children often discover some of the properties of magnetism by playing with bar magnets and metal shavings. In this way, you can easily discover that magnets have opposite poles and create magnetic fields.

What is a **magnetic field**?

Just as the gravitational field is the region around a massive object that causes the attractive force on another object with mass, a magnetic field is the region around a magnet that causes forces on magnetic materials or other magnets.

What **materials** are **attracted** to magnets?

Iron, nickel, and cobalt and most of their alloys are attracted to magnets. Other metals, like silver and gold, copper, tin, stainless steel, zinc, brass and bronze are not attracted. Non-metals are not attracted.

Iron, nickel, and cobalt are called ferromagnetic. All materials respond to magnetic fields, but most respond so weakly that the forces are hardly felt. Those that are repelled are called diamagnetic; those attracted are paramagnetic.

What **causes materials** to be **attracted** to magnets?

The ultimate cause of magnetism is electrons. When electrons are in a magnetic field the forces they experience cause them to move in tiny circles. The circling electrons create their own magnetic fields that give rise to diamagnetism. Electrons are tiny magnets themselves, with north and south poles. In most atoms these magnets are paired so their fields cancel. But, if there are an odd number of electrons, the unpaired electron produces a paramagnet. Oxygen, for example, is paramagnetic.

In ferromagnets the unpaired electrons in large groups of atoms interact with each other so that they point in the same direction. This group is called a domain. When a ferromagnet is put in a magnetic field the domains can line up with their poles facing the same direction, making the material a magnet. In most materials when the magnetic field is removed the domains revert to their former random directions and the material is no longer a magnet. For certain alloys, however, the domains remain aligned, resulting in a permanent magnet.

What materials make the **strongest** permanent **magnets**?

Traditional permanent magnets were made of an alloy of aluminum, nickel, and cobalt, called ALNICO. Ceramic and rubber magnets use ferrites, an iron oxide material. In the 1980s the automobile companies searched for materials to reduce the weight of motors in their cars. They found an alloy of cobalt and samarium, a rare earth, made strong, lightweight magnets, but were extremely brittle and expensive. Today the strongest magnets are made from a lanthanum-iron-boron (LIB) alloy. Their strength can be as much as 20 times that of alnico magnets. They are also brittle and so are coated with a plating of nickel and copper. Their price has fallen so much that they are used to hold sunglasses to eyeglass frames, in necklace clasps, and in children's toys.

How are **refrigerator magnets made**?

Examine a refrigerator magnet. It is flexible, feels like rubber, and only one surface is attracted to metals. It doesn't stick to a stainless steel door unless the stainless has been coated with steel. It's made of rubber that has been impregnated with ferrite particles and magnetized. Small pieces, each a dipole, are then pressed together under heat to bond them into one thin sheet that can be cut, folded, and bonded to other sheets. Which of the three arrangements shown below would have the properties of a refrigerator magnet as described above?

The top two wouldn't because both surfaces would act as a magnet. The top right-hand arrangement would be a very weak magnet on both faces because the alternating poles would essentially cancel each other out.

In the third drawing the sheets have been folded and then pressed together so that the poles are at only one surface, so only that surface would act like a magnet. The alternating N and S poles attract steel and stick to it. You can check this idea by taking two refrigerator magnets and holding the magnetic surfaces together, and then try sliding one over the other. You'll find that they skip as first N and S poles touch each other and attract. Then the like poles try to touch each other but repel, making the magnets skip.

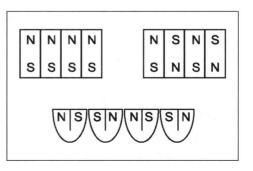

What happens when a **magnet** is cut in **two pieces**?

When a magnet is cut the atoms within the domains remain aligned. In almost

Fortunately for us, the earth generates its own magnetic field, which effectively fends off dangerous particles from the sun.

every case the cut would be between two domains, leaving aligned domains in the two halves. If you cut a domain you would create two smaller domains, each with a north pole and a south pole. So no matter where you cut the result is two magnets, each with its own north and south pole. The more domains, the stronger the magnet.

How is **Earth's magnetic field** oriented?

Because opposite poles attract, the north pole of a hanging magnet or compass must point toward a south pole. So, the south pole of Earth's magnet must be near the north geographic pole. The poles are actually far below Earth's surface, so Earth's field is not parallel to its surface.

What is the **origin** of Earth's **magnetic field**?

The source of Earth's magnetic field is its core, made of iron, so hot that it is molten. It rotates at a slightly different rate than does Earth, and this difference creates what is called a dynamo effect, generating a magnetic field. Details of how the dynamo effect works are still a matter that is under investigation.

What is **magnetic declination**?

Magnetic declination is the angular difference between north as shown by a compass and the direction to the geographic north pole, Earth's axis of rotation. Declination

> ## Has Earth's magnetic field always remained the same?
>
> The direction of magnetic poles in iron in Earth's crust demonstrates that Earth's magnetic field has reversed itself approximately nine times over the past 3.5 million years. It's not known how long the reversal lasts and what happens to the field during the reversal. The magnetic poles also wander. The magnetic north pole has moved as much as 800 kilometers toward the geographic north pole since 1831.

depends primarily on the location on Earth but, because the magnetic poles move, also on time.

Is the **magnetic field important** to **life** on Earth?

The magnetic field of Earth is very important to life on Earth because it helps deflect harmful cosmic rays and solar wind. If we were fully exposed to these charged particles the effects on us could be devastating. Communication systems would be disabled and the particles could cause damage to all living things.

How is a **compass made**?

A compass is a magnetized metallic pointer that can rotate about a low-friction pivot point. Sometimes the pointer is placed in a container of liquid to dampen the movement of the pointer. The magnetic pointer aligns itself with the north/south orientation of Earth's magnetic field, and the person using the compass can determine what direction he or she is headed by looking at the pointer.

Does a **compass sometimes point downward** along with pointing north?

For hundreds of years, navigators using compasses noticed that on occasion, the compass pointer would try to point downward in addition to pointing north. This phenomenon, which went unexplained for several hundred years, was observed by compass maker Robert Norman. He found when flying over the poles one end of the compass would point downward; he understood that the problem was that the pointer was attracted to the pole under the plane. By making the compass rotate in the vertical direction he made the first dip needle.

What is a **dip needle** and how is it similar to a compass?

A dip needle is just like a conventional compass, but instead of holding it horizontally, it is held vertically. It is a magnetic needle used for navigational purposes just like a compass, but is used predominantly when traveling around the north and south poles. Instead of measuring horizontal magnetic deflection, the dip needle measures vertical magnetic inclination. When over the equator, the magnetic field of Earth is parallel to the surface of the Earth. The closer one gets to the magnetic poles, however, the less pilots rely on compasses, and the more they rely on dip needles to tell them how close they are to the poles. The closer one gets to a pole, the more vertical the magnetic field becomes, because it's turning into the surface of Earth. Therefore, when directly over the magnetic poles, the dip needle points directly downward.

ELECTROMAGNETISM

How was the **connection** between **electricity** and **magnetism discovered**?

The close connection between electric current and magnetic fields was discovered quite by accident. In 1820, Danish physicist Hans Christian Oersted (1777–1851) gave a lecture on the heating effects of an electric current on a wire. A compass happened to be near the wire and he was surprised to see the compass rotate when the current was on. He had been looking for connections between electricity and magnetism for several years, but expected that the compass would point away from the wire. Instead he found that the compass pointed in a circle around the wire. Above the wire it pointed perpendicular to the wire; below the wire it also pointed in the perpendicular, but in the opposite direction.

What were the **implications** of **Oersted's discovery**?

The fact that moving charge in a wire could create a magnetic field created a great deal of excitement and enthusiasm in the scientific community. A week after hearing about Oersted's discovery, French physicist and mathematician André-Marie Ampère (1775–1836) gave a presentation at the French Academy of Sciences that extended Oersted's experiments and contained detailed analyses. A day later he found that two parallel current-carrying wires would either attract or repel each other depending on the relative directions of the currents. Ampère's greatest contribution, however, was the mathematical theory he created for electricity and magnetism.

British chemist and physicist Michael Faraday's (1791–1867) philosophy led him to search for connections between phenomena like electricity, magnetism, and light. In 1821 he invented what is now called a homopolar motor. One end of a wire was suspended from a support so that it could swing in any direction. The other end of the wire contacted a pool of mercury. When Faraday put current through the wire the end in the mercury traced out a circle.

The force that Faraday had observed wasn't formalized until 1891 and then by the Dutch physicist Hendrik Antoon Lorentz (1853–1928). This force, called the Lorentz

Researching magnetic fields and electric currents, Michael Faraday and, independently, the American physics teacher Joseph Henry, invented an early form of electric generator called the dynamo. Today, generators like those pictured above use the same principles to produce electrical energy from other forms of energy.

Force Law, is proportional to the current through the wire, the magnetic field, and the length of the wire. The force, which is perpendicular to both the current and the magnetic field, is strongest when the current and field are at right angles. This force is the basis of motors and many other applications.

When Faraday published his results he failed to give credit to two other important scientists and he was given assignments to work in other fields. Nevertheless he continued to do experiments on the effects of magnetic fields. For example, he found that when dense glass was put in a magnetic field the direction of polarization of light going through the glass was rotated. He spent ten years searching for ways to create a current from a magnetic field. Finally, in 1831 he tried changing the magnetic field and made the crucial discovery that an electric current is produced by a changing magnetic field. The current, a flow of charges, is produced by an electric field exerting forces on the charges. Faraday went on to invent the dynamo, an early electric generator. The American high school physics teacher Joseph Henry (1797–1878) made the discovery at almost the same time.

What **discovery** did James Clerk **Maxwell** make that **depended** on the work of **Oersted, Faraday, and Ampère**?

In an earlier chapter we have seen that charges create electric fields. In this chapter we have seen that moving charges, that is, currents, create magnetic fields and that   267 changing magnetic fields produce electric fields. In the 1860s Scottish physicist James Clerk Maxwell (1831–1879) added a crucial additional connection: changing electric fields can produce magnetic fields.

With that idea Maxwell recognized that these relationships meant that electric and magnetic fields could move through space. The fields move through space as transverse waves that are perpendicular to each other. Maxwell calculated the speed and found that it was equal to the speed of light. He published his results in 1864 and a textbook on electromagnetism in 1873. In 1881 Oliver Heaviside wrote Maxwell's famous four equations in the form they are used today.

In 1888 Heinrich Hertz (1857–1894) transmitted electromagnetic waves across his laboratory, confirming Maxwell's theoretical work.

ELECTROMAGNETIC TECHNOLOGY

Why are **electromagnets**, the kind that pick up junk cars, so strong?

An electromagnet is a coil of current-carrying wire wound on a iron core that is at the center of an iron cup. The magnetic field created by current in the wire is strengthened by the iron core. The strength of the magnetic field produced by such electromagnets creates a large force, as described by the Lorentz Force Law that allows people to more easily move large metal objects, such as steel cars, from one location to another.

What is the **difference** between a **motor** and a **generator**?

In each device, a magnet and a coil of wire are employed to change one form of energy into another form. A motor consists of multiple loops of wire placed in a magnetic field. Either the loops or the magnet can rotate. The current through the wires in the field causes a force that results in rotation and thus mechanical energy. Motors in a home are used in fans, hair dryers, and food processors. There are over a hundred motors in a modern automobile. The starter motor is the largest and most powerful.

A generator does the opposite of a motor; it changes mechanical to electrical energy, but still consists of multiple loops of wire in a magnetic field. Either the loops or the magnet can rotate. In an automobile a form of a generator, called an alternator, uses some of energy from the engine to charge the battery. Backup generators use the energy from a gasoline engine to produce enough electrical energy to keep some of the lights and appliances running in a house when the electrical power fails. Electric utilities use huge generators to provide power for a city or larger area. The generators get their energy from steam turbines. The heat required to turn water into steam can come from coal, oil, natural gas, or nuclear "burners." Wind power uses generators turned by the propeller blades.

How do earbuds use the results of electromagnetism?

An earbud contains a membrane made out of thin plastic. In the center of the membrane is a coil of wire called the voice coil. The coil fits in a cylindrical slot in a permanent magnet. The center rod of the magnet is one pole, the outside tube is the other, resulting in a magnetic field perpendicular to the wire. When there is a current through the wire the Lorentz force on the wire pushes the membrane in and out. The membrane exerts forces on the air molecules producing the longitudinal waves constituting sound. (Refer to the Sound chapter for more information.)

How are **magnetic materials** used in **computers**?

Magnets are used in the compact motors that turn the disks in the CD or DVD drive and that move the laser that reads the disk to the correct position. Motors rotate the disks in a hard drive. The arm on which the read/write head is mounted is rotated to the correct portion of the hard drive disk has a coil of wire on it in a magnetic field. When there is a current through the wire the force moves the arm to the correct position.

The disk itself is often made of aluminum coated with an extremely thin (10-20 nanometers) film of magnetic material that is divided into sub-micrometer thick regions that are perpendicular to the surface of the disk. Each region is magnetized one way to represent a "1" and another way to represent at "0." A tiny coil in the read/write head carries the current that magnetizes the regions. The state of the magnet is read using magnetorestriction where the magnetic field causes a change in resistance of a very thin wire in a coil in the read/write head.

How do **metal detectors** work?

Built into the frame of a metal detector are coils of wire that carry a current. When metal is close to the coils, the magnetic properties of the metal change the current in the coils of wire that is detected by the electronic circuits in the detector.

How do **traffic lights** at car intersections know when a **vehicle** is **present**?

Many traffic lights are triggered to change by the approach of a car. The

When you walk through a metal detector with metal anywhere on your person, that metal changes the current in the coils in the frame of the detector.

principle is similar to the metal detector, in that there are coils of current-carrying wire just below the road where the vehicles stop at the intersection. When a large enough amount of metal passes over the coil, it induces a change in the current that creates a signal in the electronic circuits that control the traffic light.

What are **MAGLEV trains**?

MAGLEV, or magnetically levitated trains, are different from conventional trains in that they use electromagnetic forces to lift the cars off the track and propel them along thin magnetic tracks. Some demonstration trains have reached speeds of 500 kilometers per hour (300 miles per hour). Although the United States has no MAGLEV train, nor an active research program in this technology, Germany and Japan have conducted a great deal of research in the field.

What are the two main forms of **MAGLEV transportation**?

The German system uses the attractive forces between electromagnets to lift the underside of the train 15 centimeters (6 inches) above its guide rail. The coils in the train and guide rail form a linear motor—like an ordinary motor that has been unrolled. The only commercial operation is a train in China that transports people 30 kilometers in slightly over seven minutes.

The Japanese have taken a slightly different approach toward MAGLEV technology. The track and train repel each other. Propulsion also uses a linear motor. Levitation works well at high speeds, but when starting and stopping traditional wheels must be used.

MAGNETIC FIELDS IN SPACE

What are the **Van Allen Belts**?

Charged electrons and protons from solar wind and cosmic rays entering Earth's magnetic field feel the Lorentz force that traps them into spiral orbits about the magnetic field lines. They create doughnut-shaped regions of charged particles called the Van Allen Belts. The belts are concentrated around the equator and become thinner as they approach the poles. The two belts are located at 3,200 kilometers and 16,000 kilometers above the surface of Earth.

Why aren't the **Van Allen Belts** present around the **north** and **south poles**?

At the equator, the magnetic field is parallel to the ground and the electrons and protons from the solar wind can become trapped around the field lines. At the poles, however, the magnetic field strengthens, the lines become closer together, and forces on the particles push them back toward the equator. Some of the most energetic particles are able to penetrate the atmosphere where they interact with oxygen and nitrogen atoms producing the natural light shows that are called auroras.

What causes the **northern lights**?

Disturbances on the sun can send large numbers of charged particles into space. When they reach Earth they disturb the Van Allen Belts, causing the belts to dump particles into Earth's atmosphere. There they interact with the gases in the atmosphere, causing them to emit light. The scientific name for this phenomena is "aurora."

Are there different names for **auroras** in the **northern** and **southern hemispheres**?

An aurora in the northern hemisphere, known as the "northern lights," is officially called aurora borealis, while an aurora in the southern hemisphere, or "southern lights," is called the aurora australis.

WHAT IS THE WORLD MADE OF?

What is **matter**?

Ancient people in many parts of the world believed that all matter was made of four elements: earth, air, water, and fire. No matter how small an amount of material you had, you could not separate an element into a combination of other materials. But if you kept dividing the amount of material into smaller and smaller pieces, what would you obtain?

Democritus, a Greek who lived around 410 B.C.E. and was the student of Leucippus of Miletus (ca. 435 B.C.E.), stated that all matter is made up of atoms and the void. Atoms are the smallest piece into which an element can be divided; they are uncuttable. They could be neither created nor destroyed, and thus were eternal. The void was empty space. This viewpoint was expanded by the first century B.C.E. by the Greek Titus Lucretius Carus (95–55 B.C.E.) in his epic poem "On the Nature of Things." Aristotle, on the other hand, regarded the atomist philosophy as pure speculation that could never be tested. He rejected the possibility of empty space and believed you could divide matter until it was infinitely small.

The Greeks were not the only ones to develop a philosophy of atomism. The Indian school of philosophy known as Valsesika, and in particular the philosopher Kanāda in the second century B.C.E., held that earth, air, fire, and water could be divided into a finite number of indivisible particles. These ideas were adopted by several other Indian schools of philosophy.

What happened to **Democritus's** and **Aristotle's ideas**?

For almost 2,000 years Aristotle's philosophy was taught in schools and accepted by educated people. In the sixteenth century doubts about Aristotle's science increased. A number of philosophers, including those whom today we would call scientists, actively opposed Aristotle's dominance of the curriculum in schools. Englishman Francis Bacon (1561–1626) developed what we today would call the scientific method in opposition to Aristotle's philosophy. In 1612 Galileo (1564–1642) published "Discourse on Floating Bodies" in which he envisioned atoms as infinitely small particles, views that he later expanded in "The Assayer" (1623) and, more completely, in "Discourses on Two New Sciences" (1638). Nevertheless, the debate over whether atoms, much too small to see, really existed or were just a successful model of matter continued for another two centuries.

How did **chemistry contribute** to the **acceptance of atoms**?

Robert Boyle (1627–1691), an Irishman who wrote *The Sceptical Chymist* in 1661, is often considered the father of chemistry. He held that matter was made of atoms or groups of atoms that were constantly moving. He urged chemists to accept only those results that could be demonstrated by experiment.

The great French chemist Antoine-Laurent de Lavoisier (1743–1794) demanded that measurements be made with precision and that scientific terms be clearly defined and carefully used. With the discovery that air contained oxygen, which is necessary to support animal life, he clearly demonstrated that air was not an element, but a mixture of oxygen and nitrogen. By showing that hydrogen, discovered by Henry Cavendish (1731–1810), when mixed with oxygen formed water, Lavoisier showed that water also was not an element, but a compound of two elements. Lavoisier's care with weighing both the reactants and products of reactions allowed him to make one of the earliest statements that mass is conserved, that is, neither created nor destroyed in chemical reactions. Lavoisier's wife was a brilliant woman who assisted her husband in many ways, including translating letters from foreign scientists. He was executed during the French Revolution for being a tax collector. In supporting the guilty verdict, the judge said "The Republic needs neither chemists nor scientists."

Lavoisier met with Joseph Priestley (1733–1804) shortly after Priestley discovered oxygen in 1774. Lavoisier did extensive work on oxygen and gave it its present name. Priestley invented carbonated water and experimented with ammonia and laughing gas (nitrous oxide) but when he opposed much of the work of chemists of his generation he was pushed to the sidelines. After writing theological books that opposed traditional Christianity in England he was forced to flee to America in 1794.

John Dalton (1766–1844) was an English chemist who, because he was a Quaker, could not obtain a position in a state-run university. For a number of years he taught at a college in Manchester for dissenters from the Church of England. His strengths were his rich imagination and clear mental pictures, but especially his astonishing physical intuition.

His first interest was meteorology. He wondered how Earth's atmosphere, consisting of gases of very different densities, could have the same composition at different altitudes. His meteorological studies led him to the conclusion that atoms were physical entities and their relative weight and number were crucial in chemical combinations. Dalton's atomic theory of chemistry was published in "A New System of Chemical Philosophy" in 1808 and 1810. Briefly, it has five parts:

- Elements are made of tiny particles called atoms that cannot be divided into smaller particles. Nor can they be created or destroyed or changed into another kind of atom.
- All atoms of an element are identical. So there are as many kinds of atoms as there are elements.
- The atoms of an element are different from those of any other element in that they have different weights.
- Atoms of one element can combine with atoms of another element to form a chemical compound (today called a molecule). A compound always has the same relative numbers and kinds of atoms.
- In a chemical reaction atoms are rearranged among the compounds; they are neither lost nor gained.

The first and last parts give an atomic basis for Lavoisier's conservation of mass that he had confirmed in many careful experiments. Two of Dalton's specifications are now known to be false. Atoms can be changed from one kind to another by the process of radioactive decay. Not all atoms of an element have the same weight. Both of these results will be discussed in the chapter "At the Heart of the Atom."

Who really **discovered oxygen**?

Three scientists are often credited with the discovery of oxygen: Carl Wilhelm Scheele, Antoine-Laurent de Lavoisier, and Joseph Priestley. The Swedish scientist Scheele discovered what he called "fire air" in 1772. The name came from the way an ember would burst into flame when immersed in the gas. Scheele wrote a book describing his work, but it took four years to be published. In the meantime, Priestley in 1774 discovered what he called "dephlogisticated air." Lavoisier, who met Priestley in 1774 and was told about Priestley's discovery, claimed to have discovered what he called "vital air." Although he was sent letters from both Scheele and Priestley describing their earlier work, he never acknowledged their receipt. Lavoisier's great contribution was to make precise studies of the role of oxygen in a variety of reactions and to give it the name oxygen.

Who introduced the **modern chemical symbols**?

Swedish chemist Jöns Jakob Berzelius (1779–1848) introduced the modern chemical notation using one or two letters to represent an element in 1813. Rather than today's notation for water, H_2O, Berzelius wrote H^2O. Berzelius did much more. He published the first accurate list of the weights of atoms.

How did **Dalton represent atoms**?

Dalton used pictographs to represent atoms. Below, in the first line, are some examples.

Dalton would represent a chemical reaction (carbon plus two oxygen to yield carbon dioxide) as shown on the second line below.

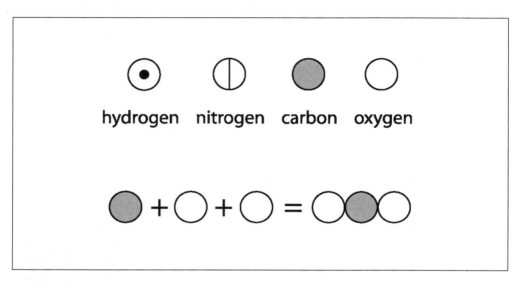

Are **atoms indivisible**?

Thanks to the invention of the electric battery in 1800 by Alessandro Volta (1745–1827) chemists had a new tool to create reactions and isolate new elements. Sir Humphry Davy was one of the most active and isolated sodium, potassium, and calcium from their salts. But it was his assistant, Michael Faraday (1791–1867), who most contributed to the discovery that the atom was not indivisible. He found that the amount of charge needed to liberate an element from a solution was proportional to the mass of the element. In modern language the amount of charge was proportional to the number of atoms liberated. It took until 1881 for the import of this result to be realized when Hermann von Helmholtz (1821–1894) pointed out that if elements are composed of atoms, then electricity could be divided into portions that could be called atoms of electricity. George Johnstone Stoney named these atoms "electrons." But what were they?

How was the "electrical atom"—the **electron—discovered**?

Further advances came not from electrolysis but from studies of gases. In the 1700s and early 1800s physicists used the vacuum pump, invented by Otto von Guericke in 1690, to reduce the pressure in glass tubes fitted with electrodes to allow electricity to pass through the tubes. In 1838 Michael Faraday passed an electric current through such a tube and noticed a strange arc-shaped light starting at the cathode (negative electrode) and ending almost at the anode (positive electrode).

When Heinrich Geissler was able to reduce the air pressure to about 1/1,000 of an atmosphere he found that the tube was filled with a glow, like the neon lamps used today. By the 1870s William Crookes was able to reduce the pressure to 1/1,000,000 of an atmosphere. As the pressure was reduced the glow gradually disappeared. Instead the glass near the anode began to glow. Without air to disrupt their passage, rays of some sort were able to travel from the cathode to the anode. At the anode end they were going so fast that they caused the glass to glow or fluoresce. By coating the glass with zinc sulfide the glow was made brighter. The mysterious invisible rays could be shown to travel in straight lines by placing metallic objects in the tube and finding that they cast sharp shadows at the anode. Because the rays came from the cathode they were called "cathode rays."

Joseph John (J.J.) Thomson (1856–1940) conducted three experiments with Crookes tubes that showed the nature of cathode rays. His first experiment used an electrode at one side of a tube, out of the direct path, which was connected to an elec-

trometer that could detect electric charge. Thomson could deflect the path of the rays using a magnet and follow their path by observing the fluorescent glow on the tube's surface. He found that the electrometer showed a negative charge, but only when the rays were deflected on to its terminal. Thus he showed that the rays consisted of a beam of negative particles.

Using a tube with the best possible vacuum, Thomson next explored the effect of an electric field on the rays. He added two parallel metal plates to the tube, connected a battery across the plates, and found that the rays were attracted toward the positive plate and away from the negative one.

In his third, and most important experiment, done in 1897, he combined the deflection of an electric field with one by a magnetic field. In doing so he could calculate the ratio of the mass to the charge of the particles. He found that this ratio was 1,800 times lower than that of a

Modern vacuum chambers like this one are used in laboratories to study atoms, molecules, nuclei, and electrons. Otto von Guericke invented the vacuum pump in 1690 to help him study electricity in a vacuum, and since then vacuums have been used to study such things as cathode rays and electrons.

277

positively charged hydrogen ion. Thus the particles must be either very light or very strongly charged. He later showed that they had the same charge as the hydrogen ion, and so they were very light. Further experiments showed that the particles had the same properties no matter what metal was used for the cathode or whether the cathode was cold or incandescent. For his work Thomson was awarded the Nobel Prize in 1906.

The charge of the electron was measured in 1909 by American physicist Robert Andrews Millikan (1868–1953). Before Millikan's experiments some physicists claimed that Thomson's results could imply that electrons had an average mass-to-charge ratio given by his experiments, but that they could have a variety of masses and charges. But Millikan showed that all electrons had the same charge, and thus the same mass.

What is the **charge** and **mass** of an **electron**?

The charge of an electron is -1.602×10^{-19} C (coulombs). Its mass is 9.11×10^{-31} kg. Both values are known to 85 parts per billion.

The electron has mass, but no size! If you direct beams of very fast, high energy particles at electrons they are deflected by the $1/r^2$ force between two charges no matter how close they come. If the electron had a finite radius then when the incoming particle penetrated the electron the deflection would be different.

What is the **structure** of the **atom**?

Thomson pictured the atom as a swarm of electrons in a positively charged sphere. This model is called the "plum pudding" model, after a then favorite English Christmas treat. Americans might picture a ball of pudding filled with raisins.

The New Zealand born physicist Ernest Rutherford (1871–1937) developed a method of testing Thomson's model. Rutherford's first scientific work was done in Montréal, Canada beginning in 1898. He and Frederick Soddy conducted a study of radioactivity and radioactive materials that won Rutherford the 1908 Nobel Prize in Chemistry. Those experiments will be described in the chapter on nuclear physics, "At the Heart of the Atom." He studied the radiation emitted by thorium and uranium, which he named alpha and beta rays. He recognized that alpha rays would be ideal probes for studying materials.

Rutherford moved to the University of Manchester in Britain in 1907 where he worked with Hans Geiger (1882–1945; later to invent the Geiger counter) on ways of detecting individual alphas. They determined that the alpha particles were doubly charged. They allowed the alphas to go through a window made of very thin mica (a mineral that could be cut into very thin slices). He noticed that the rays that penetrated the mica were deflected slightly more than they should have been by a plum-pudding atom. With Geiger and Ernest Marsden, he directed a beam of alphas on extreme-

ly thin foils of gold. To their great sur-
prise they found a significant number of
alphas were deflected into very wide
angles—some greater than 90°. Rutherford commented that it was as if a 15"
shell from a naval cannon bounced off a
sheet of tissue paper.

By 1911 Rutherford had developed
his own model of an atom. It consisted of
an extremely small positively-charged
particle (later called the nucleus) surrounded by electrons. Although the paths
of the electrons were not mentioned in

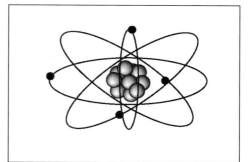

The Rutherford model, as pictured above, is used in many
common symbols, like that of a nuclear power plant or the
flag of the International Atomic Energy Agency.

the 1911 paper, it was later assumed that they orbited the nucleus as planets orbit the
sun. The nucleus was 1/10,000 as large as the atom, but had all the positive charge
and essentially all of its mass. If the atom's mass was equal to N hydrogen masses,
then the nuclear charge was about N/2. The electrons carry a total charge equal and
opposite that of the nucleus so that the atom would be electrically neutral. Rutherford's atom is mostly empty space!

Are there **problems** with **Rutherford's model**?

Electrons moving in circular orbits are experiencing centripetal acceleration. All
accelerating charges radiate energy. As a result the electrons in Rutherford's atom
would lose all their energy in a tiny fraction of a second. How then could atoms last
billions of years? Rutherford offered no answer. In addition, as the electron spiraled
into the nucleus it would create a smear of all colors of light, but hydrogen was known
to produced only specific colors, called an emission-line spectrum.

In what ways do **atoms emit** and **absorb light**?

In the mid-1800s Robert Kirchhoff published three laws that describe how materials
emit and absorb light:

1. A hot solid or a hot, dense gas produces a continuous spectrum.
2. A hot, low-density gas produces an emission-line spectrum.
3. A continuous spectrum source viewed through a cool, low-density gas produces
   an absorption-line spectrum.

What is a spectrum and what are **continuous** and **line spectra**?

Newton showed that when white light is passed through a prism it is split into a spectrum of all colors from violet through red. There were no gaps between the colors, so
the spectrum is called continuous.

279

The emission line of sodium has only two yellow lines. Hydrogen has four lines. The spectrum of iron, on the other hand, has an extremely large number of lines.

An absorption-line spectrum occurs when the low-density gas absorbs distinct colors, leaving dark gaps in the otherwise continuous spectrum. For example, the spectrum that the German physicist Joseph von Fraunhofer (1787–1826) took of the spectrum of the sun in 1814 showed 574 dark lines. In 1859 they were explained as being absorption lines from the cooler gases in the sun's atmosphere.

Fraunhofer made the best optical glass of any glassmaker of his era. He made great improvements to the achromatic lens, which refracts light of all colors the same amount. Like most glassmakers of his era, he died young, most likely from the poisonous effects of the materials used to make the glass.

How can the **emission** and **absorption** of **light** by **atoms** be **explained**?

In 1911 Niels Bohr (1885–1962), a Dane who recently had received his Ph.D. from the University of Copenhagen, joined Rutherford at Cambridge University. He quickly began work on the Rutherford model. He published his results in 1913, basing them on three postulates.

1. Electrons only move in certain allowed orbits at discrete radii and with specific energies. That is, their radii and energies are "quantized." When in these orbits their radii and energies are constant. The atoms do not emit or absorb radiation.

2. Electrons gain or lose energy when they jump from one allowed orbit to another. Then they emit or absorb light with a frequency given by $hf = E_2 - E_1$ where E_2 and E_1 are the energies of the electrons in the allowed orbits. The constant h is called Planck's constant, 6.6×10^{-34} J/Hz (joules per hertz).

3. The correspondence principle. When the electron is very far away from the nucleus classical physics must give the same answer as the new quantum physics.

A bust of Niels Bohr is displayed in Copenhagen, Denmark. Bohr developed postulates about electron behavior, including that they orbit in discrete radii and gain and lose energy in quantized amounts when they change from one orbit to another.

He later changed the third postulate from the correspondence principle to requiring that the angular momentum of the electron be quantized, that is proportional to an integer called the quantum number. The results didn't change, but the derivation of them was more straightforward. This method is presented in almost every textbook.

The two drawings below illustrate the emission of light when the electron goes from a higher-energy to a lower-energy orbit and the absorption of light when the electron's energy is increased.

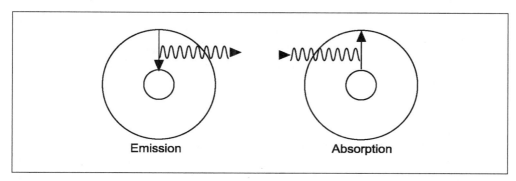

Emission   Absorption

In 1885 Johann Balmer had found a formula that accurately calculated the wavelengths of the visible hydrogen spectrum. It was purely empirical—that is, there was no physics-based explanation of it. In 1888 Janne Rydberg generalized Balmer's results to allow calculation of hydrogen emission in the ultraviolet and infrared. Bohr was able to explain Balmer's and Rydberg's formulae using the results of his postulates. The equation for the wavelength of the emitted radiation is $1/\lambda = R(1/m^2 - 1/n^2)$. The numbers m and n are the quantum numbers of the two energy levels. For example, the red line would have $m = 2$, $n = 3$. The constant R is 0.01097 nm^{-1}, making the wavelength of the red line $1/(0.01097 (1/4 - 1/9)$ nm$^{-1}) = 656.3$ nm, in excellent agreement with the experiment. Thus Bohr's model is a major advance in understanding the structure of the atom.

Is **light** a **wave** or a **particle**?

Light (and other forms of electromagnetic radiation) has the properties of both a particle and a wave. As a wave it is described by its wavelength, frequency, amplitude, and polarization. It has the ability to diffract and interfere. As a particle it has energy, momentum, and angular momentum. It has the ability to be emitted and absorbed, and to scatter off other particles, transferring energy and momentum. In some experiments it acts like a wave in part of the experiment and a particle in other parts.

For example, if you put a beam of light through a pair of narrow, closely spaced slits, the so-called Young two-slit experiment, you get an interference pattern with alternating stripes of light, where the light through the two slits constructively interferes, and darkness where the interference is destructive. If you now greatly reduce

The biochemist George Wald (1906–1997) discovered in 1958 that vision is best explained in terms of photons. The molecule $C_{20}H_{28}O$ is called retinal and is a component of both rods and cones in the retina. It can exist in two forms, one straight and one bent. When a photon strikes the molecule it changes its shape from bent to straight. The shape change creates an impulse in the nerves in the retina. The energy of the photon (and thus the wavelength of the light) that causes the transition depends on the other molecules, called opsins, in which the retinal molecule is embedded.

the intensity of the light and use a detector that can detect individual photons you will have regions where a large number of photons arrive and others where none arrive. The regions are exactly where the dark and light stripes were.

If the intensity is so low that there is only one photon in the apparatus at a time how can the dark and light regions be understood?

Particles can't split, with half going through one slit, half through the other so the two halves interfere. If you try to modify the experiment so you can tell through which slit the photon came you destroy the interference pattern.

Physicists have grappled with this mystery for decades. There are no easy explanations. Perhaps the fault lies with our language and thinking. We simply do not have the correct words or mental concepts to describe and understand how light works.

Does **light interact** with an **atom** as a **wave** or a **particle**?

A wave carries energy continuously over time; the more energy in the wave, the faster the energy is transferred. A particle, on the other hand, delivers its energy all at once. When an atom either absorbs or emits light, the transfer is almost instantaneous. Therefore light interacts with an atom like a particle.

The idea that light comes in packets of energy was first stated by Albert Einstein (1879–1955) in 1905. He called the packet a *light quantum*. The quantum was given the name photon in 1926. The photon has no mass or charge, but it does carry angular momentum. It always moves at the speed of light, c.

Each photon carries an amount of energy $E = hf$, where f is its frequency. Therefore a photon of blue light has more energy than one of red light. The energy carried by a beam of light depends both on the frequency and the number of photons per second leaving the source.

Are **emission** and **absorption** the **only ways** light interacts with an atom?

In 1917 Albert Einstein proposed a third way light could interact with an atom. If a photon with the correct energy struck an atom in the excited state then the atom would be stimulated to emit an additional photon and drop to the lower energy level. The two photons leave with the same energy (in terms of a wave—same wavelength) and in phase.

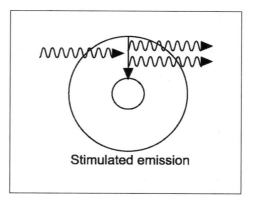

Stimulated emission

What are the **limits** of the **Bohr model**?

Today's model is totally different from Bohr's 1913 model. Bohr's model could explain only the spectra of hydrogen and helium from which an electron was removed. With some modifications is could also explain the spectra of the alkalis like lithium. From 1913 through 1926 physicists tried to extend the model, with some successes, but the lack of a physics-based explanation of the postulates led to research aimed at a model that was not based on classical ideas like Bohr's. One of the first steps was taken by the young German physicist Werner Heisenberg (1901–1976).

What were some of Niels **Bohr's accomplishments** as an exemplary **physicist** and **citizen**?

Bohr returned to the University of Copenhagen as a professor of physics in 1921. With the help of the government and the Carlsberg Beer Foundation he established the Institute of Theoretical Physics. Bohr's Institute attracted all the major theoretical physicists from around the world for short visits or extended appointments. When the Germans occupied Denmark, Bohr made a daring escape, first to Sweden, then England, then the United States. He was a consultant in the atomic bomb development effort, but after the war he tried to get President Harry Truman (1884–1972) and British leader Winston Churchill (1874–1965) to agree to share the secrets of the bomb with all countries, including the Soviet Union. They both rejected Bohr's proposal. But after the Soviets developed the bomb, Bohr's ideas helped found the United Nations' International Atomic Energy Agency. Until his death in 1962 he worked to reduce the threat of a nuclear war. In the centenary of his birth Denmark issued a postal stamp showing Bohr and his wife, Margarethe.

What is the **Heisenberg Uncertainty Principle**?

Heisenberg, together with Max Born and Pascal Jordan, tried a totally different approach using mathematical matrices. As part of their work Heisenberg developed a principle that demonstrates that in the atomic world our knowledge is limited. The uncertainty principle is written as $\Delta x \times \Delta p \geq h/4\pi$. In words, the uncertainty of a particle's position times the uncertainty in its momentum is never less than Planck's constant divided by 4π. If it has a precise location, then its momentum, and thus its speed (measured at the same time), must be imprecise. Planck's constant is extremely small, and so the uncertainty principle is important only for objects the size of atoms or smaller. The position and momentum of a baseball, for example, can both be precisely known at the same time.

The uncertainty principle shows why Bohr's electron orbits cannot exist. If you know the radius of the circle precisely, then it must have some velocity along the radius—smearing out its orbit. The uncertainty principle also exists in a form linking energy and time. In this form it says that if an electron is in a state that lasts for only a short time, then its energy is not precisely defined.

How did **probability** rather than **certainty** enter into the **model** for the **atom**?

In 1926 the Austrian physicist Erwin Schroedinger (1887–1961) published an equation for a wave function that describes the probability of finding an electron at a particular position. It agrees with Bohr's model in that the most probable radius for an electron is that given by the Bohr model, and the energy of the electron is the same as Bohr calculated, but its results are fundamentally different.

The solution of Schroedinger's equation can be shown as a probability cloud that shows the most probable locations for the electron.

The $n = 1$ state of the hydrogen atom is small and spherical. There are two $n = 2$ states. The s state has angular momentum zero and another spherically symmetric cloud. The p state, with angular momentum 1 has two most probable locations, the top and bottom. The $n = 3$ state has three possible angular momenta: 0, 1, and 2. The d state, with angular momentum = 2, has four angular regions with high probability.

Red light is emitted when the electron goes from the $n = 3$ "p" state to the $n = 2$ "s" state. The lifetime of the higher-energy state is very short, less than one billionth of a second, so when many atoms emit the light, the energy they emit is spread out— the energy from any particular atom cannot be precisely predicted.

Are **electrons waves** or **particles**?

The electrons in an atom are not confined to one region of space, but are spread out. They are acting more like waves than particles. In Louis-de Broglie's (1892–1987) 1924 doctoral thesis he proposed that electrons behave like waves with a wavelength given by $\lambda = h/mv$, where h is Planck's constant, and m and v the mass and velocity of the electron. The thesis was forwarded to Einstein, who enthusiastically endorsed the idea and recommended that the thesis be approved. De Broglie was awarded the Nobel Prize in

1929 for this work. The de Broglie wavelength is associated with any particle, although for an object the size of a baseball it is much smaller than the diameter of a nucleus!

The de Broglie wavelength of a particle determines its wave-like properties. Just as light photons interfere with themselves in a two-slit experiment, so do particles. The interference of electrons, atoms, and even molecules as large as C_{60} (so-called Bucky-balls) has been observed, and the measurements fit de Broglie's wavelength perfectly.

So, both matter and light can act like either a particle or a wave. This phenomenon is given the name "wave–particle duality."

Where do **X rays come from**?

X rays are electromagnetic waves of very short wavelength. Alternatively, they are very high-energy photons. They are emitted by atoms with many electrons, such as those high in the periodic table. The more electrons, the greater the charge of the nucleus, and the higher the energy of the electrons that are close to the nucleus (with $n = 1$ or 2). Therefore, when the atom is disturbed, as it is in an X-ray tube when the anode is bombarded by high-energy electrons, one of the $n = 1$ electrons can be kicked out. When an $n = 2$ electron loses energy and takes the place of the kicked-out electron, an X ray is emitted.

What is **quantum mechanics**?

The field of study of atoms by themselves, as well as in molecules, liquids, and solids in which the wave nature is important is called quantum mechanics. The atoms obey the Schroedinger equation, but for any atom more complicated than hydrogen the equation can be solved only by a computer. Using computers to solve equations, run complicated models, and simulate experiments is an important sub-division of the study of physics called computational physics, which now has equal status with theoretical and experimental physics.

Some of the recent accomplishments of quantum mechanics are in the areas of condensed matter physics, or the study of solids, and atomic physics. All of the integrated circuits used in mp3 players, televisions, cameras, and automobiles are designed using quantum mechanics. They are all based on diodes and transistors where quantum mechanics describes and explains their properties and guides materials scientists to select appropriate materials for their construction. One of the most exciting new materials is graphene, a film of carbon where the atoms are in hexagon-shaped arrays, but the film is only one atom thick. Graphene is being investigated as a way of constructing extremely tiny transistors, and thus powerful integrated circuits. It may also be used as a sensor to detect single atoms or molecules of selected materials or as a transparent conductor for touch-screen computer displays. Its discoverers won the 2010 Nobel Prize.

Lasers are now commonly used in industry, surgery, and communications technology, among other uses. Theodore Maiman demonstrated the first laser in 1960, but Gordon Gould was awarded the patent.

From 1924 to 1925 Einstein and the Indian physicist Satyendra Nath Bose (1894–1974) predicted that if an atomic gas were cooled enough the atoms could form a new state of matter that would exhibit quantum effects on a macroscopic scale. The first experimental confirmation of this prediction occurred in 1995 when Eric Cornell (1962–) and Carl Wieman (1952–) at the University of Colorado cooled a gas of rubidium atoms to about 1/6 of a millionth of a Kelvin. They and Wolfgang Ketterle (1957–) at MIT shared the 2001 Nobel Prize for their work. While this new state of matter, called a Bose-Einstein Condensate, has no existing applications, physicists are using it to improve their knowledge of how atoms interact at very low temperatures and to explore possible future applications to atomic clocks and computers.

How was the **laser invented**?

One of the useful applications of quantum mechanics is in each CD or DVD player in your home, or in the laser pointer. It's the device that uses stimulated emission of light predicted by Einstein so many years ago. In 1953 Charles Townes (1915–) of Columbia University (and later MIT) developed and patented the first application that he called a "maser" for microwave amplification by stimulated emission of radiation. It used a beam of ammonia molecules and later led to the hydrogen maser, now used as an extremely accurate clock. In 1958 Townes and Arthur Schawlow described how molecules could be used to extend the maser concept to optical frequencies. After

> ## What did Einstein mean by "God does not play dice with the universe"?
>
> As quantum mechanics was developed in the 1920s and 1930s its statistical nature became more evident. Einstein was convinced that there were variables, that could not be seen, but that actually controlled the outcomes. Einstein conducted many debates with Niels Bohr and Max Born about the statistical, or probabilistic, interpretation of quantum mechanics. In 1935 Einstein, Boris Podolsky, and Nathan Rosen wrote a famous paper that explored what they considered to be the incomplete description of the world provided by quantum mechanics. They proposed an experiment that has now been done and confirmed the predictions of quantum mechanics.
>
> In a 1926 letter to Max Born, Einstein wrote: "Quantum mechanics is certainly imposing. But an inner voice tells me that it is not yet the real thing. The theory says a lot, but does not really bring us any closer to the secret of the 'old one'. I, at any rate, am convinced that He does not throw dice."

their paper was published a number of physicists at university and industrial labs rapidly tried to apply these ideas to working devices.

In May 1960, Theodore Maiman (1927–2007) at Hughes Aircraft Company demonstrated an "optical maser" that used a ruby crystal with a flash lamp (similar to the camera flash lamp) to put the chromium atoms in the ruby into their excited states. Maiman was involved in a court fight over the validity of his patent for the laser with Gordon Gould (1920–2005), who worked with Townes at Columbia. In 1973 Gould was awarded the patent rights. Maiman won number of awards, but never the Nobel Prize.

While the laser was first described as a "solution looking for a problem" over the past fifty years lasers have become a multi-million dollar industry. Lasers have been constructed using gases, as in the familiar helium-neon, or HeNe laser; the carbon-dioxide laser used to cut fabrics and metals; the argon-ion laser used in surgery; the ultraviolet excimer laser used for eye surgery; and the free-electron laser used in research. Lasers made of tiny semiconducting crystals are used in CD and DVD players and laser pointers, as well as optical fiber communications equipment that brings video and the internet almost to your front door. Lasers have revolutionized research in physics, chemistry, and biology.

What is a **laser tweezer**?

When a laser is sent through a microscope it creates a tiny spot of very intense light in the material on the slide. A tiny plastic sphere will interact with the light in such a

287

way that it is pulled into the center of the light. The light can be moved around the slide, dragging the sphere with it. That is the essence of a "laser tweezer." The sphere can be chemically attached to the end of a long molecule, such as DNA. When the other end of the DNA is similarly fastened to the surface of the slide, the sphere can be used to stretch the DNA, straightening it out, and measuring properties such as the force needed to stretch it. Proteins, enzymes, and other polymers can be used in place of the DNA and the forces they exert similarly measured. In addition, the tweezers can be used to sort cells, moving them to specific locations on the slide.

We'll explore other mysterious consequences of quantum mechanics in the last chapter, "Unanswered Questions."

AT THE HEART OF THE ATOM

What makes up the **nucleus**?

Ernest Rutherford (1871–1937) had determined that the charge on the alpha particle, actually the helium nucleus, was two proton charges, but its mass was four proton masses. Dmitri Mendeleev (1834–1907), in building the periodic table, had arranged the elements in order of their mass and had arbitrarily assigned them sequential numbers that he called the atomic number. The English physicist Henry Moseley (1887–1915) measured the wavelengths of X rays emitted when metals were struck by other X rays. He was able to provide a physical basis for the atomic number and, in doing so, greatly strengthened the case for Rutherford's nuclear model.

But, as is the case for helium, if the atomic number, and thus the nuclear charge, was about half the atomic mass number, what made up the additional mass of the nucleus? The first proposal was that the missing particle was a combination of a proton and an electron, which would have the correct mass and charge. But the Heisenberg uncertainty principle showed that if an electron were confined to the size of a proton its energy would be larger than ever observed. In addition, by the late 1920s the angular momentum of the nitrogen nucleus, with charge 7 and mass 14 had been measured. The result could not be obtained from a combination of 14 protons and 7 electrons.

In 1931 two German physicists found that when energetic alpha particles struck light elements a very penetrating, neutral radiation was produced. The next year Marie Curie's daughter, Irène Joliot-Curie (1897–1956), and her husband, Frederick, found that if this radiation struck paraffin protons were ejected, suggesting that the radiation was actually a neutral particle with a mass near that of a proton. The next year James Chadwick (1891–1974), working in Manchester, England, experimentally confirmed the suggestion. The particle was named a neutron, combining "neutral" with the ending of the word proton. He was awarded the Nobel Prize for his discovery in 1935.

The neutron has a mass slightly larger than that of the proton. While neutrons are stable in non-radioactive nuclei, if they are free from the nucleus they decay with a half-life of about 10 minutes. Neutrons are used extensively in creating nuclear reactions and are necessary for nuclear fission, which will be discussed later in this chapter.

Do all **elements** have a **fixed number** of **protons** and **neutrons**?

The number of protons in the nucleus of an atom determines what element it is, so the number of protons is fixed. But, the number of neutrons can vary. Nuclei with the same number of protons but different numbers of neutrons are called isotopes. For example, carbon, with six protons, can have five, six, seven, or eight neutrons. The isotopes are labeled carbon-11, carbon-12, carbon-13, and carbon-14. A more compact notation is ^{11}C, ^{12}C, ^{13}C, and ^{14}C. Chemical properties in general do not depend on the isotope. There are at least 3,100 isotopes of all the elements presently known.

How do the numbers of **protons** and **neutrons** in **nuclei compare**?

For lighter elements the number of protons and neutrons are approximately equal. For example, the most common isotope of helium with two protons is ^4He, so it has 4 − 2 = 2 neutrons. Oxygen, with eight protons, has eight neutrons in its most common isotope, ^{16}O.

In elements heavier than calcium (20 protons, 20 neutrons) the number of neutrons is larger than the number of protons. In uranium-238 (92 protons, 146 neutrons) the ratio is almost 3 neutrons for 2 protons.

What **holds** a **nucleus together**?

Protons, all being charged positively, will repel each other. So there must be an attractive force that holds the nucleus together. The strong nuclear force acts between protons and protons, protons and neutrons, and neutrons and neutrons, all with the same strength. Because the strong force acts the same on protons and neutrons, the two particles are frequently lumped together under the name nucleon. While the repulsive electromagnetic force acts over long distances, the strong nuclear force only acts between nucleons that are in contact. Nucleons have angular momentum, or spin, and the strong force depends on the relative orientation of the spins. The force is stronger if the spins are in opposite directions.

How were the **details** of the **strong force determined**?

Examining the kinds of isotopes that exist gives clues about the nature of the strong force. The most stable nucleus, helium-4, has two neutrons having spins in opposite directions and two protons also having spins in opposite directions. A majority of non-radioactive isotopes have even numbers of neutrons and even number of protons, allowing them to form pairs with opposite spins. Most of the other isotopes have either an even number of protons and an odd number of neutrons or the opposite. Isotopes with both an odd number of protons and neutrons are extremely rare. So, the strength of the strong force clearly depends on the spins of the nucleons.

The mass of a nucleus is less than the sum of the masses of the protons plus the sum of masses of the neutrons. The larger the mass difference, the stronger the forces holding the nucleus together and the more energy needed to pull the nucleus apart. Studies of the mass of the nuclei can thus be used to gain further insight into the strong force. Maria Goeppert-Mayer (1906–1972) was a German-born American physicist who explained why nuclei with certain numbers of protons and/or neutrons, called magic numbers, were extremely stable. Her theory showed that the nuclear force depended on both the spin of the nucleon and its orbital angular momentum. The magic numbers are 2, 8, 20, 28, 50, 82, and 126. Thus helium-4, the most stable nucleus has a magic number of protons and a magic number of neutrons, and so is called doubly magic. Oxygen-16, calcium-40 and calcium-48, and lead-208, which is the heaviest stable nuclide, are also doubly magic. But, are all nuclei stable?

What **rays** do **radioactive materials** emit?

Among the scientists who explored radioactivity immediately after Becquerel's discovery was Ernst Rutherford, then at McGill University in Montréal, Canada. He and Frederick Soddy found that uranium and thorium emitted two different kinds of rays. One could be stopped by paper, the other required metal about a centimeter thick. Rutherford named them alpha and beta rays after the first two letters in the Greek alphabet. In 1907 he named the even more penetrating rays produced by radium gamma rays.

In 1900 Pierre (1859–1906) and Marie Curie (1867–1934), using an electroscope, found that beta rays are negative particles. Becquerel used the same kind of apparatus J.J. Thomson (1856–1940) had used (see the chapter "What Is the World Made Of?") to measure the ratio of mass to charge of an electron to determine that betas typically travel at about half the speed of light and are identical to electrons. In late 1907 Rutherford demonstrated that alpha rays were helium atoms with the two electrons removed. (He had not yet discovered the nucleus. Today we say that an alpha particle is the helium-4 nucleus.) Gamma rays were later found to be very high energy photons.

Henri Becquerel earned the 1903 Nobel Prize for his discovery of radioactivity.

Alpha, beta, and gammas all are hazardous to our health. Alphas are blocked by skin, but if a radioactive material, such as the gas radium, is inhaled, the alpha particles can cause damage to the lungs. Betas can penetrate skin and tissue and, if they strike a cell, can cause mutations to the DNA. Gammas can cause mutations and kill cells; they are used in cancer radiation therapy.

What contributions did **Marie** and **Pierre Curie** make?

Marie Skłodowska Curie, born and raised in Poland, but working in France, used an electrometer to measure the ionization of the air caused by radioactive minerals. Because the amount of radioactivity produced by a uranium compound depended only on the amount of uranium present, she concluded that the atom itself must be the source. She found that the uranium-containing mineral pitchblende was more radioactive than the uranium itself and concluded that the mineral must contain a small quantity of another element that was more radioactive than the uranium. Her husband, Pierre, stopped his own work and joined Marie in searching for the element.

They started by grinding up 100 grams of pitchblende, but by the time they had found the element they had processed tons of the mineral. In July 1898 they announced the discovery of an element they named "polonium" in honor of Poland, where she had been born. In December of the same year they announced they had found an even more radioactive element that they named radium. It took until 1902 for them to separate one-tenth of a gram of radium chloride from a ton of pitchblende. In 1910 Marie announced that she had obtained pure metallic radium.

Pierre and Marie shared the 1903 physics Nobel Prize with Becquerel for their work on radioactivity. Marie won the 1911 chemistry Nobel Prize for her discovery of polonium and radium.

The dangers of radioactivity were unknown when the Curies did their work, but both were affected by it. The radon that was produced when they were processing the pitchblende would have caused lung cancer. But, the leaky windows in their workplace and their frequent bicycle rides in the country spared them. Pierre died in 1906 when he slipped on a wet street and was run over by a horse-drawn wagon. Marie's death in 1934 was due to anemia, known now to be frequently caused by radiation.

What is a **half life**?

When will a particular radioactive nucleus decay? It's impossible to know. All that can be known is the average time between formation and decay. And we know that the number of decays will be proportional to the number of nuclei present. Now, suppose we start with a large number of nuclei. In a given time interval, say one second, a certain number will decay. At the end of that second the number of nuclei that haven't decayed will be smaller, so in the next second there will be fewer decays. At some time there will be only half as many decays in a second as there were originally. That time is called the half life. One half life later there will be only 1/4 as many decays as at the beginning. After another half life there will be half again as many, or 1/8 the number of initial decays.

How is **radioactive decay** used to **determine dates**?

Radioactive carbon-14 (^{14}C) has a half life of 5,730 years. It is produced in the atmosphere when cosmic ray neutrons strike nitrogen atoms. The ^{14}C then reacts with oxygen to form carbon dioxide in the atmosphere and dissolves into the oceans. Plants take up the atmospheric CO_2 in respiration and animals ingest it. Therefore all living things exchange both $^{12}CO_2$ and radioactive $^{14}CO_2$ throughout their lives. When they die the exchange stops and the fraction of ^{14}C to ^{12}C in their bodies is fixed. As time goes on the amount of ^{14}C decreases. Objects up to 50,000 years old can be dated by radioactive carbon as long as correction figures, agreed to by international agencies, are applied. Willard F. Libby (1908–1980), an American chemist, won the Nobel Prize in chemistry in 1960 for his development of carbon-14 dating in 1949.

There are several other dating techniques that use the half lives of radioactive isotopes in rocks including urani-

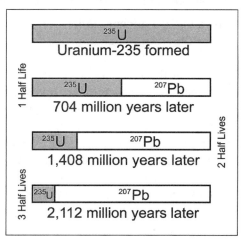

Radioactive uranium-235 decays into lead (Pb-207) very gradually. The half life is 704 million years!

293

um/thorium and rubidium/strontium ratios that extend dating techniques to hundreds of millions of years.

What **happens** to an **element** when it **decays** by emitting an **alpha particle**?

The nucleus that results from a radioactive decay is called the daughter. The number of nucleons in a radioactive decay does not change, and in an alpha decay the number of protons of the original must equal the number of protons in the daughter plus the number of protons in the alpha. The same is true for the number of neutrons. So, for example, when uranium-238, with 92 protons and 146 neutrons emits an alpha, the daughter has 90 protons and 144 neutrons. It is thorium-234. The alpha is emitted with a specific energy. Only very heavy nuclei decay by alpha decay.

Are there **uses** for **alpha decay**?

The decay of uranium and thorium produces all the helium that exists on Earth. Most of the helium is mixed with natural gas and is extracted from gas wells. It is expensive to separate and store the helium, but, given the increasing needs for this resource for cooling superconducting magnets used in hospital MRI machines, it is an effort that must be made.

Alpha-emitting radioactive elements are used in smoke detectors. The charged alpha particles leaving the element are collected on a metal plate where they produce a small electric current. Smoke scatters the alphas, reducing the current, and triggering the alarm.

How did the study of **beta decay** result in the **discovery** of a **new particle**?

In beta decay a nucleus emits an electron. The electron isn't in the nucleus originally, but results from the change of a neutron into a proton. Studies of the energy of the emitted electron showed that, instead of having a single energy like an alpha has, the energies of the electrons were spread from near zero to a maximum energy. Early investigators recognized that this suggested that energy was not conserved in beta decay. Austrian physicist Wolfgang Pauli (1900–1958) proposed in 1930 that a second particle was emitted along with the electron. This particle had to be neutral and zero, or extremely small mass, and he named it the neutrino, or "little neutral one." The neutrino wasn't detected experimentally until 1956. We now know that the neutrino emitted in beta decay is actually an anti-neutrino. The question of its mass will be discussed in the chapter on "Unanswered Questions."

When a nucleus undergoes beta decay the number of protons increases by one as the number of neutrons decrease by one. So, for example, carbon-14 (6 protons and 8 neutrons) becomes nitrogen-14 (7 protons and 7 neutrons) with the emission of the beta (electron) and an anti-neutrino.

What **nuclear force** is responsible for **beta decay**?

The weak nuclear force is the cause of beta decay. It is 10^{-13} (one-ten trillionth) as strong as the strong force. As a result the half life of most beta-decaying nuclei is long.

In 1968 Abdus Salam, Sheldon Glashow, and Steven Weinberg showed that the electromagnetic force and the weak nuclear force were actually different aspects of a single force, the electroweak force. How can these two forces that are so different in strength and range be unified? It only happens at extremely high energies (corresponding to a temperature of 10^{15} K), such as occurred in the first few moments after the Big Bang or in an accelerator. They received the 1979 Nobel Prize for their work.

What is the **origin** of **gamma rays**?

Gamma rays, high-energy photons or short-wavelength electromagnetic waves, are emitted from the nucleus along with an alpha or beta decay. When an alpha or beta decay produces a daughter nucleus that nucleus is often in an excited state. One or more gammas are emitted as the nucleus settles down to its lowest energy, or ground state. Gammas are like high-energy X rays, but are emitted from the nucleus, not the electrons, of an atom.

What is **antimatter**?

In 1932 Carl Anderson (1905–1991) was studying particles produced when cosmic rays struck lead sheets in a cloud chamber that was in a magnetic field. He found low-mass particles that curved the opposite direction from electrons, showing that they had positive charge. He later confirmed the existence of these particles by using a laboratory source of high-energy gamma rays. This positive electron was named the positron and was the first form of antimatter found. Anderson shared the 1936 Nobel Prize for his discovery.

When a gamma ray with sufficient energy strikes matter it can produce an electron–positron pair. Energy is converted into particles with mass. The minimal amount of gamma-ray energy needed is given by Einstein's famous equation, $E_{gamma} = m_{electron}c^2 + m_{positron}c^2$. The uncharged gamma produces a negatively charged electron and a positively charged positron, so electric charge is conserved.

Positrons are also emitted in radioactive decay of isotopes that have a deficit of neutrons. For example, stable carbon exists as ^{12}C or ^{13}C; six protons and either six or seven neutrons. As was discussed above, ^{14}C decays by emitting an electron. One of the neutrons changes to a proton with the emission of the electron and anti-neutrino. On the other hand, ^{11}C, with only five neutrons, is a positron emitter. One of the protons changes to a neutron with the emission of a positron and a neutrino.

When a positron strikes matter the positron and an electron annihilate each other, producing two or three gammas. Particles with mass are converted to energy.

Again, Einstein's equation can be used to find the total energy of the gammas: $E_{gammas} = m_{electron}c^2 + m_{positron}c^2$.

Antiprotons are created in accelerators, where the accelerated particle is slammed into a metal target, emitting gammas that create, in this case, proton–antiproton pairs.

How is **antimatter** used in **medicine**?

A three-dimensional image that shows biological activity in a person can be made using PET, or Positron Emission Tomography. PET uses a short-lived positron-emitting isotope, typically ^{11}C, ^{13}N, ^{15}O, or ^{18}F. The isotope is chemically attached to a molecule that is involved in the activity that is to be studied. Fluid containing that molecule is injected into the person and after enough time passes for the molecule to reach its target, the person is put into the PET machine. The positron that is emitted by the decaying nucleus strikes an electron and decays into two gammas that are simultaneously emitted back-to-back, that is, 180° apart. Detectors record the arrival of the two gammas, and computers extrapolate them back to the location of the gamma emission. When a sufficient number of events are recorded a three-dimensional image of the region where the biologically-active molecule accumulated can be made. PET scans can be combined with CT or MRI scans to pair information on the anatomy with biological activity for diagnostic purposes.

What **elements beyond uranium** have been discovered?

Atomic #	Symbol	Name	Longest-lived Isotope	Year Discovered	Location*	Formation Reaction
93	Np	Neptunium	^{237}Np 2.1×10^6 y	1940	Berkeley	n + U
94	Pu	Plutonium	^{244}Pu 8.0×10^7 y	1940	Berkeley	α + U

An aerial view of CERN, just northwest of Geneva, Switzerland. The Large Hadron Collider ring covers 22.4 square miles (large circle in the photo). The smaller ring at the lower right, which is the Proton Synchotron that is used to accelerate protons that are then fed into the LHC, has a circumference of 2.7 square miles.

297

Atomic #	Symbol	Name	Longest-lived Isotope	Year Discovered	Location*	Formation Reaction
95	Am	Americium	^{243}Am 7370 y	1944	Berkeley	n + U
96	Cm	Curium	^{247}Cm 1.6×10^7 y	1944	Berkeley	α + Pu
97	Bk	Berkelium	^{247}Bk 1380 y	1949	Berkeley	α + Am
98	Cf	Californium	^{251}Cf 898 y	1950	Berkeley	α + Cm
99	Es	Einsteinium	^{252}Es 742 d	1952	Berkeley	N + U
100	Fm	Fermium	^{257}Fm 101 d	1952	Berkeley	O + U
101	Md	Mendelevium	^{258}Md 52 d	1955	Berkeley	α + Es
102	No	Nobelium	^{257}No 58 m	1956	Dubna	C + Cm
103	Lr	Lawrencium	^{262}Lr 3.6 h	1962	Dubna/ Berkeley	B + Cf
104	Rf	Rutherfordium	^{267}Rf 1.3 h	1966	Dubna/ Berkeley	Ne + Am
105	Db	Dubium	^{268}Db 28 h	1968	Dubna/ Berkeley	Ne + Am
106	Sg	Seaborgium	^{271}Sb 1.9 m	1985	Dubna/ GSI	Cr + Pb
107	Bh	Bohrium	^{270}Bh 61 s	1981	GSI	Bi + Cr
108	Hs	Hassium	^{269}Hs 10 s	1984	GSI	Fe + Pb
109	Mt	Meitnerium	^{278}Mt 8 s	1982	GSI	Fe + Bi
110	Ds	Darmstadtium	^{281}Ds 10 s	1994	GSI	Ni + Pb
111	Rg	Roentgenium	^{281}Rg 20 s	1994	GSI	Bi + Ni
112	Cn	Copernicium	^{285}Cn 30 s	1996	GSI	Ca + U
113	Uut	Ununtrium	^{286}Uut 20 s	2003	Dubna/ Berkeley	Ca + Am
114	Uuq	Ununquadium	^{289}Uuq 2.6 s	1999	Dubna	Ca + Pu
115	Uup	Ununpentium	^{289}Uup 0.22 s	2004	Dubna/ Berkeley	Ca + Am
116	Uuh	Ununhexium	^{293}Uuh 0.060 s	2000	Dubna	Ca + Cm
117	Uus	Ununseptium	^{294}Uus 0.036 s	2010	Dubna	Ca + Cm
118	Uuo	Ununoctium	^{294}Uuo 0.89 ms	2002	Dubna	Ca + Cf

*Berkeley = University of California at Berkeley, Lawrence Radiation Laboratory, Lawrence Livermore Laboratory; GSI = Heavy Ion Accelerator at Darmstadt, Germany; Dubna = Joint Institute for Nuclear Research, Dubna (near Moscow), Russia.

Elements 113 to 118 have temporary names until the International Union of Pure and Applied Chemistry decides on the name. The number of atoms detected in the decays of elements 113–118 vary between a dozen down to even one! In the search for element 117 a total of 10^{19} Ca ions bombarded the Californium target and resulted in three atoms of element 117.

Elements up through Californium (98) have been produced in milligram or gram quantities. They have been used as target materials for the accelerators that bombard them with ions in the search for heavier elements. The estimated cost of producing enough Einsteinium (99) to be used as a target is $50 million!

What is the **Island of Stability**?

All the isotopes of elements beyond lead are radioactive. Some have lifetimes of tens of millions of years, while others are fractions of a second. Nuclear physicists and chemists know that isotopes that have magic numbers of neutrons or protons are more stable than others. They have proposed that very heavy isotopes with a neutron number around 180 and a proton number around 110 should be more stable than those with fewer or greater neutrons and protons. A glance at the table of elements beyond uranium shows that elements around 110 have longer lifetimes than others. But researchers have not yet been able to create isotopes with enough neutrons to reach this island.

What is **nuclear fission**?

Radioactive nuclei can decay in another way—they can split in two. This process is called spontaneous fission. Fission can also be produced artificially by bombarding nuclei with low-energy neutrons. In fact, fission was discovered in this manner.

How was **nuclear fission discovered**?

Italian physicist Enrico Fermi (1901–1954) was appointed professor at the University of Rome at the age of twenty-four. Among many projects in which he and his group were involved, perhaps none was more important than his studies of the reactions produced when slow neutrons struck nuclei. Fermi had discovered in 1934 that slowing neutrons by passing them through paraffin greatly increased this ability to produce nuclear reactions. He did systematic studies of the results of bombarding a series of materials with slow neutrons.

The German chemist Otto Hahn (1879–1968) had a distinguished career that included inventing the field of radiochemistry in 1905. Using chemical techniques and measurements of half lives to study the results of nuclear reactions he had discovered dozens of isotopes and at least one element. Three times he was nominated for the Nobel Prize. Since 1907 he had collaborated with the Austrian physicist Lisa Meitner (1878–1968). The teamwork between a physicist and a chemist was a great advantage.

Hahn and Meitner, together with Hahn's young assistant Fritz Strassman (1902–1980), employed Fermi's slow neutron techniques to create nuclear reactions, and thus more isotopes. When, in 1938, they tried bombarding uranium with neutrons they expected to create new elements beyond uranium in the periodic table. But they kept finding the element barium in the bombarded uranium.

Starting in 1933, the Nazi regime forced people of Jewish origin out of all laboratories and universities. Meitner, who had Jewish parents but had converted to Protestantism in 1908, was protected because she was Austrian. But, when Austria was incorporated into Germany, she lost that protection. In July 1938, she took the train from Berlin to the Netherlands. Thanks to the intervention of two Dutch physicists she was allowed to leave Germany, but with no possessions. She soon moved to Sweden and kept up her collaboration with Hahn by mail.

On December 17, 1938, Hahn and Strassman submitted their findings for publication but admitted that they had no explanation for the appearance of barium. Meitner and her nephew Otto Frisch utilized Niels Bohr's "liquid drop" model of the nucleus and Einstein's $E = mc^2$ equation to propose that the nucleus had split into two, releasing both extra neutrons and a large amount of energy. The Meitner-Frisch paper was submitted a few days after Hahn's. Frisch returned to his laboratory in England and confirmed Hahn's result in January 1939. Hahn won the Nobel prize in chemistry for his work in 1944, but Meitner never did.

Did physicists recognize the **military uses** of **fission**?

Lisa Meitner had recognized that extra neutrons could produce a chain reaction that would produce a very large amount of energy. In early 1939 physicists from many countries attempted to create such chain reactions by slowing down the released neutrons. Among these were Enrico Fermi and a Hungarian-born physicist Leó Szilárd. They saw signs that such a reaction had occurred.

In August 1939 Szilárd drafted a letter to President Franklin D. Roosevelt (1882–1945) that the German results could lead to in an extremely powerful new weapon. To give his letter more weight he convinced Einstein to sign the letter. It worked. Roosevelt directed the government to support fission research and created the Uranium Committee. While there were several important studies during the next three years, it was the British who made the breakthrough finding that the rare isotope uranium-235

It was not long after scientists figured out how to create nuclear fusion that the Nazis began conducting research on how to turn this science into a powerful bomb. Albert Einstein and physicist Leó Szilárd urged U.S. President Franklin D. Roosevelt start a program to complete a nuclear warhead before the Germans did.

What was the Manhattan Project?

In 1942 the "Manhattan Project" was started to produce nuclear weapons. It was named after the Manhattan Engineering District in New York City from which it was run. The chief was U.S. Army General Leslie Groves who appointed physicist J. Robert Oppenheimer (1904–1967) as scientific director. Its first success was at the University of Chicago where, in December 1942, Enrico Fermi's uranium reactor created the first self-sustained nuclear chain reaction.

could be used in a weapon. The Americans were informed but ignored the results until a personal visit by one of the British team members convinced the Uranium Committee of the need for action. The United States then established a new office that could authorize large-scale engineering projects.

How can **uranium** be used in a **weapon**?

The British discovery that naturally occurring uranium, a mixture of uranium-238 and only 0.7% uranium-235, would have to be highly enriched in uranium-235 created a need for enrichment plants. One method chosen had been developed in California. Uranium metal would be evaporated in a vacuum. The atoms went through a narrow slit and then into a region with a strong magnetic field. Because of their mass difference the two isotopes followed slightly different paths. The atoms condensed on the surfaces of separate containers. Dozens of giant machines, called calutrons, were built in a plant in Oak Ridge, Tennessee, chosen because abundant electricity was available from the nearby hydroelectric plants. Not enough copper was available to wind the coils for the magnets so 70,000,000 pounds of silver bullion were borrowed from the U.S. Treasury to be formed into wires for the machines.

Somewhat enriched uranium from the calutrons was then combined with fluorine to produce the gas U_6. Because of the mass difference of the two isotopes $^{235}U_6$ would diffuse through porous membranes slightly faster (about 0.5%) than its more massive counterpart. Thousands of separations were needed to produce weapons-grade uranium (85 to 90% ^{235}U). The plant at Oak Ridge built to accomplish this gaseous diffusion had an area of 2 million square feet, employed 12,000 workers, and cost the equivalent of $6.2 billion in 1999 dollars. At one time it consumed 17% of all the electricity produced in the United States—more than New York City!

Why was **plutonium used** in **bombs**?

Plutonium is not found in nature, but it is produced in reactors by bombarding uranium-238 with neutrons. Plutonium-239 (^{239}Pu) can be fissioned by slow neutrons, and

Today ultracentrifuges are used for uranium enrichment. A centrifuge is routinely used in medical labs to separate materials of different density. The test tubes are spun rapidly and the denser materials move away from the center of rotation because it requires more centripetal force to pull them toward the center. A gas ultracentrifuge uses a rapidly rotating drum to separate the UF_6 with the two isotopes. Gas centrifuges supply about 54% of the enriched uranium today. Each centrifuge is a more effective separator than a stage in a gaseous diffusion plant and requires only 6% of the electrical energy of gaseous diffusion.

Because of the relatively small size and lower energy needs of a centrifuge-based separation plant, there are serious concerns about the use of these plants to produce weapons-grade uranium. Pakistan used such a plant for its bomb, and there are indications that North Korea and Iran have similar plants.

so could be used in weapons. In December 1942, Hanford, Oregon, was chosen as a site for reactors that would produce plutonium. Hanford was selected because it was isolated but also on the Columbia River, which afforded a source of cooling water.

What was done in **Los Alamos**?

In September 1942, General Groves and Robert Oppenheimer chose Los Alamos, New Mexico, as the site for the top-secret laboratory at which weapons would be developed. Thirty-five miles northwest of Santa Fe, it was almost totally isolated and the site was occupied only by a school. During World War II hastily erected housing held Nobel Prize winning scientists, younger scientists and engineers recruited into the project, wives and children, and soldiers.

After determining the "critical mass," the minimal amount of enriched uranium needed to create a bomb, they designed and built the uranium-based weapon called "Little Boy." The uranium was divided into two halves and placed in a cannon-like container. An explosive charge drove the two masses together, forming a large enough mass of uranium to sustain a rapid chain reaction and explode. This weapon was never tested. It contained 64 kilograms (141 pounds) of uranium, about 2.5 times the critical mass. Less than 1 kilograms (2 pounds) of the uranium fissioned. Only 0.6 grams (0.001 pounds) was converted into energy, but the result was the equivalent of 15,000 tons of TNT. Little Boy was dropped on Hiroshima, Japan, on August 6, 1945. Over 100,000 people were killed in the blast, resulting fires, and effects of radiation.

The second task of Los Alamos was to design and build a weapon using plutonium. Originally they had expected to use the cannon-type method used with the uranium bomb, but when the plutonium was produced by the Hanford reactors it was found to contain ^{240}Pu, and another design had to be developed. They arranged a sub-critical plutonium mass in the shape of a sphere and used specially designed explosive charges to simultaneously compress the plutonium, increasing its density above the critical point. Scientists were uncertain that the design would work, so they decided to test the device first.

When was the **first "atomic device"** exploded?

"The gadget" was a test version of the plutonium bomb. It was installed on the top of a 30-meter (100-foot) tower in the New Mexico desert at a location 35 miles southeast of Socorro, New Mexico, on the White Sands Proving Ground. The explosion, called "Trinity," occurred on July 16, 1945. The energy yield was about 20,000 tons of TNT, more than twice what had been expected. The implosion-type bomb was much safer and more effective than the cannon-style "Little Boy" design and has been used for all other nuclear weapons.

How was the **plutonium bomb used**?

The plutonium bomb, named "Fat Man" for its shape, was dropped on the city of Nagasaki, Japan, on August 9, 1945. It contained 6.4 kilograms (14 pounds) of plutonium-239. About 20% fissioned, and less than one gram was converted into energy with the equivalent of 21,000 tons of TNT. As many as 80,000 people were killed in the attack.

How many **nations** have **nuclear weapons**?

The Soviet Union exploded a nuclear device in 1949. It was similar to "Fat Man" and was built using information delivered to that country by spies. China, Britain, and France developed nuclear weapons in the 1950s.

South Africa had nuclear weapons but abandoned the program. Israel is suspected of having weapons, but has never admitted it. India and Pakistan have both tested nuclear weapons. Iran and North Korea are suspected of developing nuclear weapons.

The A-Bomb Arch in Peace Park is a memorial to those who died when a nuclear warhead was dropped on Hiroshima in 1945. The United States also dropped the bomb on Nagasaki, Japan. The war in the Pacific ended soon afterwards with Japan's surrender.

303

By 1953 there had been 50 above-ground tests of nuclear weapons that created radioactive fall-out, contaminating milk and animals. These effects alerted the public to the danger of such testing. The Cold War, however, created an atmosphere in which treaties could not be negotiated. In 1963 a partial test ban treaty was signed, prohibiting tests in the atmosphere, underwater, and in space. In 1968 the nuclear non-proliferation treaty was signed. Non-nuclear weapon states were prohibited from building or acquiring nuclear weapons. Many nations have signed the treaty, although some major states did not on the basis that it makes no effort to curb development by states that already have such weapons. In 1996 the comprehensive test-ban treaty was adopted by over two-thirds of the members of the United Nations general assembly. The United States has signed but rejected ratification in 1999. Nevertheless, 337 facilities around the world monitor compliance with the treaty. They send data to a center in Vienna for analysis and distribution to the states that have signed the treaty.

How is **nuclear fission** used **peacefully**?

Nuclear reactors produce electric power. The energy from uranium fission heats water that circulates through the reactor. The heated water produces steam that turns turbines connected to generators. As is the case with all electric power plants, only about 1/3 of the energy produced by the reactor is converted into electrical energy. The remaining energy heats local rivers, lakes, or the atmosphere. In the United States there are 104 reactors that provide about 20% of the electricity used by our country.

What are the **strengths** and **weaknesses** of **nuclear power**?

Electric power is produced primarily by plants using hydrocarbon fuels—coal, oil, and natural gas. These are usually called fossil fuels and are no longer being created; when they are used up these resources are gone. Nuclear power plants can reduce our reliance on such fuels. Uranium, however, is also a fossil fuel. Coal mining has significant environmental costs. Oil is used mostly for transportation. Natural gas is relatively clean and is used mostly for home and industrial heating. Another advantage of nuclear power is the reduction in greenhouse gases, primarily carbon dioxide.

One major disadvantage of nuclear power is the cost of the plant and the extremely long time scale associated with obtaining approval and constructing the facility. Costs are difficult to calculate precisely, but nuclear power and off-shore wind farms are the two most expensive methods of generating electricity, while oil from the Middle East is the cheapest. As a result of these uncertainties, factors other than costs are increasingly important.

Other disadvantages include the production of nuclear waste that poses long-term dangers to people due to its intense radioactivity. No long-term storage plans have been approved, although underground storage in salt deposits is the most likely method.

Are nuclear power plants safe?

There have been two major accidents in nuclear power plants. One was the failure of water cooling at the Three-Mile Island plant in Pennsylvania in 1979 that resulted in a partial melt-down of the reactor core. While there was a significant release of radioactive gases, studies have shown that there has been no increase in cancer that can be traced to the accident. Nevertheless, this accident has been cited as the major reason the growth in the use of nuclear power has essentially stopped in the United States.

The accident that caused the greatest loss of life and contamination was the explosion at Chernobyl—then in the Soviet Union, now in Ukraine—in 1986. The graphite-moderated reactor had a power spike during a test of the cooling system that led to explosions and fire. Radioactive material, four-hundred times the amount released by the Hiroshima bomb, was spread over a huge area of the Soviet Union and Western Europe. The estimates of the number of deaths range from 50 direct deaths to thousands due to cancers caused by the radiation.

Will **more nuclear power plants** be built?

Concerns over the build up of greenhouse gases and the environmental problems caused by coal mining has led to renewed interest in building new nuclear power plants. Advocates of construction point out that if a standardized plant could be designed, then licensing delays could be reduced and design costs minimized. In addition, several new types of plants have been suggested and undergone small-scale testing. These promise to be simpler and safer than traditional designs. While recycling nuclear fuel is an attractive option, the plutonium in used fuel rods raises issues of nuclear weapon proliferation. The problem of long-term storage of nuclear wastes still needs to be solved.

What is **nuclear fusion**?

Nuclear fusion is the opposite of fission. Two nuclei join, or fuse together, forming a more massive nucleus. The mass of the resultant nucleus is less than that of the reacting nuclei, so energy is released. Because the reacting nuclei are both positively charged, there is a large repulsive force between them. To overcome this force the reacting nuclei must have very high energy. Fusion was first observed by the British physicist Mark Oliphant (1901–2000) in 1932.

A typical fusion reaction involves two isotopes of hydrogen: ^{2}H, or deuterium, and ^{3}H, or tritium. They fuse to produce ^{4}He, releasing a neutron. The energy released is more than a million times larger than that released when an electron combines with a proton to produce a hydrogen atom.

305

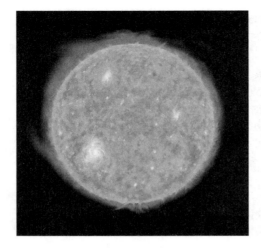

Nuclear fusion occurs naturally in stars like our sun, but reproducing a fusion reaction in the laboratory has proven to be very difficult because it takes a lot of energy to fuse nuclei together.

What is the **closest fusion reactor**?

The sun! In stars like the sun the principal reaction is called the proton–proton cycle that was first described by the German-American physicist Hans Bethe (1906–2005) in 1939. In the first step two protons fuse into a deuterium (^2H) nucleus. The deuterium has a proton and a neutron, so the second proton changed into a neutron, releasing a positron and a neutrino. The positron annihilates with an electron, producing two gamma rays. In the second stage the deuterium fuses with another proton to produce helium-3 (^3He) plus a gamma ray. In the sun the third stage is primarily a fusion between two ^3He nuclei producing ^4He plus two protons. So the net reaction is an input of four protons and an output of two ^4He plus six gammas and two neutrinos, as well as a lot of energy due to the loss of 0.7% of the mass of the protons. In order to accomplish these reactions the protons must be moving with a large amount of kinetic energy, the equivalent of a temperature of about 10 million kelvins.

What are **fusion weapons**?

Even before the nuclear bomb was completed, some physicists at Los Alamos started work on what they called the "super," a weapon based on fusion. After the war was over there were heated discussions, both scientific and political, about the wisdom of developing a new, even more destructive weapon. Due to tensions in the Cold War, both the United States and the Soviet Union embarked on programs to create these weapons, informally called hydrogen bombs. The United States tested such a weapon in 1952; the Soviet Union in 1955. Britain, China, and France are known to have tested fusion bombs.

Some aspects of fusion weapons are known while others are still secret. What is known is that the bomb starts with a uranium or plutonium implosion bomb with deuterium and tritium gases inside its core. The neutrons released by the fusion increase the fission of plutonium or uranium, boosting the efficiency of the fission device. The energy released is transferred to a lithium-dihydride fusion fuel in a process that remains secret. That undergoes fusion, and the energy released causes fission in a surrounding layer of natural uranium. The Soviet Union exploded a fusion weapon that released the equivalent of 50 million tons of TNT. The most recent effort on weapons development is to decrease the size of the weapons so they can be deployed in smaller delivery rockets.

Can **fusion** be used for **peaceful purposes**?

The energy released in a fusion reaction could be captured and converted into electrical energy. The difficulty in creating a fusion reactor is confining the reactants, typically deuterium and tritium nuclei, at temperatures needed for fusion. There are two approaches: magnetic and inertial confinement. Most effort has gone into magnetic confinement where the positively charged nuclei are trapped in evacuated regions containing strong magnetic fields that keep the nuclei from colliding with the metallic walls of the equipment. The swarms of charged nuclei, a plasma, must then be raised to very high temperatures and held at that temperature long enough for fusion to take place. So there are three variables: number of nuclei, their energy (or temperature), and the confinement time. The product of these three variables determines whether or not fusion will occur.

What is the latest **fusion reactor research project**?

The ITER reactor project is international in scope, including the United States, Russia, the European Union, India, Japan, China, and South Korea. The plan is to spend 10 years in construction and 20 years in operation. No electrical energy would be generated. That would require a more advanced reactor. Using 0.5 grams of deuterium-tritium fuel, it is designed to produce 500 megawatts of energy over 1,000 seconds. For the first time in a fusion reactor, more energy would be produced than needed to run the reactor. In May 2010, contracts were signed to construct the first buildings in the project that are sited in the south of France. Fusion reactions are scheduled to begin in 2018.

How would **lasers produce fusion**?

The second method of producing fusion is called inertial confinement. Tiny glass, plastic, or metallic spheres containing about 10 milligrams of deuterium-tritium fuel mixture at high pressure are bombarded on all sides by extremely powerful lasers. The lasers vaporize the sphere, driving a strong inward shock wave that compresses the fuel enough to create fusion.

The National Ignition Facility, located at Lawrence Livermore Laboratory in California, has been under construction since 1997. The design goal is to create a 2 megajoule flash of ultraviolet energy that hits the target from 192 separate beamlets. All of the beamlets must strike the target within a few picoseconds. In January 2010 the laser beams delivered 700 kilojoules to a test sphere, heating the gas inside to 3.3 million kelvins. It takes hours for the lasers to cool enough to deliver another pulse. The engineering goal is to deliver one pulse each five hours. Before deuterium-tritium fusion fuel can be used neutron shields have to be built to protect the lasers. There are no accepted engineering plans to convert the energy released to electrical energy. One proposed method is to surround the bead with a blanket of liquefied lithium that would be further heated by the energy released.

When are **fusion reactors** expected to **generate electricity**?

No one knows, but the prediction of scientists working on fusion projects that has been made every year since the 1960s is "about forty years from now."

UNANSWERED QUESTIONS

BEYOND THE PROTON, NEUTRON, AND ELECTRON

What is the **"Standard Model"**?

The description of all the particles described below is called the Standard Model. A model is something like a theory in that it is intended to explain observations. But it is not as complete as a theory. The Standard Model works very well, but there are still many questions about it. Some of these will be explained below.

Why do physicists believe that there are **particles inside protons** and **neutrons**?

Ernest Rutherford (1871–1937) scattered alpha particles off gold atoms and discovered that the atom was not filled with a uniform positive material. In the same way, scattering of protons at high energies off hydrogen nuclei showed that the proton is not composed of uniform positive material. Rather, it is composed of three much smaller charged particles called quarks. The neutron is also composed of three quarks.

How are **particles accelerated** to very high energies?

The two highest energy particle accelerators are the Tevatron at Fermilab, near Chicago, and the Large Hadron Collider (LHC) at CERN (European Organization for Nuclear Research) near Geneva, Switzerland. Both machines accelerate and store protons moving in circular paths in a metallic tube, called the beam line, from which all but minute amounts of air have been evacuated. Beams of particles circulate through the tube in both the clockwise and counter-clockwise direction. The beams intersect at several places around the circle so that the particles can collide head-on. Many superconducting magnets bend the paths of the particles into the circle while other magnets focus the particles into small beams. Electric fields created in high-power vacuum tubes provide the accelerating forces on the particles.

In particle accelerators energies are measured in electron volts (eV). One eV is the energy a particle obtains by being accelerated by a 1 volt potential difference. Modern accelerators give particles energies of hundreds of MeV (mega electron volts) or GeV (billions of electron volts). The Tevatron creates collisions between protons and antiprotons, each giving an energy of almost 1,000 GeV resulting in an energy of 2 GeV when the two beams collide. In the LHC both beams are protons. The design energy is 7 GeV in each beam for a total collision energy of 14 GeV.

An inside look at the ATLAS (A Toroidal LHC ApparatuS) at CERN in Switzlerland.

How are the results of **collisions** in **accelerators detected**?

A collision between two protons or a proton and anti-proton can produce a large number of particles of different kinds. The task of a detector is to find the direction in which each particle is going, find its charge, momentum, and energy, and identify it. Detectors are huge devices. The CDF (Collider Detector at Fermilab) weighs 5,000 tons and is 12 meters (39 feet) in each dimension. It is operated by a collaboration of about 600 physicists from 60 universities and organizations. ATLAS (A Toroidal LHC ApparatuS) at LHC is even larger. It weighs 7,000 tons, is 44 meters (144 feet) long and 25 meters (82 feet) in diameter; 2,000 physicists are involved in building and operating this experiment.

In both detectors position detectors locate the particles close to where they are produced in the beam line. Somewhat farther away from the beam line a strong magnet deflects the paths of charged particles. Outside the magnet are additional position detectors and finally detectors that measure the energy of the particles. The curved paths charged particles take in the magnet are used to find their charge and momentum.

Each collision can produce billions of pieces of data, only a small number of which are interesting. Electronic circuits and fast computers use data from the position detectors to decide whether the set of data are interesting enough to be recorded. ATLAS produces 100 megabytes of data each second. The entire LHC produces 4 gigabytes of data each second, enough to fill a DVD disk.

> ## Why have physicists given these particles the name "quark" and what do they mean by flavor and color?
>
> The American physicist Murray Gell-Man (1929–) has written that he gave the particles the name "quark" from the phrase "Three quarks for Muster Mark" in *Finnegan's Wake* by James Joyce. Flavor and color have nothing to do with their traditional meanings. The names up and down are associated with a property of protons and neutrons central to an earlier theory. While some people believe that using ordinary words in a very technical context suggests that physicists are not serious about the particles, others believe that their use suggests a whimsical approach to physics that invites people into the discussion rather than erecting barriers to keep others out.

What are the **properties** of **quarks**?

The quarks are fractionally charged, that is their charges are either 2/3 or –1/3 the charge of the proton. They have been given the names "up" and "down" respectively. The proton is composed of two up quarks and one down quark while the neutron is composed of one up and two down quarks. You can check to see that the charges of the quarks add up correctly. Up and down quarks are said to be two different quark "flavors." Because the quarks have half-integral spin (like electrons and neutrinos) there cannot be two identical quarks that have the same quantum properties. Quarks have additional property called color charge that can be red, green, or blue. A proton or neutron must have quarks with one of each of the colors so that they add to white, just like ordinary colors do.

Are there **anti-quarks** like there are antineutrinos and anti-electrons (positrons)?

Yes, there are anti-up quarks with charge –2/3 and anti-down quarks with charge 1/3. An antiproton is composed of two anti-up and one anti-down quarks. Anti-quarks have color charges that are the complementary colors to red, green, and blue: cyan (anti-red), magneta (anti-green), and yellow (anti-blue).

An up (or down) quark and an anti-up (or anti-down) quark will annihilate each other, resulting in energy being released.

What **holds the quarks together** in a proton or neutron?

Eight different massless force carriers called gluons exert the force that holds quarks together. Gluons are members of the boson family. Bosons have integral spin while quarks, electrons, and neutrinos have half-integral spins and are called fermions. Bosons can be created or destroyed while fermions can be created only if an anti-fermion is also created. A single fermion cannot be destroyed unless an anti-fermion is simultaneously destroyed.

Can **quarks** be **free**, or are they always bound in the proton or neutron?

There have been many searches for free quarks but they have never been observed. The theory that describes the interactions between quarks and gluons, quantum chromodynamics (QCD for short), says that quarks can never break free because the more that the gluons are stretched, the stronger the force they exert. In this way they act like springs whose force also increases the more they are stretched.

Do **electrons** and **neutrinos interact** with **quarks** and **gluons**?

Electrons and neutrinos are called leptons, meaning a lightweight particle. They don't participate in the strong interaction, so they don't interact directly with quarks and gluons. They are, however, involved in the weak interaction through two additional force carrier particles, the W and Z bosons that can interact with quarks.

Beta decay involves neutrons, protons, electrons, and neutrinos. How is it **explained using quarks**?

Beta decay involves two steps. First the down quark changes to an up quark with the emission of a W$^-$ boson. Then, almost instantaneously the W$^-$ changes into an electron and anti-neutrino. Physicists diagram the two beta decay processes this way:

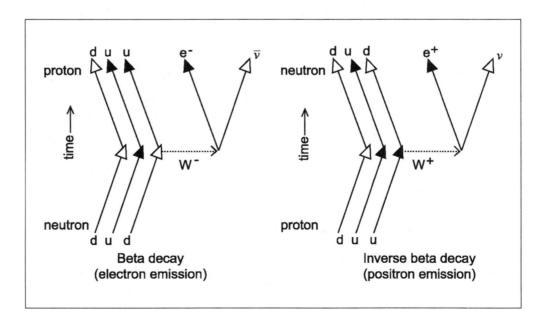

Beta decay
(electron emission)

Inverse beta decay
(positron emission)

When a nucleus undergoes beta decay the number of neutrons goes down by one, the number of protons goes up by one, and an electron and anti-neutrino are emitted. In terms of quarks and leptons one of the down quarks in the neutron is changed to an up quark in the proton, and a lepton (electron) and anti-lepton (anti-neutrino) are emitted. The change from down to up quark is a flavor change and that can occur only in beta decay. When a positron is emitted in inverse beta decay, an up quark in the proton is changed to a down quark in the neutrino and a lepton (neutrino) and anti-lepton (positron) are emitted.

Are **protons** and **neutrons** the only particles containing **up** and **down quarks**?

In the 1950s new particles with masses in between those of the electron and the proton were found. They were given the name "mesotrons" where "meso" meant intermediate. This was soon shortened to meson. The most commonly produced meson in particle accelerators is the pi-meson or pion. Pions can be positively or negatively charged or uncharged.

In the quark model mesons are composed of a quark and an anti-quark. The quarks that make up the three pions are shown below.

Pion	π^+	π^-	π°
Quarks	$u\bar{d}$	$d\bar{u}$	$1\sqrt{2}\,(d\bar{d} + u\bar{u})$
Quark charge	$+2/3 + 1/3$	$-1/3 - 2/3$	$0.707(-1/3 + 1/3 + 2/3 - 2/3)$

It is also possible to create particles with different combinations of three up and down quarks. These particles have higher masses than protons and neutrons and can only be created by colliding protons and neutrons together with very high energies. Examples are the combinations uuu with charge +2 and the ddd with charge −1. These particles decay via the strong interaction. The uuu decays when its extra energy creates a $d\bar{d}$ pair. The d quark replaces one u quark in the uuu particle, resulting in a proton and the $\bar{d}$ combines with the extra u quark to form a π^+.

Are there yet **more quarks** and **leptons**?

There are two more generations of both quarks and leptons. In the second generation the quarks are called strange and charm. The strange quark was named after particles that contain these quarks. The particles decayed in very different ways and were therefore called strange particles. The charm quark was found in a particle that had a very long, or "charmed" life. The leptons are the muon and muon neutrino. The muon was first called the mu-meson before its true identity was discovered. In the third generation the quarks are called bottom and top. The leptons are called tau and the tau neutrino. The table shows their properties.

Generation	Quarks		Leptons		Anti-Quarks		Anti-Leptons	
	Charge +2/3	Charge −1/3			Charge −2/3	Charge +1/3		
1st	up	down	electron	electron neutrino	Anti- up	Anti- down	positron	Anti- electron neutrino
	u	d	e	ν	$\bar{u}$	$\bar{d}$	e^+	$\bar{\nu}$
2nd	charm	strange	muon	muon neutrino	Anti- charm	Anti- strange	Anti- muon	Anti- muon neutrino
	c	s	μ	ν_μ	$\bar{c}$	$\bar{s}$	$\bar{\mu}$	$\bar{\nu}_\mu$
3rd	top	bottom	tau	tau neutrino	Anti- top	Anti- bottom	Anti- tau	Anti-tau neutrino
	t	b	τ	ν_τ	$\bar{t}$	$\bar{b}$	$\bar{\tau}$	$\bar{\nu}_\tau$

What **role** do the **quarks** and **leptons beyond the first generation** quarks and leptons **play** in the universe?

The second and third generation quarks and leptons require large amounts of energy to be created. Particle accelerators like the Tevatron and the LHC have been built to create and study them. They can also be created when very high-energy cosmic rays strike Earth's atmosphere. Some astrophysicists believe that stars exist that contain the strange quark. These particles also existed at times shortly after the Big Bang when temperatures in the universe were extremely high. Thus studying high energy collisions using accelerators is a way of studying what the universe was like when it was very, very young.

What are the **masses** of the **quarks** and **leptons**?

Because quarks can't be separated their masses can only be found by using one or more theories to calculate them. The masses of the charged leptons can be measured to high precision. In the Standard Model neutrinos are all massless, but recent experimental results demand that they have mass.

Masses are given in a unit that is energy divided by the speed of light squared based on Einstein's $E = mc^2$.

Quarks	Mass	Leptons	Mass
u	$2.01 \text{ MeV}/c^2$	e	$0.511 \text{ MeV}/c^2$
d	$4.79 \text{ MeV}/c^2$	ν	$<2.2 \text{ eV}/c^2$
c	$1.27 \text{ GeV}/c^2$	μ	$105.7 \text{ MeV}/c^2$
s	$92.4 \text{ MeV}/c^2$	ν_μ	$<0.17 \text{ MeV}/c^2$
t	$171.2 \text{ GeV}/c^2$	τ	$1.777 \text{ GeV}/c^2$
b	$4.2 \text{ GeV}/c^2$	ν_τ	$<15.5 \text{ MeV}/c^2$

> ## Is there a fourth (or fifth) generation of quarks and neutrinos?
>
> **W**hile there have been many experiments searching for additional generations that have proved negative and many theoretical arguments against more, this is still an open question.

What is the **origin** of the **masses** of the **quarks** and **leptons**?

Peter Higgs (1929–), a British physicist, proposed a mechanism that gives rise to the masses of the quarks, leptons, and the two weak-interaction bosons, the W and Z. The mechanism successfully predicted the mass of the Z boson, but it requires the existence of a particle, called the Higgs boson. Despite many searches, the Higgs boson has not yet been discovered. Physicists hope that the Large Hadron Collider (LHC) that started operation in 2010 will accelerate protons to high enough energies that the Higgs will be found.

Do the **masses** of the **quarks add up** to the **mass** of the **proton**?

The mass of two up quarks and one down quark sum to 8.81 MeV/c^2, and gluons are massless, but the proton's mass is 938 MeV/c^2. How can that be? The remainder of the mass is due to the high energies of the quarks and gluons within the proton.

What are **force carriers**?

The photon is called the force carrier for the electromagnetic interaction. That means that, for example, the attractive force between two oppositely charged objects is described mathematically not by an electric field, but by an exchange of (invisible) photons. The photon has neither mass nor charge, but does have angular momentum. The gluon carries the color force which, in combination with quarks, creates the strong force. The gluon also has neither mass nor charge. The weak nuclear force is carried by the massive W^+, W^-, and Z° particles. Gravity is carried by a massless particle called the graviton.

How are **neutrinos detected**?

Neutrinos hardly interact at all with matter. Some 100 trillion pass through your body every second! Yet the chance that one interacts in your body over your lifetime is only one in four. For this reason neutrino detectors must be huge and can expect to detect only an extremely small fraction of the neutrinos striking them. Most detectors are large cavities filled with extremely pure water or mineral oil. Neutrino interactions result in flashes of light that are seen by sensitive phototubes. The detectors have identified neutrinos from the sun, from cosmic rays, a supernova, and from reactors and particle detectors. Some experiments have used beams of neutrinos that are aimed through Earth by a reactor or accelerator at a distant detector.

What is the **evidence** that **neutrinos** have **mass**?

The number of neutrinos detected from the sun is only a fraction of what should be detected if our understanding of the nuclear reactions in the sun is correct. A proposed solution to this problem is that neutrinos change from one flavor to another on their travel from the sun to Earth. This phenomenon is called neutrino oscillation. It can only occur if the neutrinos have a small mass. Most experiments that support neutrino oscillation measure the lack of neutrinos of one flavor that they interpret as an oscillation into an undetected flavor. Recently an experiment has detected tau neutrinos in a muon-neutrino beam from an accelerator at CERN and a detector, 732 kilometers away, in the Gran Sasso tunnel in Italy. This flavor change supports neutrino oscillation.

How does **gravity** fit into the **Standard Model**?

It does not fit. Gravity, explained by Einstein's General Theory of Relativity, is extremely successful on the macroscopic scale. But, all attempts to create a quantum gravity have failed. So have attempts to fit gravity into the Standard Model.

What are **possible replacements** or **additions** to the **Standard Model**?

One proposed replacement is called supersymmetry. Each boson in the standard model has a supersymmetric fermion partner particle (or sparticle). And each fermion has a supersymmetric boson sparticle. They are called sleptons, squarks, glunios, and photinos. These sparticles must be much more massive than the Standard Model particles (at least 480 billion electron volts). At this time there is no direct evidence of their existence, although the LHC might be able to reach high enough energies to find them.

A second replacement is string theory. String theory requires a ten-dimensional space-time rather than the normal four-dimensional space-time (three spatial dimensions and one time dimension). A string is a tiny closed loop (like a rubber band). In some theories the loop can break into an open length (like a piece of yarn) but in others it cannot. Strings vibrate at characteristic frequencies, like guitar strings. Each vibration frequency corresponds to an elementary particle. If a string theory is to include fermions (quarks and leptons) supersymmetry must exist.

The great advantage of string theory is that the graviton, the particle that carries the gravitational force, is naturally accommodated. But, if string theory is to be a theory of quantum gravity then the size of a string must be the size of the so-called Planck length, 1.6×10^{-35} m! That is an unimaginably small length, 10^{-20} as large as a proton.

String theories, however, contain symmetries that make their effects felt at larger distances. At this time there is no experiment that could test string theory. If supersymmetric particles are found, then string theory may garner more support.

In summary, what are some of the **unanswered questions** in **particle physics**?

What is the origin of mass? Does the Higgs particle exist? How can the mass of physical particles be completely explained based on the masses of the quarks? Are there additional generations of quarks and leptons? What are the masses of the neutrinos? Could a neutrino be its own antiparticle? How can the gravitational interaction be incorporated into particle

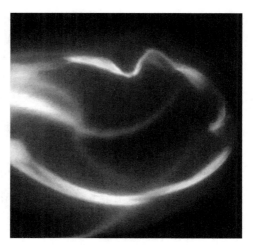

Too small for most people to even visualize in their minds, strings are theoretical objects that exist in ten dimensions and are much tinier than quarks. It is impossible to prove that strings exist using any observational methods known to humankind, but string theory helps explain how matter, energy, and forces like gravity might work in the universe.

physics theories? Are supersymmetry and string theory able to replace the Standard Model? Perhaps you will be able to find the answers to these and other questions.

ENTANGLEMENT, TELEPORTATION, AND QUANTUM COMPUTING

What is **entanglement**?

Consider what happens when a positron and electron annihilate. Two gammas (high energy photons) are produced that go off in opposite directions, 180° apart. They can be detected many meters from their source and have opposite spins, but which gamma has which spin is a random choice. That is, for each gamma there is a 50/50 chance that it will be in a particular direction. Suppose you find that the spin of one gamma is pointing up. But, the moment you detect the spin of that gamma, the spin of the other gamma must be pointing down. The result of one measurement determines the results of the other. The two detectors would measure the spins of their gammas at the same time, so there is no way that one gamma could communicate with the other gamma.

Physicists say that the spins of the two photons are entangled and that the spin state of each photon is the superposition of the two possible spin directions. When the

317

spin is measured the "wavefunction collapses" and gives a definitive result. Albert Einstein (1879–1955) called results like this "Spooky action at a distance." Others have called it quantum weirdness.

Similar results can be obtained with atoms or ions, in which case the photons are light quanta and may be transported through space (or the air) or by optical fibers. For example, if an atom is excited by the absorption of a photon it can emit two photons that are entangled the same way the gammas are in the example above.

What is **quantum teleportation** and how does it use entanglement?

First, what it is not. It is not "Beam me up, Scotty" where a person disappears at one place and appears at another. It involves only information transfer by either photons or atoms.

Teleportation involves an atom whose state is to be communicated from a sender, Alice, to a receiver, Bob. It also involves a source of two entangled photons.

Alice first sets or measures the state of the atom. This is the information she will send to Bob. At the same time she measures the polarization of one of the two photons. This polarization information is the encryption key that will allow Bob to decode the information. Because the two photons are entangled the polarization of the transmitted photon is fixed. Bob now has the key that allows him to decode the information Alice sent. In practice, of course, the encryption key isn't the polarization of a single photon, but many.

The longest distance information has been teleported is 16 kilometers through the air in Beijing.

What is a **qubit**?

Qubit is short for quantum bit, the basic piece of information for a quantum computer. A qubit can be the state of an atom, ion, photon, or an electron. As was the case of the polarization of the two gammas, the qubit doesn't have a value of 0 or 1, but is in a state that is a superposition of 0 and 1, so it can have any value between those two numbers. If you try to measure the value of the qubit you will get either 0 or 1, so it will act like a bit in a normal computer. The trick is to use one of two entangled photons (or atoms) to determine the qubit's state without collapsing it into either 0 or 1.

Quantum computers are by nature parallel machines that can work on many problems at once. At this time computers with 12 qubits have been demonstrated, so there is a long way to go before such a device is practical. Still, progress is being made. A company has demonstrated a way of using the kind of chip lasers used in CD players to create entangled photons, greatly reducing the size, complexity, and cost of a quantum computer.

Will **quantum teleportation** be able to be extended to **longer distances** with **error-free results**?

At the present the encryption codes cannot be broken. Will that continue to be true? Will quantum computers become practical? At present their projected use is to factor large numbers, a task used to create encryption keys to keep electronic messages secure. What additional problems will they be able to solve that traditional computers cannot?

THE STRUCTURE
AND END OF THE UNIVERSE

What **makes up** the **universe**?

If you said stars and galaxies, gas clouds, black holes, photons, neutrinos, etc. you're only partially correct. These known materials constitute only 5% of the mass-energy in universe! Dark matter comprises another 23%. Dark matter does not interact with electromagnetic radiation but does interact with gravity. Thus it is invisible and we can see right through it. The remaining 72% is made up of dark energy.

Why do **astronomers believe** there is **dark matter** in the universe?

The first hints that there is more to the universe than visible stars and galaxies came in 1933 when the Swiss-American astronomer Fred Zwicky (1898–1974) studied the motion of galaxies in the Coma cluster and estimated that in order to account for their rapid motion there must be 400 times as much mass as could be accounted for. In the 1970s the American astronomer Vera Rubin (1938–) studied the rotation of galaxies and found that at least half of the matter in galaxies must be invisible—dark matter.

Dark matter is also in evidence in gravitational lensing, where the gravitational interaction of mass with light causes the light from very distant galaxies to be bent by nearer galaxies. The result is a distortion of shape and position in the images of the distant galaxies as if they are being seen through a lens. In one set of colliding galaxies the center of mass, determined by the lensing effect, is separated from the center of optical and X-ray brightness, probably by the effect of the collisions.

Radio and infrared emissions from the initial Big Bang that formed the universe is called the 3 kelvin cosmic microwave background radiation. Understanding the details of this radiation also requires large amounts of dark matter.

What makes up **dark matter**?

Many candidates have been proposed, but there is no agreement at this point. Neutrinos with mass could make up a small portion, as could brown dwarf stars that have

319

This image from the Hubble Space Telescope shows a dark ring surrounding the CL0024+17 region of galaxies. Scientists speculate that the distorted images of these galaxies is a result of dark matter in the ring, which is five hundred million light years wide.

failed to ignite. More likely candidates are the supersymmetric particles described above. Much discussion recently has focused on WIMPs, Weakly Interacting Massive Particles. If they exist they would be passing through Earth, and so several experiments are underway to detect them. Early results from one experiment claimed to have detected two particles. Other experiments look for indirect evidence from gamma rays or anti-matter. One result claimed to have found an excess of positrons, but not anti-protons. The composition of dark matter is one of the most active areas of investigation in astrophysics.

What is **dark energy**?

For the first ten billion years of the existence of the universe the attractive forces of gravity on matter—both ordinary and dark—slowed the expansion of the universe. But beginning about five billion years ago the universe began to expand at an increasing rate. Two studies of supernovae have documented this acceleration. The cause of this expansion has been called dark energy, but the nature of dark energy is totally unknown. It interacts only via gravity and is very dilute. One possibility is that Einstein's General Relativity has to be modified by the addition of a cosmological constant. Einstein himself considered such a constant, then discarded it, calling it his biggest blunder. The problem with such a constant is that particle physics estimates the value (in proper units) of 1, but the value required to explain the extra expansion is 10^{-120}. This huge discrepancy cannot be explained.

What will happen to the **ever expanding universe**?

If the acceleration of the universe continues in many billion years galaxies beyond the Milky Way and its companions will become invisible because their apparent velocity as seen from Earth would exceed the speed of light. In some models dark energy would rip apart all galaxies and solar systems, and eventually become so strong that it could tear apart atoms and even nuclei. The universe would end in what has been called the "Big Rip." In other models gravity would again take over, causing the universe to contract and end in the "Big Crunch." More measurements of acceleration are needed to decide between these two ultimate ends.

BIBLIOGRAPHY

Physics of Sports

Adair, Robert. *The Physics of Baseball*. New York: Harper-Collins, 2002. ISBN 0-06-08436-7.

Armenti, Angelo, editor. *The Physics of Sports*. College Park, MD: American Institute of Physics, 1992. ISBN 0-88318-946-1.

Fontanella, John J. *The Physics of Basketball*. Baltimore: The Johns Hopkins University Press, 2006. ISBN 0-8018-8513-2.

Gay, Timothy. *The Physics of Football: Discover the Science of Bone-Crunching Hits, Soaring Field Goals, and Awe-Inspiring Passes*. New York: HarperCollins, 2005. ISBN 10-06-0862634-7.

Haché, Alain. *The Physics of Hockey*. Baltimore: The Johns Hopkins University Press, 2002. ISBN 0-8018-7071-2.

Jorgensen, Theodore P. *The Physics of Golf*. New York: Springer Science, 1999. ISBN 0-387-98691-X.

Leslie-Pelecky, Diandra. *The Physics of NASCAR: How to Make Steel + Gas + Rubber = Speed*. New York: Dutton Penguin Group, 2008. ISBN 978-0-525-95053.

Lorenz, Ralph D. *Spinning Flight*. New York: Springer Science, 2006. ISBN 978-0387-30779-4.

History of Physics

Cropper, William H. *Great Physicists: The Life and Times of Leading Physicists from Galileo to Hawking*. Oxford: Oxford University Press, 2001. ISBN 019-517324-4.

Hakim, Joy. *The Story of Science: Aristotle Leads the Way*. Washington, DC: Smithsonian Books, 2004. ISBN 1-58834-160-7.

Hakim, Joy. *The Story of Science: Newton at the Center*. Washington, DC: Smithsonian Books, 2004. ISBN 1-58834-161-5.

Hakim, Joy. *The Story of Science: Einstein Adds a New Dimension.* Washington, DC: Smithsonian Books, 2007. ISBN 1-58834-162-4.

Holton, Gerald, and Brush, Stephen G. *Physics, the Human Adventure: From Copernicus to Einstein and Beyond.* Piscataway, NJ: Rutgers University Press, 2005. ISBN 0-8175-2908-5.

Physics of Light, Music, and the Arts

Falk, David, Dieter Brill, and David Stork. *Seeing the Light: Optics in Nature, Photography, Color, Vision, and Holography.* New York: Harper & Row, 1986. ISBN 0-476-60385-6.

Fletcher, Neville, and Thomas Rossing. *Physics of Musical Instruments.* New York: Springer Science, 1998. ISBN 378-0387-98374-5.

Laws, Kenneth. *Physics and the Art of Dance: Understanding Movement.* Oxford: Oxford University Press, 2002. ISBN 0-19-514482-1.

Rogers, Tom. *Insultingly Stupid Movie Physics: Hollywood's Best Mistakes, Goofs, and Flat-Out Destructions of the Basic Laws of the Universe.* Naperville, IL: Sourcebooks, 2007. ISBN 978-1-4022-1033-4.

Rossing, Thomas. *Light Science: Physics and the Visual Arts.* New York: Springer Verlag, 1999. ISBN 387-98827-0.

Sundburg, Johan. *The Science of the Singing Voice.* DeKalb, IL: Northern Illinois University Press, 1987. ISBN 087-58012-0-X.

General Physics with a Unique Approach

Nitta, Hideo. *The Manga Guide to Physics.* San Francisco: No Starch Press, 2009. ISBN 978-1-59327-196-1.

Walker, Jearl. *The Flying Circus of Physics.* New York: John Wiley and Sons, 2007. ISBN 978-0-471-76273-7.

21st-Century Physics

Carroll, Sean. *From Eternity to Here: The Quest for the Ultimate Theory of Time.* New York: Dutton/Penguin, 2010. ISBN 978-05259-5133-9.

Greene, Brian. *The Elegant Universe: Superstrings, Hidden Dimensions, and the Quest for the Ultimate Theory.* New York: Random House, 2000. ISBN 0-375-70811-1.

Kaku, Misho. *Physics of the Impossible: A Scientific Exploration into the World of Phasers, Force Fields, Teleportation, and Time Travel.* New York: Doubleday, 2008. ISBN 978-0-307-27882-1.

Krauss, Lawrence. *The Physics of Star Trek.* New York: Basic Books, 2007. ISBN 978-0-465-00204-7.

How Things Work

Bloomfield, Louis A. *How Things Work: The Physics of Everyday Life.* New York: John Wiley, 2010. ISBN 978-0-470-22399-4.

Macaulay, David. *The New Way Things Work.* New York: Houghton Mifflin, 1998. ISBN 0-395-93847-3.

Online Resources*

American Physical Society's Website: *http://www.physicscentral.com.*

"Learn How Your World Works": *http://www.compadre.org.* A digital library of resources for physics and astronomy communities. See especially "Physics to go."

How things work: *http://www.howeverythingworks.org.* Associated with the book of this name. Questions and answers.

How things work: *http://www.howstuffworks.com.* Associated with Discovery magazine.

Recent research results in physics: *http://physics.aps.org.*

Recent research results in the sciences: *http://www.insidescience.org/research* Less technical than the site above.

Physics news and research results: *http://www.physicstoday.org.*

Physics news and research results with a more European prospective: *http://physics world.com.*

*All last accessed on September 30, 2010.

SYMBOLS

Symbol	Meaning
a	acceleration
A	ampere
AC	alternating current
AM	amplitude modulation
b	bottom quark
B	magnetic field strength
Btu	British thermal unit
c	centi (10^{-2})
C	capacitor
c	charm quark
C	coulomb
c	speed of light
°C	degrees celsius
cal	calorie
Cal	food calorie (kcal)
cd	candela
d	deci (10^{-1})
d	distance
d	down quark
da	deka (10^1)

Symbol	Meaning
dB	decibel
DC	direct current
E	electric field
E	energy
e	electron
EM	electromagnetic
emf	electro-motive force
eV	electron volt
f	femto (10^{-15})
F	force
f	frequency
°F	degrees fahrenheit
FM	frequency modulation
G	giga (10^9)
g	gram
g	gravitational field strength
G	universal gravitational constant
h	hecto (10^2)
h	Planck's constant
Hz	hertz (cycles per second)
i	ac electric current
I	dc electric current
I	impulse
I	moment of inertia
J	joule
k	generic symbol for constant
K	kelvins
k	kilo (10^3)
KE	kinetic energy
kg	kilogram

Symbol	Meaning
kWh	kilowatt hour
l	length
lum	lumen
M, m	mass
M	mega (10^6)
m	meter
m	milli (10^{-3})
MA	mechanical advantage
n	nano (10^{-9})
n	index of refraction
n	neutron
N	newton
N	normal force
n	generic integer number
p	momentum
P	peta (10^{15})
p	pico (10^{-12})
P	power
P	pressure
Pa	pascal
PE	potential energy
psi	pounds per square inch
q	charge
R	resistance
S	entropy
s	second
s	strange quark
T	temperature
T	tera (10^{12})
t	time

t	top quark
u	up quark
V	potential difference (volt)
v	velocity
V	volt
W	charged boson (weak force)
x	direction in Cartesian coordinate system
X, x	generic unkmown
y	direction in Cartesian coordinate system
z	direction in Cartesian coordinate system
Z	neutral boson (weak force)

Greek	Letter	Meaning
α	alpha	alpha particle
α	alpha	angular acceleration
β	beta	beta particle
γ	gamma	gamma ray
γ	gamma	special relativity factor
Δ	Delta	change in, or uncertainty in
θ	theta	angle
μ	mu	micro (10^{-6})
υ	nu	frequency
π	pi	pi (3.14167…)
τ	tau	torque
Ω	Omega	ohm
ω	omega	angular velocity

GLOSSARY

Absolute zero—Temperature at which molecular motion is at a minimum.

Acceleration (*a*)—Change in velocity divided by the time over which the change occurs (vector).

Accuracy—How correct or how close to the accepted result or standard a measurement or calculation has been.

Acoustics—Study of the ways musical instruments produce sounds, the design of concert halls, using ultrasound images.

Active noice cancellation (ANC)—Device that creates a waveform that is the opposite of the noise so that the noise is cancelled.

Aerodynamics—An aspect of fluid dynamics dealing with the movement of air.

Air drag—Friction caused by an object moving through air. Drag depends on velocity, area, shape, and the density of the air.

Alloy—A metal that is a mix of two or more different metals.

Alpha decay—Type of radioactive decay where nucleus emits an alpha particle. Result is nucleus with atomic number reduced by two and mass number reduced by four.

Alpha particles—Helium atoms with two electrons removed emitted with high energy by some radioactive nuclei.

Alternating-current (AC)—Polarity of the voltage source and thus current changes back and forth at a regular rate.

Ampere (A)—Unit of measure of electric current. Equal to one coulomb of charge passing through a wire divided by one second.

Amplitude—The distance on a wave from the midpoint to the point of maximum displacement (crest or compression).

Amplitude modulation (AM)—EM wave amplitude changes to represent transmitted information.

Aneroid barometer—Device to measure gas pressure in which the elastic top of an extremely low-pressure drum is bent by the pressure.

Angular momentum—Equal to the product of its moment of inertia and its angular velocity.

Angular velocity (ω)—Equivalent of velocity for rotational motion.

Anode—Positive electrode or terminal on, e.g. a battery, electrolytic cell, or cathode ray tube.

Antenna—Used to transmit or receive electromagnetic radio waves.

Antinode—Locations in a standing wave where there is the largest amplitude caused by constructive interference.

Apparent weightlessness—Occurs when an object is in free fall. Force on object in direction opposite gravity is zero.

Archimedes' Principle—An object immersed in a fluid will experience a buoyant force equal to the weight of the displaced fluid.

Arrow of time—The forward direction of time is the one in which entropy increases or remains the same.

Astrophysics—Study of how astronomical bodies, such as planets, stars, and galaxies, interact with one another.

Atmospheric physics—Study of the atmosphere of Earth and other planets, especially effects of global warming and climate change.

Atomic and molecular physics—Study of single atoms and molecules that are made up of these atoms.

Atomic mass number—Number of protons and neutrons in nucleus.

Atomic number—Number of protons in the nucleus and thus the nuclear charge.

Atoms—Smallest piece into which an element can be divided and retain its properties. Uncuttable; can be neither created nor destroyed.

Audion—A vacuum-tube amplifier invented in 1906 by Lee DeForest.

Barometer—Device to measure gas pressure.

Bernoulli effect—One of three causes of lift. Due to difference in air pressure on upper and lower surfaces of wing.

Beta decay—Type of radioactive decay where nucleus emits a beta particle (electron) from the change of a neutron into a proton. Antineutrino also emitted. Result is nucleus with same atomic mass number but atomic number increased by one.

Beta particles—Electrons emitted at high energy by some radioactive nuclei.

Biophysics—Study of the physical interactions of biological molecules.

Black—The absence or the absorption of all light.

Block and tackle—Simple machine that is a combination of fixed and movable pulleys.

Bohr model—A nuclear model but with electrons moving in only certain allowed orbits at discrete radii and with specific energies. When in these orbits their radii and energies are constant. The atoms do not emit or absorb radiation. Electrons gain or lose energy when they jump from one allowed orbit to another. Then they emit or absorb light with a frequency given by $hf = E_2 - E_1$ where E_2 and E_1 are the energies of the electrons in the allowed orbits.

Boiling point—Temperature at which pressure of the water vapor equals the atmospheric pressure.

Bose-Einstein Condensate—Quantum effects on a macroscopic scale exhibited by extremely cold gas of atoms.

Boson—Particles like photons and gluons with integral spin.

Brass—An alloy that is typically 80% to 90% copper with zinc.

British Thermal Unit (Btu)—Unit of energy often used to measure heat in homes and industries.

Brittle materials—Usually break when under tension rather than compression.

Bronze—An alloy of copper and tin.

Buoyant force—The reduction in weight equal to the net upward force of a fluid.

Calorie (cal)—A unit of energy that is used both to measure both thermal energy and heat.

Candela (cd)—Unit of measurement of luminous intensity.

Capacitor—Device that consists of two conducting plates with opposite charge separated by an insulator to store electric charge.

Carbon dating—Method of dating objects by measuring relative amounts of radioactive and stable carbon in them.

Carnot efficiency—Largest efficiency that a heat machine can obtain. $e = (T_{hot} - T_{cold})/T_{hot}$.

Cathode—Negative electrode or terminal on, e.g. a battery, electrolytic cell, or cathode ray tube.

Cathode rays—Electron beam in an evacuated glass tube.

Celsius scale—Temperature scale on which the freezing point of water is 0° C and the boiling point 100° C.

Center of gravity—Depends on how weight of extended object is distributed. The object's motion is the same as that of a point object in which the gravitational force that acts on its center of gravity.

Center of mass—Because the force of gravity is proportional to the mass, the center of mass is at the same location as the center of gravity.

centi—10^{-2}

Centrifugal force—A pseudo-force that exists only in rotating reference frames. Apparent force outward from center of rotation.

Centripetal acceleration—Acceleration due to a centripetal force toward the center of a circle.

Centripetal force—Force that keeps an object moving in a circle. Must be exerted by an external agent.

Charge of electron—$e = -1.602 \times 10^{-19}$ C.

Charging by contact—Results when a neutral object touches a charged object.

Charging by induction—Results when a conductor is first polarized by a charged object that doesn't touch it. Then one polarity of charge is removed from conductor.

Chemical energy—Potential energy of a system due to its chemical composition. A chemical change can cause this energy to be changed into or out of other forms. Animal bodies and batteries are two common objects that have chemical energy.

Chemical physics—Study of the physical causes of chemical reactions between atoms and molecules and how light can be used to understand and cause these reactions.

334   **Chromatic aberration**—Defect of a lens that introduces colors into images.

Circuit—Circular path through which charge can flow. Needs a source of potential difference, wires and a device with resistance.

Circuit, open—Circuit that is open so no charges flow.

Code division multiple access (CDMA)—Method of sending cell phone calls in which three calls are packed together with six more calls at two other frequencies.

Coefficient of friction μ—Friction force divided by normal force.

Coefficient of kinetic friction μ_k—Coefficient when there is relative motion between the two surfaces. Smaller than static friction.

Coefficient of static friction μ_s—Coefficient when there is no relative motion between the two surfaces.

Color blindness—An inability to see some colors due to an inherited condition.

Color mixing, additive—Combining blue, green, and red light to form other colors.

Color mixing, subtractive—Combining dyes, pigments, or other objects that absorb and reflect light.

Colorimetry—Measurement of the intensity of particular wavelengths of light.

Colors, complementary—Pair of one secondary and one primary color that form what is close to white light.

Colors, primary—Blue, green, and red of light; cyan, magenta, and yellow of pigments.

Colors, secondary—Cyan, magenta, and yellow of light; blue, green, and red of pigments.

Compass—Magnetized pointed metallic pointer that can rotate about a low-friction pivot point.

Composite materials—Modern composite materials use carbon fibers with very high tensile strength but low compressive strength to increase the tensile strength of brittle materials.

Compression—Region of higher pressure in a longitudinal wave.

Compressive force—Force pushing an object together.

Condensation—Change from vapor to liquid on a cold object.

Condensed matter physics—Studies the physical and electrical properties of solid materials.

Conduction, heat—Occurs when two objects are in contact, like when you put your hand in hot water.

335

Conductor, electrical—A material that allows electrons to move easily through it.

Cone—Cone-shaped nerve cells on the retina that can distinguish fine details in images. Located predominantly around the center of the retina called the fovea; responsible for color vision.

Conservation of energy—The energy of the system doesn't change; energy is neither created nor destroyed. The energy put into a system equals the energy change in the system plus the energy leaving the system. True as long as no objects are added to or removed from a system, and as long as there are no interactions between the system and the rest of the world.

Conservation of mass—Mass is neither created nor destroyed in chemical reactions.

Conservation of momentum—The momentum of the system is constant when external forces are zero.

Constant in Coulomb's law—$k = 9.0 \times 10^9$ Nm²/C².

Convection—The motion of a fluid, usually air or water, due to a difference in temperature. An efficient means of heat transfer.

Coordinate system—A set of axes used to locate an object. Usually defined at the reference location. Cartesian coordinates are three directions at right angles to each other, usually the x, y, and z directions.

Coriolis force—A pseudo-force that exists only in rotating reference frames. Apparent force in direction of rotation.

Cornea—A transparent membrane that contains the fluid in the eye.

Cosmology—Study of the formation, properties, and future of the universe, galaxies, and stars.

Coulomb—Equal to the charge of 6.24×10^{18} electrons (negative) or protons (positive).

Coulomb's law—Describes the strength of the electrical force between two charged objects. $F = k \, (q_1 \, q_2/r^2)$.

Crest—The highest point of a transverse wave.

Critical angle—Angle at which refracted light angle is 90°.

Critical mass—Minimal mass of enriched uranium or plutonium needed to create a bomb.

Current, electrical—The flow of electric charge.

Dark energy—Said to be cause of accelerating expansion of universe. Comprises 72% of mass and energy of universe. Its nature is totally unknown.

Dark matter—Matter in universe that does not interact with electromagnetic radiation but does interact with gravity. Comprises 23% of mass of universe. Its composition is unknown.

de Broglie wavelength—Wavelength associated with moving particle $\lambda = h/mv$.

deci (d)—10^{-1}

Decibel (dB)—Unit of relative intensity of sound.

deka (da)—10^1

Diamagnetic—Caused by circling electrons that create their own magnetic field. Material is repelled by magnetic fields.

Difference tone—Frequencies produced as a result of two different frequencies mixing with each other.

Diffraction—Light or other wave bends around the edge of the hole, through a narrow slit, or around a small object.

Dip needle—Like a conventional compass, but the needle rotates in vertical plane.

Direct-current (DC)—Charges travel only in one direction.

Dispersion—Light of different colors are refracted into different directions caused by wavelength dependence of refractive index.

Displacement—Distance and direction of an object from reference location (vector).

Distance—The separation between object's position and the reference location (scalar).

Domain—Group of atoms in a ferromagnet whose spins are aligned in the same direction.

Doppler effect—Change in frequency of a wave that results from an object's changing position relative to an observer.

Drag—Force that opposes the motion of an object through a fluid.

Earth's mass—$m_{\text{Earth}} = 5.9736 \times 10^{24}$ kg.

Efficiency, energy—Useful energy output divided by energy input.

Elastic collision—In a collision the kinetic energy of colliding objects is the same before and after the collision.

Elastic deformation—When the deforming force is no longer applied object returns to its original shape (see plastic deformation).

Elastic potential energy—The energy in the squeezed spring or stretched band.

Electric field—Region that surrounds a charged object. Another charged object placed in that field will experience a force.

Electricity, resinous—Produced when plastic, amber, or sealing wax is rubbed.

Electricity, vitreous—Produced when glass or precious stones are rubbed.

Electromagnet—Coil of current-carrying wire wound on a iron core that produces a magnetic field that depends on the current.

Electromagnetic spectrum—The wide range of electromagnetic (EM) waves from low to high frequency.

Electromagnetic wave (EM)—Two coupled transverse waves, one an oscillating electric field, the other a corresponding magnetic field perpendicular to it and perpendicular to motion of wave.

Electromagnetism and optics—Studies how electric and magnetic forces interact with matter.

Electromotive force (*emf*)—Potential difference resulting from separation of charges.

Electron volt (eV)—Energy particle with charge of an electron or proton obtains by being accelerated by a 1 volt potential difference.

Electroscope—Device used to measure electrical charge on an object.

Electrostatics—Study of the causes of the attractive and repulsive forces, called static electricity, that result when objects made of two different materials are rubbed together.

Elementary particles and fields—Study of the interactions and properties of the particles from which all matter is built.

Energy transfer—Requires a source, whose energy is reduced, a means of transferring the energy, and an energy receiver, whose energy is increased.

Entanglement—An effect in an atom or nucleus when two or three photons are emitted. The spin of one photon depends on that of the other, or is entangled with it. Measuring one spin tells you what the direction of the other is.

Entropy—A measure of the dispersal of energy. The greater the energy is dispersed the larger the entropy. As a consequence of the second law of thermodynamics is that the entropy of the system and the environment can never decrease.

Evaporation—Change from liquid to gas at a temperature below the boiling point.

Extremely high frequency (EHF) waves—Frequency 30 Ghz–300 Ghz.

Extremely low frequency (ELF) waves—Frequency less than 30 kHz.

Fahrenheit scale—Temperature scale on which the freezing point of water is 32° F and the boiling point 212° F.

Faraday cage—Cage or metal grating that can shield electrical charge by gathering charges on outer shell of the cage.

Faraday's law—An electric field is produced by a changing magnetic field. The field can cause a current in a wire.

Farsightedness—Also called hyperopia. Allows only objects far from the eye to be focused on retina. Images from close objects are behind the retina so they are blurry.

femto (f)—10^{-15}

Fermion—Particles like quarks, electrons, and neutrinos with half-integral spins.

Ferromagnetic—Iron, nickel, and cobalt. Unpaired electrons in large groups of atoms interact with each other so that they point in the same direction. Strongly attracted by magnetic fields.

First law of thermodynamics—A restatement of the conservation of energy: net heat input equals net work plus change in thermal energy.

First-class lever—Simple machine with pivot at one end or bar, the input force applied at the other end, the effort force between the two. *MA* greater than one.

Fission—Form of radioactive decay where nucleus splits in two. Extra neutrons and a large amount of energy is released.

Flavor—Generation of quark (up/down, strange/charm, top/bottom) or lepton (electron/electron neutrino, muon/muon neutrino, tau/tau neutrino).

Fluid—A liquid or gas that can flow and assume the shape of its container.

Fluid dynamics—Study of fluids in motion.

Fluid statics—Study of fluids in a state of rest.

Focal length—Measures the strength of a lens or mirror. Distance from lens or mirror to focal point.

Focal point—Location where light rays from a very distant object are reflected from a mirror or refracted by a lens to come together.

339

Force (F)—A push or a pull on an object by an external agent (vector).

Fourier theorem—Any repetitive waveform can be constructed from a series of waves of specific frequencies, a fundamental and higher harmonics.

Frequency (f)—How many cycles of an oscillation occur per second.

Frequency modulation (FM)—EM wave frequency changes to represent transmitted information.

Friction force—A force due to an interaction between an object and the surface on which it rests. Always in the direction opposite the motion. Depends on the force pushing the surfaces together. In most cases is larger if the surface is rougher. Generally doesn't increase if the speed of motion increases. $F_{friction} = \mu N$.

Fuse—Protects household electrical circuits are protected by fuses or circuit breakers. Opens when current exceeds a predetermined limit.

Fusion—Joining of two nuclei to form a more massive nucleus.

Galvanic cell—Combination of two different metals separated by a conducting solution that produces a potential difference and thus a continuous flow of charge.

Gamma ray—High energy photons emitted by some radioactive nuclei.

Gas—State of matter in which the atoms are about ten times further apart than liquids or solids and forces between them are very weak. A fluid that assumes the shape of its container.

Gear—Toothed wheels that transmit torque between two shafts. The smaller gear in a pair is called a pinion and the larger one a gear.

General relativity—Study of the descriptions and explanations of the causes and effects of gravity.

Generation (nG), cell phone—1G analog voice phone calls. 2G digital signals with more simultaneous users; 3G able to receive television-like video, video images; 4G very high-speed networks.

Generator—Converts mechanical to electrical energy. An electric field is created at the ends of multiple loops of wire on the armature rotating in a magnetic field. The field causes a current in an attached circuit.

Geometrical optics—Deals with the path that light takes when it encounters mirrors and lenses and the uses of these devices.

Geophysics—Study of the forces and energy found within Earth including tectonic plates, earthquakes, volcanic activity, and oceanography.

giga (G)—10^9.

Gigahertz (Ghz)—One thousand million or one billion hertz.

Global Positioning System (GPS)—Consists of 24 satellites in 12-hour orbits that broadcast their location and the time the signal was sent; a control system that keeps the satellites in their correct orbits, sends correction signals for their clocks as well as updates to their navigation systems; and a receiver.

Gluon—Eight different massless particles that carry the color force that holds quarks together.

Gold alloys—24-karat gold is pure gold. 18-karet gold has 18 parts gold and 6 parts other metals. 10-karet gold has 10 parts gold and 14 parts other metals.

Gravitational field energy—Energy stored in the gravitational field when an object moves away from the (larger) object that is attracting it.

Gravitational field strength—Ratio of gravitational force to mass of attracted object ($g = 9.8 \, N/kg$).

Gravitational mass—Mass defined by the effect of gravity on object. Gravitational force divided by gravitational field strength ($m = F_{gravitation}/g$).

Graviton—Particle that carries the gravitational force.

Greenhouse gas—Gases transparent to light and short-wavelength infrared radiation, but are opaque to the long-wavelength infrared emitted by warm objects. Examples are carbon dioxide, methane, and water vapor.

Ground fault interrupter (GFI)—Detects difference in current into and out of circuit to protect against accidental grounding, typically through water.

Half life—Time when there are half as many decays in a second as there were originally.

Harmonic—Mode of vibration that is a whole-number multiple of the fundamental mode. The first harmonic is the fundamental frequency. The second harmonic is twice its frequency, etc.

Heat—Energy transfer that results from a difference in temperature is called heat. Heat always flows from the hotter (energy source) to the cooler object (energy receiver). Heat is transferred by conduction, convection, and radiation.

hecto (h)—10^2.

Hertz (Hz)—Unit of measurement of frequency. Equals one cycle per second.

Higgs particle—Hypothesized particle associated with the mechanism that gives rise to the masses of the quarks, leptons, and the two weak-interaction bosons, the W and Z.

High frequency (HF) wave—Frequency 3 Mhz–30 Mhz.

Horsepower (hp)—Unit of power. 1 hp = 746 W = 0.746 kW.

Hue—Related to the wavelength of a color.

Hydraulics—Use of a liquid in a device, usually a machine such as a pump, lift, or shock absorber.

Image, light—Point at which light rays converge.

Image, real—Light rays do converge. Can be projected on a screen.

Image, virtual—Light rays do not converge, but your eye believes that rays came from a single point. Cannot be projected on a screen.

Impedance—Opposition to wave motion exerted by a medium.

Impedance-matching device—Placed between two media to create a smooth impedance transition to maximize energy transfer by the wave through the interface. A transformer.

Impulse—The product of force and the time the force is applied.

Inclined plane—Simple machine used to help lift an object.

Induced drag—A consequence of the lift generated by the wing. Induced drag causes vortices.

Inelastic confinement—In a collision the kinetic energy of colliding objects is smaller after the collision than before it.

Inertial confinement—Method of producing nuclear fusion by using lasers to compress tiny amounts of a deuterium-tritium fuel mixture.

Inertial mass—Mass defined as the net force divided by its resultant acceleration ($m = F_{net}/a$).

Inertial reference frame—A reference frame in which there is no acceleration.

Infrasonic—Frequencies below human hearing, 20 Hz.

Inner ear—Consists of the cochlea and three semicircular canals that sense the body's motions and give rise to a sense of balance.

Instantaneous speed (v)—Limiting value of distance divided by time interval when the time interval is reduced to zero.

Insulator, electrical—A material in which electrons are strongly bound to their nuclei and thus cannot move through the material.

Interference, constructive—Two waves both either positive or negative that add producing a larger wave amplitude.

Interference, destructive—Two waves, one positive, the other negative that produce a smaller wave amplitude, even zero.

Interfernce, light—Light waves from two closely-spaced slits or two surfaces of a thin film that produces regions of bright and dark regions.

Isotrope—Nuclei with the same number of protons but different numbers of neutrons.

ITER project—International effort to produce electricity from nuclear fusion.

Jelly doughnut (JD)—Informal energy unit. Energy in a medium-sized jelly-filled doughnut. 1 JD = 239 kcal.

Kelvin temperature scale—Based on absolute zero being 0 K. A kelvin (K) is equal to a degree on the Celsius scale. 0°C is equal to 273.15 kelvins.

Kepler's first law—Planetary orbit is an ellipse with the sun at one focus.

Kepler's second law—In equal times planets sweep out equal areas.

Kepler's third law—The square of the period of the planet is proportional to the cube of the radius of the orbit.

kilo (k)—10^3

Kilocalorie (kcal or Cal)—The food calorie used to measure energy content in foods.

Kilogram (kg)—The mass of a platinum-iridium cylinder that is permanently kept near Paris.

Kilohertz (kHz)—One thousand hertz.

Kilowatt (kW)—1,000 watts or 1,000 joules per second.

Kilowatt-hour (kWh)—Unit of energy used to measure electrical energy.

Kinetic energy (*K* or *KE*)—energy of motion ($K = 1/2mv^2$).

Laminar flow—Flow of slow-moving fluid or object moving slowly through a fluid. Each thin film of water moves slightly faster than the one closer to the surface.

Large Hadron Collider (LHC)—Very high synchrotron at CERN near Geneva, Switzerland.

Laser—Light source using stimulated emission of radiation invented in 1960. Creates tiny spot of very intense, monochromatic light.

Latent heat—Energy needed to change phase of matter.

Latent heat of fusion—Energy involved in change from solid to liquid phase.

Latent heat of vaporization—Energy involved in change from liquid to gas.

Law—Summary of many observations that describes phenomena.

Law of reflection—Angle of incident light is equal to angle of reflected light.

Lens, converging—Has at least one convex side. Its shape causes the entering light rays to converge, that is, come closer together.

Lens, diverging—Has at least one concave side. The shape of the lens causes the entering light rays to spread apart when they leave the lens.

Lepton—Light-weight particles like electrons and neutrinos.

Lepton generation—Each generation has a charged and uncharged particle. In first electron and neutrino, in second muon and muon neutrino, in third tau and tau neutrino.

Lever—Simple machine consisting of a rigid bar that pivots. Three different classes.

Leyden jar—Device consisting of an insulating jar with conductors on inner and outer surfaces. Stores charges. Modern version is a capacitor.

Lift—Upward force on an airplane wing caused by the air moving past it.

Light—An electromagnetic wave to which human eyes respond.

Light detector—Converts light energy to another form of energy.

Light-year—Distance that light travels in one year. 9.4605×10^{12} km or about 6 trillion miles.

Limits of human vision—Lower and upper boundaries of light. 4×10^{14} Hz, or 700 nm and 7.9×10^{14} Hz, or 400 nm.

Liquid—State of matter in which the forces keep the atoms close together, but they're free to move. A liquid is a fluid that flows freely and assumes shape of its container.

Load—An engineering term for force.

Longitudinal wave—Wave and its energy move in the same direction as direction of oscillations.

Lorentz force law—Force on a current-carrying wire in a magnetic field. Proportional to the current, the magnetic field, and the length of the wire. Force is perpendicular to both the current and the magnetic field and is strongest when the current and field are at right angles.

Loudness—Measure of how a listener responds to intensity of sound. Measured in sones.

Low frequency (LF) waves—Frequency 30 kHz–300 kHz.

Lumen (lum)—Unit of measurement of luminous power.

Luminous intensity—Eye response to intensity of light electromagnetic waves.

Luminous power—Eye response to power of light electromagnetic waves.

Lunar eclipse—Occurs when Earth casts its shadow on the moon. Earth must be directly between the sun and the moon.

Mach number—Ratio of a velocity to the speed of sound.

MAGLEV—Magnetically levitated trains that use electromagnetic forces to lift the cars off the track and propel them along thin magnetic tracks.

Magnetic declination—Angular difference between north as shown by a compass and the direction to the geographic north pole, Earth's axis of rotation.

Magnetic field—Region around a magnet that causes forces on magnetic materials or other magnets.

Magnetic pole—Region of magnetic strength at ends of magnets. Always come in North-South pairs.

Magnus force—Force on a spinning ball that causes its path to curve.

Mass (*m*)—A property of the object, the net force exerted on object divided by its acceleration (scalar).

Mass energy—Potential energy the result of the mass of an object first postulated by Einstein.

Mathematical physics—Study of how physical processes can best be described by mathematics.

Maxwell's equations—Four equations that describe the nature of electric and magnetic fields.

Mechanical advantage (*MA*)—Output force divided by the input force ($F_{output} / F_{input} = MA$).

345

Mechanics—Study of the effect of forces on the motion and energy of physical objects including fluids and granular particles.

Medical physics—Study of how physical processes can be used to produce images of the inside of humans, as well as the use of radiation and high-energy particles in treating diseases such as cancer.

Medium frequency (MF) waves—Frequency 300 kHz–3 Mhz.

mega (M)—10^6

Megahertz (Mhz)—One million hertz.

Meson—Intermediate mass particles made up of a quark and antiquark.

Meter (m)—The distance light travels in 1/299,792,458 seconds.

micro (μ)—10^{-6}.

Microwave oven—Uses 2.4 GHz EM waves to cook food.

Microwaves—Electromagnetic waves with frequencies above about 3 Ghz.

Middle ear—Between outer and inner ear. Includes the ear drum, the hammer, anvil, and stirrup, the three smallest bones in the human body, and the oval window on the inner ear.

milli (m)—10^{-3}.

Mirage—Occurs when there is a temperature difference between the air directly above the surface, which is hot and thus less dense, and the cooler, denser, air a few meters above the surface.

Mirror, concave—Curved inward so that the incident rays of light are reflected and can be brought together. Can form real or virtual images.

Mirror, convex—Curved outward. The reflected light spreads out rather than converging at a point.

Mirror, plane—Flat mirror. Forms virtual images.

Moment of inertia (*I*)—Equivalent of mass for rotational motion. Depends on mass and how far it is from axis of rotation. The further the mass is from the axis, the larger the moment of inertia.

Momentum—The product of mass and velocity $p = mv$ (vector).

Monopole—Isolated North or South poles. Predicted by some theories, but never found.

Motor—Converts electrical to mechanical energy. Current through multiple loops of wires on the armature in a magnetic field causes a force that results in rotation.

nano (n)—10^{-9}.

Natural oscillation frequency—A property of an object that can vibrate. Depends on the mass and force that restores object to equilibrium position.

Nearsightedness—Also called myopia. Allows only objects relatively near the eye to be focused on retina.. Images from distant objects are in front of the retina, so they are blurry.

Neutrino oscillation—Change of neutrinos from one flavor to another that is a proposed solution to deficit of neutrinos emitted by sun.

Neutron—Uncharged particle in nucleus with mass slightly larger than that of the proton.

Newton's first law—If there is no net force on an object, then if it was a rest it will remain at rest. If it was moving, it will continue to move at the same speed and in the same direction.

Newton's second law—If a net force acts on an object, it will accelerate in the direction of the force. The acceleration will be directly proportional to the net force and inversely proportional to the mass. That is, $a = F_{net}/m$ or $F_{net} = ma$.

Newton's third law—When two objects, A and B, interact the force of A on B is equal in magnitude but opposite in direction of the force of B on A.

Newton (N)—Unit of measure of force. In the English system, force is measured in pounds (lbs).

NEXRAD—Next-generation weather radar, relies on the Doppler Effect to calculate the position and the velocity of precipitation.

Node—Locations in a standing wave where there is no amplitude caused by destructive interference.

Noise—Sound intensity at a wide variety of different wavelengths.

Noise, white—Roughly equal sound intensities at all frequencies.

Normal force (N)—Force pushing two surfaces together.

Nuclear mass defect—Difference between mass of a nucleus and sum of the masses of the protons plus the sum of masses of the neutrons. Larger the defect, the larger the energy needed to pull nucleus apart or energy released when nucleus is assemble.

Nuclear model—Extremely small positively charged particle, the nucleus, surrounded by electrons.

Nuclear physics—Study of the properties of the nuclei of atoms and the protons and neutrons of which they are composed.

Nuclear reactor—Method of producing electric power where energy released by nuclear fission is used to boil water producing steam that turns generators.

Nucleon—Generic name of protons and neutrons.

Object, light—Point from which light leaves into many directions.

Opaque—An object that allows no light through it.

Optical fiber—Strands of glass that use the principle of total internal reflection to transmit information near the speed of light.

Optics—Area of study within physics that deals with the properties of and applications of light.

Overtone—Mode of vibration are is not a whole-number multiple of the fundamental frequency.

Parabola—Path taken by an object acted upon by gravity that is given an initial horizontal velocity.

Parallel circuit—Circuit allows the charges to flow through different branches.

Paramagnet materials—Caused by unpaired electrons in atoms. Material is attracted by magnetic fields.

Parasitic drag—Force when an airplane wing, automobile, or any other object moves through a fluid.

Pascal (Pa)—Unit of measure of pressure. One pascal is one newton per square meter.

Pascal's principle—A fluid, like water exerts the same pressure in all directions.

Penumbra—Area of a shadow where light from only part of the source is blocked.

Period—The time it takes for a an oscillating object or wave to complete one full vibration; the inverse of the frequency.

Permanent magnet—Domains remain aligned after the polarizing magnetic field is removed, resulting in a permanent magnet.

Perpetual motion—Forbidden by the second law of thermodynamics.

peta (P)—10^{15}.

Pewter—An alloy of tin with copper, bismuth, and antimony.

Photon—Packet of light energy light quantum by Einstein in 1905. Given name photon in 1926. Has no mass or charge, but carries angular momentum. Always moves at the speed of light.

Photon energy—$E = hf$, where f is frequency.

Physical optics—Division of optics that depends on the wave nature of light, polarization, diffraction, interference, and the spectral analysis of light waves.

Physics—The study of the structure of the natural world that seeks to explain natural phenomena in terms of a comprehensive theoretical structure in mathematical form; from the Greek *physis*, meaning nature.

Physics education research—Study of how people learn physics and how best to teach them.

pico (p)—10^{-12}.

Pinna—Outer ear. A cartilage flap that forms a transformer to match the impedance of the sound wave in air to that at the end of the ear canal.

Pitch—Like loudness, a description of how the ear and brain interpret sound frequency.

Planck's constant—$h = 6.6 \times 10^{-34}$ J/Hz.

Plasma—State of matter consisting of electrically charged particles.

Plasma physics—Study of the properties of large numbers of electrically charged atoms.

Plastic deformation—Object's shape changed after deforming force is removed (see elastic deformation).

Plum pudding model—Swarm of electrons in a positively charged sphere.

Pneumatics—Use of a compressed gas rather than a liquid in a hydraulic machine.

Polarized wave—Transverse wave, like light. with amplitude in one direction.

Position—The location of an object. Requires a reference location (vector).

Positron—Antimatter form of electron with same mass but positive charge. Emitted in decay of radioactive nucleus or as result of collision of high-energy gamma ray with the nucleus.

349

Positron emission tomography (PET)—Method of forming three-dimensional image of human body. Uses a short-lived positron-emitting isotope inserted into a biologically active molecule.

Potential difference—Difference in electric potential or energy difference divided by charge, also called voltage.

Pounds per square inch (psi)—Measure of pressure in English system.

Power (P)—The change in energy divided by the time taken.

Power, electrical—Given by $P = I \times V$ (power equals current times voltage).

Precession—The rotation of the axis of rotation of a rotating object due to a torque on the object.

Precision—How well the results can be reproduced.

Pressure (P)—Force divided by area over which force is exerted.

Pressure of air—At sea level about 101 kPa.

Principle of equivalence—A central tenant of Einstein's General Theory of Relativity. The laws of physics in an accelerating reference frame or a gravitating frame are indistinguishable. You can't tell the difference between falling or being acted on by gravity. Requires that inertial mass equals gravitational mass.

Principle of relativity—The laws of physics are the same in two coordinate systems moving at constant velocity with respect to each other.

Proton—Positively charged particle in nucleus.

Pulley—Simple machine. A fixed, or unmovable, single pulley can be considered a wheel and axle where both have the same radius. The mechanical advantage is one, but the pulley changes the direction of the force. If the pulley is allowed to move you achieve a mechanical advantage of two.

Pulse width modulation (PWM)—Method of digital modulation in which a narrow pulse represents a 0, a wider pulse represents 1.

Quad—Unit of energy used for measuring energy use by nations. One quad is one thousand billion BTUs.

Quantum chromodynamics (QCD)—The theory that describes the interactions between quarks and gluons.

Quantum mechanics—Study of atoms by themselves, as well as in molecules, liquids, and solids in which the wave nature is important.

Quark—Particles that make up protons and neutrons and other short-lived particles. Fractionally charged, that is their charges are either 2/3 or –1/3 the charge of the proton and are called "up" and "down" respectively.

Quark generations—Each generation has two quarks: up and down in first, strange and charm in second, top and bottom in third.

Qubit—Quantum bit, the basic piece of information for a quantum computer.

R-value—Measure of resistance of material to heat transfer. R is the inverse of conduction: $R = 1/U$.

Radar—An acronym for "RAdio Detection And Ranging."

Radiation—A means of heat transfer when infrared waves areemitted by hotter objects and absorbed by colder ones.

Radio astronomy—Uses VHF, UHF, and microwave portions of the electromagnetic spectrum to study planets, stars, and galaxies.

Radioactive nucleus—Unstable nucleus that emits particles in decay to another nucleus.

Rainbow—Spectrum of light formed when sunlight interacts with water droplets. Angle between entering and leaving is 40° for blue light, 42° for red.

Rainbow, secondary—Reversed color spectrum caused by an additional reflection of the light inside water droplet.

Ramp—Example of an inclined plane $MA = F_{ourput} / F_{input}$ = (length of ramp) / (height of ramp).

Rarefaction—Region of lower pressure in a longitudinal wave.

Reflection—Light or other wave bounces off a surface or interface between two media.

Reflection, diffuse—Light rays reflect into different, random directions, as from paper.

Reflection, specular—Light rays reflect into a single direction, as from a mirror.

Refraction—Bending of light as it goes from one medium to another.

Refractive index (n)—Ratio of speed of light in a vacuum to that in a medium.

Resistance (R)—Friction that electrons encounter due to collisions with the atoms in a wire. Causes electric charges to lose electrical energy. Equal to potential difference divided by current.

Resistor—Device used in electrical and electronic circuits to put a definite resistance in a circuit.

Resonance—Occurs when an external oscillating force is exerted on an object that can vibrate. When the frequency of the external force equals the natural frequency.

Retina—Screen on back of eye. Has cells that detect light and a neural network that does preliminary processing of the information received.

Reverberation time—Time for a sound to diminish by a factor of 60 dB, that is, to 1/1,000,000th of its original intensity.

Rod—Rod-shaped nerve cells on the retina that sensitive to low light levels. Responsible for a general image over a large area, but not fine details.

Rolling friction—The result of deformation of either the rolling object or the surface.

Rotational inertia—See moment of inertia.

Rotational kinetic energy—Equivalent of kinetic energy for rotational motion.

Rydberg formula—Wavelength of the emitted radiation from hydrogen atom is given by $1/\lambda = R(1/m^2 - 1/n^2)$. The numbers m and n are the quantum numbers of the two energy levels. For example, the red line would be $m = 2, n = 3$. The Rydberg constant $R = 0.01097$ nm^{-1}.

Saturation—Extent to which other wavelengths of light are present in a particular color.

Scalar quantity—A quantity that has only magnitude (size).

Schroedinger's equation—Equation of the wavefunction of a particle. Wavefunction is related to probability of finding particle at a specific location.

Screw—Simple machine like an inclined plane wrapped around an axle.

Second (s)—The time it takes for 9,192,631,770 periods of microwave radiation that result from the transfer of the cesium-133 atom between lower-energy and higher-energy states.

Second law of thermodynamics—It is impossible to convert heat completely into work in a cyclic heat engine; there is always some heat output. Also: Heat doesn't flow from cold to hot without work input.

Second-class lever—Simple machine with effort force exerted at one end or bar, the input force applied at the other end, and the pivot between the two. *MA* less than, equal to, or greater than one.

Series circuit—Electrical devices arranged in a single line allowing only one path for the charges.

Shadow—Areas of darkness created by an opaque object blocking light.

Shear force—Force exerted that parallel to a plane.

Simple machine—Device that matches human capabilities to do work to tasks that need to be done. There are four major groups, the lever, the wheel and axle, the pulley, and the inclined plane.

Snell's law of refraction—Describes how light behaves at a boundary between two different media.

Solar eclipse—Occurs when the moon casts its shadow on Earth. The moon must be directly between the sun and Earth.

Solid—State of matter. Solids retain their shape because strong forces hold the atoms in their places.

Sonar—Acronym for "SOund Navigation And Ranging," a method of using sound waves to determine the distance between an object and a sound transmitter.

Sonic boom—Caused by a shock wave produced when an aircraft reaches the speed of sound.

Sound intensity—Power in a sound wave divided by the area it covers.

Sound wave—Created by some type of mechanical vibration or oscillation that forces the surrounding medium to vibrate.

Sparticle—A supersymmetric partner particle. The sparticle for a boson is a fermion; the sparticle for a fermion is a boson. Including sleptons, squarks, glunios, and photinos, sparticles are theoretical particles that are (or have been) proposed as part of an alternative to the Standard Model of sub-nuclear particles.

Special relativity—Study of the descriptions and explanations of the motion of objects moving near the speed of light.

Specific heat capacity—Amount of energy needed to increase the temperature of one kilogram of a substance by one degree Celsius.

Spectrum, absorption-line—Continuous spectrum with dark gaps produced by a continuous spectrum source viewed through a cool, low-density gas.

Spectrum, continuous—Light of all wavelengths within a range produced by a hot solid or a hot, dense gas.

Spectrum, emission-line—Light of a few, distinct wavelengths produced by a hot, low-density gas.

Speed (v)—Distance moved divided by the time needed to move (scalar).

Speed of light in a vacuum (c)—= 299,792,458 km/s.

Speed of sound—In air about 340 m/s or 760 mph (miles per hour) at 20° C. Increases by 0.6 meters per second for every degree Celsius increase in temperature ($v = (331$ m/s) $(1 + 0.6\,T)$ where T is measured in Celsius.

Sphygmomanometer—Device used to measure blood pressure.

Stainless steel—A non-rusting alloy of iron and chromium and nickel. May also contain silicon, molybdenum, and magnesium.

Standard model—Presently accepted description and explanation of sub-nuclear particles and their interactions.

Standing wave—Two waves of same frequency moving in opposite directions produce a standing wave that appears to stand still.

Static—Not moving. All the forces and torques acting on a body must sum to zero the net force on the body is zero so the object does not move or rotate.

Steel—An alloy of iron and a small amount of carbon. Other metals may be added.

Stimulated emission—Atom in excited state struck by photon with correct energy is stimulated to emit another photon with same energy and drop to lower energy level.

Streamlines—Lines that represent the flow of a fluid around an object or through another fluid.

String theory—Proposed theory to replace the standard model. Each particle is a string, a tiny closed loop. Requires a ten-dimensional spacetime rather than the normal four-dimensional one.

Strong nuclear force—Force within nucleus that acts between protons and protons, protons and neutrons, and neutrons and neutrons, all with the same strength.

Sublimation—Change of state from solid directly to gas.

Superconductor—Allow electrical current to travel without resistance, and therefore no voltage drop across them or energy loss.

Superhigh frequency (SHF) waves—Frequency 3 Ghz–30 Ghz.

Superposition—Two or more waves at the same location pass through each other without interaction. The amplitudes of the two waves add or subtract.

Supersymmetry—Proposed theory to solve difficulties with the standard model in which each boson has a supersymmetric fermion partner particle (or sparticle) andeach fermion has a supersymmetric boson sparticle.

Suspension bridge—Bridge using cables from which wires are attached that suspend the roadbed.

Sustainable energy source—Primarily wind, water, and solar energy, ultimately receive their energy from the sun, and therefore will be available for billions of years.

Synthesizer—Electronic device that generates, alters, and combines a variety of waveforms to produce complex sounds.

Système International (SI)—The International System of Units. Based on the meter-kilogram-second (MKS) or metric system.

Teleportation—Communicating the state of an atom involving entangled photons.

Telescope, reflecting—Uses a concave mirror to gather light and focus it. Uses a secondary mirror to direct light to eyepiece.

Telescope, refractor—Uses one lens to gather, refract, and focus light toward an eyepiece that contains one or more lenses that create an image that they eye can see.

Temperature—A quantitative measure of hotness. The more thermal energy an object has the higher its temperature.

Tensile strength—Maximum tension force that can be applied without breaking object.

Tension force—Force that pulls an object apart. See tensile strength.

Tera (T)—10^{12}

Terminal velocity—Constant velocity attained by a falling object affected by drag in a gas or liquid.

Tevatron—Very high-energy synchrotron at Fermilab, near Chicago, Illinois.

Theory—An explanation of a large number of observations.

Thermal energy—The sum of the kinetic and elastic energy of the atoms and molecules.

Thermal physics—The study of objects warm and cold, and how they interact with each other.

Thermodynamics and statistical mechanics—Studies how temperature affects matter and how heat is transferred. Thermodynamics deals with macroscopic objects; statistical mechanics concerns the atomic and molecular motions of larger numbers of particles.

Thermograph—Picture that shows the temperature of every location in the picture.

Thermometer—Device to measure temperature.

Thermostat—A device that is part of a system that maintains a constant temperature.

Third law of thermodynamics—Absolute zero can never be reached.

Third-class lever—Simple machine with pivot at one end or bar, the effort force exerted at the other end, the input force between the two (*MA* less than one).

Threshold of hearing—Minimum sound intensity that can be heard (0 dB).

Timbre—Quality of sound. Sound spectrum produced by an instrument, characterized by relative amplitudes or harmonics.

Time division multiple access (TDMA)—Method of sending cell phone calls that splits up three compressed calls and sends them together.

Time standard—The cesium-133 atomic clock is the standard for the second.

Torque (τ)—Equivalent of force for rotational motion. If forces is at a right angle to the radius then torque equals the force times the distance from the axis of rotation.

Torr—Measure of pressure using measurement of height of mercury in a glass tube that would create a pressure that balances the pressure of the fluid. Used to be called millimeters of mercury (mmHg).

Torsional force—Force exerted to twist an object.

Torsional wave—Material is not only displaced vertically, but also twists in a wave-like fashion.

Total internal reflection—Occurs when a ray of light in a medium with a higher index of refraction strikes the interface between that medium and one with a lower index of refraction.

Transformer—Impedance matching device.

Translucent—Media that allow light to pass through, but that bend closely-spaced rays into different directions.

Transparent—Media that allows light to pass through the material. Rays of light are either not bent or closely-spaced rays are bent together.

Transverse wave—Wave and its energy move perpendicular to the direction of oscillations.

Trough—The lowest point of a transverse wave.

Tsunami—Series of waves in water caused by underwater earthquakes and volcanic eruptions.

Turbulent—Flow of fast-moving fluid or object moving fast through a fluid. Causes usually circular or helical motion of fluid.

Ultrahigh frequency (UHF) waves—Frequency 300 Mhz–3 Ghz.

Ultrasonic sound—Frequencies above human hearing, 20kHz.

Ultrasound—A method of looking inside a person's body to examine tissue and liquid-based organs and systems without physically entering the body.

Umbra—Area of the shadow where all the light from the source has been blocked.

Uncertainty principle—Limitation to simultaneous knowledge of position and momentum of particle: $\Delta x \times \Delta p \geq h/4\pi$. Equivalent equation for energy and time.

Universal gravitational constant—Proportionality between gravitational force between two objects and GMm/r^2.

Van Allen belt—Doughnut-shaped regions of charged particles around Earth. Created by charged electrons and protons from solar wind and cosmic rays trapped by Earth's magnetic field.

Vad de Graaff generator—Device using static electricity to produce very large potential differences.

Vector quantity—A quantity that has both a magnitude and a direction.

Velocity (*v*)—Displacement divided by the time required to make the change (vector).

Very high frequency (VHF) waves—Frequency 30 Mhz–300 Mhz.

Very large array (VLA)—Group of 27 radio telescopes spaced as much as 13 miles apart from each other in Socorro, New Mexico.

Vocal cords—Or vocal folds. Vibrating source of the human voice.

Volt (V)—Unit of measurement of potential difference or voltage.

Voltaic pile—Or battery in which chemical energy is converted into increased energy of electric charges.

Watt (W)—Unit of power. One watt is one joule (J) per second (s).

357

Wave—A traveling disturbance that moves energy from one location to another without transferring matter.

Wave velocity—Depends upon the material or medium in which it is traveling.

Wavelength—The distance from one point on the wave to the next identical point; the length of the wave.

Weak nuclear force—Cause of beta decay. Weaker than strong nuclear force. Shown to be an aspect of the electroweak force. Carried by the massive W^+, W^-, and Z° particles.

Wedge—Simple machine used to force two objects apart. Examples are the knife, hatchet or axe, and plow.

Weight—Force of gravity on an object ($F = mg$).

Wheel and axle—Simple machine. If the radius of the axle is a and the radius of the wheel is w, then if the input force is applied to the wheel, the output force is given by $F_{output} = F_{input}\ (w/a)$.

White light—The combination of all the colors in the visible spectrum.

Work—Energy transfer by mechanical means.

X ray—Electromagnetic wave of very short wavelength or very high-energy photons. Emitted by an atom with many electrons that is disturbed by the collision with a high-energy particle.

Young's modulus—Applied pressure (called stress) divided by the ratio of the change in length to the original length (called strain).

Zeroth law of thermodynamics—If objects A and B are in equilibrium and B and C are in equilibrium, then A and C are also in equilibrium.

γ—In Einstein's Theory of Special Relativity a ratio that depends on the speed of an object. It is 1 when the speed is zero and infinite when the object moves at the speed of light. Affects distance, time, momentum, and energy.

Index

Note: (ill.) indicates photos and illustrations.

3-D, mechanics of, 226
9/11 terrorist attacks, 116

A

Abhimanyu, 202
Abrikosov, Alexei A., 14
absolute zero, 123
absorption, light, 279–81, 281 (ill.), 283
absorption, sound, 178, 178 (ill.)
AC (alternating-current) circuits, 255–56
acceleration, 32–36, 40–41, 45, 57, 58 (ill.), 309–10
accuracy, difference between precision and, 6
Acoustical Thermometry of Ocean Climate (ATOC), 166
acoustics, 3, 177–84, 186
active noise cancellation (ANC), 185–86
aerodynamics, 109–13
air as a conductor, 241
air conditioners, mechanics of, 134, 134 (ill.)
air drag, 37–38, 45–46, 46 (ill.), 110
airplanes
  airships, use of, 106–7
  controls, difference between automobile and, 113
  fastest, 116

safety during lightning storms, 243–44, 244 (ill.)
  stealth, 161–62
  wings, creation of lift, 110
  wings, importance of angles, 115
Akashi Kaikyo, 91
al-Baghdaadi, 9
al-Biruni, 9, 187
al-Din, Taqi, 187
Alferov, Zhores I., 15
Alfven, Hannes, 17
al-Haitham, Ibn (Alhazen), 9, 187, 210
al-Khazini, 9
alpha decay, 294–95
altitude, relationship between sound barrier and, 114–15
altitude of blimps, 105–6, 106 (ill.)
Alvarez, Luis W., 18
AM (amplitude modulation), 148–49, 149 (ill.), 151
ambulance, word printed backwards on, 205, 206 (ill.)
Americans with Disabilities Act, 74
Ampere, Andre-Marie, 247, 266–67
amplitude modulation (AM), 148–49, 149 (ill.), 151
amplitude of waves, 139–40

analog signals, 147–48, 147–48 (ill.), 150
ANC (active noise cancellation), 185–86
ancient culture
  clocks in, 5, 6 (ill.)
  computers, 80
  electricity, 232–33
  ideas about light, 187–88
  lenses, 209–10
  mirrors, 204
  motion, perception of, 29
  solar eclipses, 202
Anderson, Carl David, 20, 295
Anderson, Philip W., 17
Angels and Demons (Brown), 296
angle, critical, 212
angle of incidence, 205
angles of airplane wings, importance of, 115
angular momentum, 58–59
animals, hearing bandwidths of, 170 (ill.)
animals, sight of nocturnal, 225–26
antennas, transmission of signals using, 145–46
antimatter, 295–96
anti-noise headphones, mechanics of, 185–86
anti-quarks, 311
Appleton, Sir Edward Victor, 20
applications of physics, 3

359

Archimedes, 8, 77–78, 80–81, 103–4
Archimedes' Principle, 104
architectural acoustics, 177–80
Arecibo Telescope, 163
Aristarchus of Samos, 9
Aristophanes, 209
Aristotle
    author of first physics book, 1
    bust of, 2 (ill.)
    contributions to field, 8–10, 29
    description of gears, 81
    description of matter, 273–74
    discovery of magnetism, 261
    ideas about light, 187
    laws of motion, 33
    observation over measurement, belief in, 3
Arjun, 202
arm, blood pressure taken from the, 98–99
Armstrong, Edwin Howard, 149
arrays, radio telescope, 164, 164 (ill.)
astronomical methods to measure speed of light, 191–92
astronomical objects, gravitational fields of, 41 (ill.), 41–42, 51
astronomy, radar's connection to, 162–63
astronomy, radio, 163–64
astrophysics, 3
athletes' use of physics, 8, 59, 101
ATLAS (A Toroidal LHC ApparatuS), 310
atmosphere, Earth's, 196–97, 208–9
atmospheric physics, 3
atmospheric pressure, 49, 99–101
ATOC (Acoustical Thermometry of Ocean Climate), 166

atomic bombs, 13, 301–3
atomic clocks, 4
atomic devices, 303
atomic physics, 2
atoms, 274–76, 276 (ill.), 278–84, 279 (ill.), 281 (ill.), 283 (ill.), 289–308
attractive electrical forces, 234
Audion, 144
auditoriums, dead spots in, 154
auroras, 271
Avempace, 29
axis of rotation, 59, 59 (ill.)
axles, wheels and, 76–78, 77 (ill.)

B

Babinet, Jacques, 213
Bacon, Francis, 274
ball, path of a thrown, 48–50
balloons, release of helium, 107
balloons, static electricity of, 235
balloons, submerging in water, 100
Balmer, Johann, 281
bandwidths, animals' hearing, 170 (ill.)
Bardeen, John, 17, 19, 249
Barkla, Charles Glover, 21
barometers, 100, 100 (ill.)
barrier, sound, 113–15, 114 (ill.), 167
basics of physics, 1–23
Basov, Nicolay Gennadiyevich, 18
battery, creating out of a lemon, 246 (ill.)
beaches, best surfing, 141–42
becoming a physicist, 7
Becquerel, Antoine Henri, 22–23, 291 (ill.), 291–92
Bednorz, J. Georg, 16, 249
Bell X-1, 114, 116
bendability of material, factors in, 87, 90, 90 (ill.)
bends, 99–100

Bernoulli, Daniel, 109
Bernoulli's Principle, 109–10, 112
Berossus, 5
Berzelius, Jons Jakob, 275
beta decay, 294–95, 312 (ill.), 312–13
Bethe, Hans Albrecht, 18, 306
Big Bang, 160, 295, 314, 319
Binnig, Gerd, 16
biophysics, 3
birds on power lines, safety of, 251, 251 (ill.)
black, definition of, 217–19
Blackett, Lord Patrick Maynard Stuart, 19
blimps, altitude of, 105–6, 106 (ill.)
blindness, color, 224
Bloch, Felix, 19
Bloembergen, Nicolaas, 16
blood pressure, 98–99
blue, why is the ocean, 220
blue, why is the sky, 220
blue shift, 160
blurriness of objects seen, 224
body, atmospheric pressure on the, 101
body, resistance to electricity in the human, 250
Bohr, Aage, 17
Bohr, Margarethe, 283
Bohr, Niels, 21, 280 (ill.), 280–81, 283–84, 287, 300
Bohr model, 283–84
Boltzmann, Ludwig, 136
A-Bomb Arch, 303 (ill.)
bombs, atomic, 13, 301–3
Book of Optics (al-Haitham), 9
Book of the Devil Valley Master, 261
Born, Max, 19, 283, 287
Bose, Satyendra Nath, 286
Bose-Einstein Condensate, 286
Boston Symphony Hall, 179
Bothe, Walther, 19
Boyle, Robert, 166, 274

Boyle, Willard S., 14
Bradley, James, 191–92
Bragg, Sir William Henry, 21
Bragg, Sir William Lawrence, 21
Brahe, Tycho, 51–52
Brattain, Walter Houser, 19
Braun, Carl Ferdinand, 22
breaking crystal, resonance a cause of, 157, 157 (ill.)
breaking of waves, 141, 141 (ill.)
bridges, physics of, 90 (ill.), 90–94, 157–58
Bridgman, Percy Williams, 20
brightness of light, measuring, 193
British Thermal Unit (BTU), 67, 127
Brockhouse, Bertram N., 15
Brown, Dan, 296
Brownian motion, 12
BTU (British Thermal Unit), 67, 127
bubbles, iridescence of, 215, 215 (ill.)
building, tallest in world, 93
buoyancy, 101–7
Burj Dubai (Dubai Tower), 93
Buys Ballot, 160

C

calories, 127
camera lens, creation of image by, 212
cameras, mechanics of, 227 (ill.), 227–28
cameras, pinhole, 210, 211 (ill.)
capacitors, 238, 238 (ill.), 241–42
carbon dating, 293 (ill.), 293–94
careers in physics, 7
Carnot, Sadi, 132–33, 135–36
cars
    controls of different from airplane, 113

day/night rearview mirrors, mechanics of, 205
design using physics, 55–56
efficiency of, 135
electric, 71
objects viewed in side-view mirrors of, 207
safety during lightning storms, 243
spots in windows, 194
Carus, Titus Lucretius, 273
Cavendish, Henry, 274
CDF (Collider Detector at Fermilab), 310
CDMA (code division multiple access), 152
cell phones, mechanics of, 151 (ill.), 151–52
Celsius, Anders, 120
Celsius temperature scale, 120
center of gravity, 83–85
centripetal force, 50–51
CERN (European Organization for Nuclear Research), 297 (ill.), 309, 316
CFLs (compact fluorescent lamps), 189
Chadwick, Sir James, 20, 289
Chamberlain, Owen, 19
Chandrasekhar, Subramanyan, 16
charge of electrons, 278
charges, electrical, 233–37, 239–42, 247–48
Charpak, Georges, 15
chemical physics, 3
chemical symbols, 275
chemistry, relationship to discovery of atoms, 274–75
Cherenkov, Pavel Alexseyevich, 19
Chu, Steven, 15
Churchill, Winston, 283
circuits, electrical, 253–57
circular motion, 50–51
circular rainbows, 221
cities, increase of wind in, 110
Citigroup building, 92–93

Clausius, Rudolf, 135–36
clepsydra, 5
clocks, 4–5, 6 (ill.), 81–82
clothing as insulation, 129–30
clouds, formation of, 127
clouds as capacitors, 241–42
CN Tower, 94
Cockcroft, Sir John Douglas, 19
code division multiple access (CDMA), 152
Cohen-Tannoudji, Claude, 15
cold, collapse of containers in the, 101
Collider Detector at Fermilab (CDF), 310
collisions in accelerators, detection of, 310
Collodon, Daniel, 213
color, study of, 215–22
color as factor in temperature of objects, 130
color blindness, 224
color shifts, 160
colorimetry, 219
colors, Newton's theory of, 188
communicating with electromagnetic waves, 144–46, 146 (ill.)
communication, microwaves used for, 152
compact fluorescent lamps (CFLs), 189
compasses, 265–66
complementary colors, 217
composite materials, 88–89, 89 (ill.)
Compton, Arthur Holly, 21
computer, ancient, 80
computers, avoiding static buildup in, 235, 236 (ill.)
computers, use of magnetism in, 269
concave mirrors, uses of, 206
concert hall, optimal shape of, 178–80
Concorde, 116
condensation, 126–27
condensed matter physics, 2
conduction, 129–30

conductors of electricity, 235, 241, 247–50
cones, 223
conservation of energy, 63–64, 66–67
conservation of momentum, 56–57, 64, 66–67
containers, collapse of in the cold, 101
continuous spectra, 279–80
controlling temperature, 122, 122 (ill.)
convection currents, 129
converging lenses, 210
convex mirrors, uses of, 207
cooling process, evaporation as a, 128
Cooper, Leon N., 17, 249
Copernicus, Nicolas, 9–10
Coriolis force, 51
Cornell, Eric A., 15, 286
cosmic year, 30
Coulomb, Charles, 236
Coulomb's Law, 236–37
critical angle, 212
Cronin, James W., 16
Crookes, William, 277
crystal, resonance a cause of breaking, 157, 157 (ill.)
Ctesibius, 5
Curie, Marie, 22–23, 289, 291–93
Curie, Pierre, 22–23, 291–93
current electricity, 246–47, 250, 250 (ill.), 254–56
curve balls, 111

D

da Vinci, Leonardo, 33
Dalen, Nils Gustaf, 22
Dalton, John, 224, 274–76
dams, thickness of, 98
danger of looking at a solar eclipse, 201–02
danger of noise pollution, 184
danger of operating electrical devices in water, 258–59

danger of short circuits at home, 254
dangerous levels of electrical current, 250
dark, seeing in the, 225–26
dark energy, 320
dark matter, 319–20, 320 (ill.)
darkness of Earth during solar eclipse, 198
Davis Jr., Raymond, 14
Davisson, Clinton Joseph, 20
Davy, Sir Humphrey, 246, 252, 276
day/night rearview mirrors, mechanics of, 205
DC (direct-current) circuits, 254–56
de Broglie, Prince Louis-Victor, 20, 284–85
de Gennes, Pierre-Gilles, 15
dead loads, 94
dead spots in auditoriums, 154
deaths from lightning, 243
decay, radioactive, 293 (ill.), 293–95, 312 (ill.), 312–13
decibels, 174–75, 174–75 (ill.)
declination, magnetic, 264–65
definition of physics, 1
DeForest, Lee, 144
Dehmelt, Hans G., 16
Democritus, 273
density of objects, 102, 104 (ill.), 104–5
Descartes, Rene, 187–88
detection of light, 189–90
detectors, metal, 269, 269 (ill.)
Dialogue Concerning the Two Chief World Systems (Galileo), 10
diamonds, sparkle of, 212
Diesel, Rudolf, 133
difference, potential, 247
difference tones, 183–84
diffraction of light, 214
digital signals, 147–48, 147–48 (ill.)

dimensions, displacement in multiple, 26, 26 (ill.)
dimples, inclusion of in golf balls, 112
dip needles, 265–66
Dirac, Paul Adrien Maurice, 20
discus, throwing into the wind, 111–12
displacement, difference between distance and, 25
displacement, number of dimensions of, 26, 26 (ill.)
distance
    connection to wave amplitude, 139–40
    converting latitude and longitude to, 27
    determining of lightning, 168
    difference between displacement and, 25
    relationship between blurriness of objects and, 224
    relationship between sound intensity and, 173–76, 176 (ill.)
    relationship between time, velocity and, 44
    travel time of light of a specific, 192, 192 (ill.)
    velocity's effect on, 30
diverging lenses, 210
divisibility of atoms, 276
division of magnets, 263–64
door grating, function of microwave, 152–53
doorknobs, static shocks from touching, 235–36
Doppler, Johann Christian, 159
Doppler Effect, 159–61
Doppler shift, 29
Dorsey, N.E., 191
drag, air, 37–38, 45–46, 46 (ill.), 110
dryer, using a microwave as a, 153
Dubai Tower (Burj Dubai), 93
Dufay, Charles-Francois, 233

E

E = mc², 12

earbuds, use of electromagnetism in, 269

ears, cause of ringing in, 175

ears, frequency limits of human, 169–70

ears, pressure in, 97

Earth, atmosphere of relating to viewing stars, 208–9

Earth, darkness of during solar eclipse, 198

Earth, gravitational field of, 38–42

Earth, magnetic field of, 264 (ill.), 264–66

Earth, radiation in atmosphere, 196–97

Earth in moon's shadow during eclipse, 199

eclipses, 197–202, 199 (ill.), 200 (ill.)

Edison, Thomas Alva, 252, 255

education research, physics, 2

eels, electric, 251

efficiency, energy, 67, 135

Einstein, Albert
  becoming a physicist, 7
  comparison to Galileo, 49
  comparison to Maxwell, 143
  entanglement, 318
  equation of, 314
  and gamma rays, 295–96
  ideas about light, 282–84
  and momentum, 57
  most influential modern scientist, 11–13
  Nobel Prize winner, 21
  nuclear fission, 300
  photo of, 12 (ill.)
  study of quantum mechanics, 286–87
  theory of gravity, 52
  theory of relativity, 30–32, 39, 41–43, 53, 316

electric cars, 71

electric force, 47, 234, 236

electricity, physics of, 231–59, 308

electromagnetic spectrum, 143, 149–50

electromagnetic waves, 142–52

electromagnetism, 2, 266–70

electromotive force (emf), 247

electrons, 242, 276–78, 284–85, 312

electroscopes, 237

electrostatics, 231–37, 232 (ill.)

elements, properties of, 290, 296 (ill.), 298 (ill.), 298–99

elevation, athletes' training at high, 101

emission of light, 189, 279–81, 281 (ill.), 283, 283 (ill.)

Empedocles, 187

energy, 59–82, 117–18, 125–26, 128, 135, 242, 320

enrichment of uranium, 302

entanglement, 317–18

entertainment, electricity as a form of, 233

entropy, 135–36

Erathosthenes, 9

Esaki, Leo, 17

Essen, Louis, 191

etymology of physics, 1

Euclid, 11, 187

Europe, voltage system in, 258

European Organization for Nuclear Research (CERN), 297 (ill.), 309, 316

evaporation of liquids, 127–28

expansion of the universe, 321

eye, perception by related to photons, 282

eye, similarity between a camera and the, 227, 227 (ill.)

eyesight, 216, 222–26

F

Fahrenheit, Daniel Gabriel, 119

Fahrenheit temperature scale, 119

famous physicists, 7–13

Faraday, Michael, 39, 237, 240 (ill.), 241, 266–67, 276

Faraday Cage, 241, 243–44

farsightedness, 225, 225 (ill.)

"Fat Man" bomb, 303

FCC (Federal Communications Commission), 147

federal standards for hearing protection, 175

Fermi, Enrico, 20, 299–301

Fert, Albert, 14

Feynman, Richard P., 18

fiber optics, 213 (ill.), 213–14

fields, electric, 237

fields, magnetic, 262, 264 (ill.), 264–66, 270–71

fields, study of, 2

Finnegan's Wake (Joyce), 311

first law of thermodynamics, 131 (ill.), 131–32

Fischer, Avery, 179

fission, nuclear, 299–301, 304

Fitch, Val L., 16

Fitzgerald, Ella, 157

Fizeau, Hippolyte, 160, 190–91

floating and sinking, 101–7

flow of charges, 247–48

flow of fluids, 108–9

flow of heat, 129–30

fluid dynamics, 107–9

fluids, electric charges in, 233–34

fluids, study of, 95–116

Flying Boy experiment, 233

FM (frequency modulation), 148–51, 149 (ill.)

focal length of lenses, 210

focus on objects, lens shape, 223

forces
  buoyant, 102

carriers, 315
electrical, 47, 234, 236
exertion of, 86–87
nuclear, 295
physics of, 33–53
rotational, 58
strong, 290–91
on structures, 89
supporting, 86
Foucault, Leon, 190–91
Fourier, Jean Baptist, 183
Fourier Theorem, 183
Fowler, William A., 16
Franck, James, 21
Frank, Il'ja Mikhailovich, 19
Franklin, Benjamin, 158, 233, 238–39, 239 (ill.), 245–46
Fraunhofer, Joseph von, 280
free electrons, 242
free quarks, 312
freezing point of water, 119–20
freon, 134
frequency
    difference between pitch and, 180
    of eclipses, 199
    of lightning striking the ground, 242
    limits of the human ear, 169–70
    of a tone, 180–81
    of waves, 138–39, 139 (ill.), 146, 146 (ill.)
frequency modulation (FM), 148–51, 149 (ill.)
Fresnel, Augustin-Jean, 188
friction, 36–38, 63
Friedman, Jerome I., 16
Frisch, Otto, 300
Froome, K.D., 191
fuels, energy in common, 68, 68 (ill.)
fusion, nuclear, 300 (ill.), 305–8, 306 (ill.)

G

Gabor, Dennis, 17
galaxies, movement of, 160

Galileo Galilei
    contributions to field, 9–10
    description of matter, 274
    illustration of, 10 (ill.)
    invention of the telescope, 228
    inventor of the thermometer, 119
    Principle of Relativity, 49
    study of fluids, 100
    study of motion, 45
    study of sound, 177
    study of speed of light, 190
Galvani, Luigi, 246
gamma rays, 295
gas, objects floating in, 103, 103 (ill.), 105–6, 106 (ill.)
gas pressure, similarity to liquid pressure, 99
gasoline, iridescence of, 215
gears, 80–82, 81 (ill.)
Geiger, Hans, 278
Geim, Andre, 14
Geissler, Heinrich, 277
Gell-Mann, Murray, 18, 311
General Conference on Weights and Measures, 3–4
General Theory of Relativity, 12, 41–42
generation devices, 150
generators, difference between motors and, 268
generators, efficiency of electrical, 135
generators, power, 267 (ill.)
generators, Van de Graaff, 239–41, 251
geometrical optics, 202–3
geophysics, 3
George III, King (Count Rumford), 66–67, 117
GFI (Ground Fault Interrupter), 258
Giacconi, Riccardo, 15
Giaever, Ivar, 17
Gibbs, Josiah Willard, 136
Giffard, Henry, 107
Gilbert, William, 233, 261
Ginzburg, Vitaly L., 14

Glaser, Donald A., 18
Glashow, Sheldon L., 17, 295
glasses, polarized, 194–95
Glauber, Roy J., 14
Global Positioning Systems (GPS), 4, 26–27
gluons, 312
gnomons, 5
Goeppert-Mayer, Maria, 18, 23, 291
golf balls, inclusion of dimples in, 112
Gordon-Smith, A.C., 191
Gould, Gordon, 287
GPS (Global Positioning Systems), 4, 26–27
grating, function of microwave door, 152–53
gravitational force, 40–43, 51
gravitational mass, 40
gravity, 38–47, 39 (ill.), 47 (ill.), 316
gravity, center of, 83–85, 94, 94 (ill.)
Gray, Stephen, 233
Grimaldi, Francesco, 188
Gross, David J., 14
ground, frequency of lightning striking the, 242
ground as a capacitor, 241–42
Ground Fault Interrupter (GFI), 258
grounding wires, 257 (ill.), 258
Groves, Leslie, 301–2
Grunberg, Peter, 14
Guillaume, Charles Edouard, 21

H

Hahn, Otto, 299–300
half lives, 293, 293 (ill.)
Hall, John L., 14
hammers, wobble of, 83–84
hand behind their back, electricians working with a, 251
Hansch, Theodor W., 14
harmonics, 183

headphones, mechanics of anti-noise, 185–86

hearing, mechanics of, 169–70

hearing protection, 175–76, 176 (ill.)

heat, difference between thermal energy and, 65–66

heat, physics of, 128–30

heat, relationship between objects and, 117–18

heat capacity, 125 (ill.), 125–26

Heaviside, Oliver, 268

Heisenberg, Werner, 20, 283

Heisenberg Uncertainty Principle, 283–84, 289

helium, buoyancy of, 106–7

Henry, Joseph, 267

Hero of Alexandria, 187

Hertz, Gustav, 21

Hertz, Heinrich, 143, 188, 268

Hess, Victor Franz, 20

Hewish, Antony, 17

Higgs, Peter, 315

highest possible temperature, 123

Hindenburg, 106

Hipparchus, 9

hippopotamus, density of a, 105

Hitler, Adolf, 13

Hofstadter, Robert, 18

holes, purpose of outlet, 257 (ill.), 257–58

holiday lights, circuitry of, 256

homes, circuitry in, 257

homes, methods of heating, 128

Hooft, Gerardus T., 15

horsepower, 69, 69 (ill.)

hourglasses, 5

Hubble Space Telescope, 229–30, 230 (ill.)

hue, 219

hula hoop, center of gravity of, 94, 94 (ill.)

Hulse, Russell A., 15

humidity, relating to sky color, 219

Huygens, Christiaan, 66, 188, 192

hydraulics, 107–8, 108 (ill.)

hydrogen, buoyancy of, 106–7

hydrostatics, 104

I

Ibn Bajja, 29

ideas about light, early, 187–88

ideas about light, modern, 188

images, 206, 212

impedance, 158–59

impulse, 56–57

inclined planes, 74–75

index of refraction, 208 (ill.), 208–9

Indian Point Energy Center, 254 (ill.)

indigo, 215–16

inertia, 34

infrared radiation, 196–97

infrasonics, 172 (ill.), 172–73

injuries from lightning, 243

instantaneous speed, 28–29

Institute for Advanced Study, 13

Institute of Theoretical Physics, 283

instruments, mechanics of wind, 181–82

instruments, standing waves in musical, 155 (ill.), 155–56

insulation, clothing as, 129–30

insulators of electrical charges, 236

intensity of light, 193

intensity of sound, 173–76

interaction of objects, momentum and, 56

interference of light, 214

International Astronomical Union, 42

International Atomic Energy Agency, 279, 283

International System of Units (SI), 3–4

International Union of Pure and Applied Chemistry, 298

inverse-square law, 173–74

iridescence, differences in, 215, 215 (ill.)

Island of Stability, 299

J

Jansen, Sacharias, 228

Jayadrath, 202

JD (jelly doughnut), 67

Jensen, J. Hans D., 18

jobs that use physics, 8

Joliot-Curie, Frederick, 289

Joliot-Curie, Irene, 289

Jordan, Pascal, 283

Josephson, Brian D., 17

Joule, James Prescott, 66–67, 127

Joyce, James, 311

K

Kamerlingh-Onnes, Heike, 21

Kanada, 273

Kant, Immanuel, 11

Kao, Charles K., 14

Kapitsa, Pyotr Leonidovich, 17

Kastler, Alfred, 18

Kelvin, Lord (William Thompson), 121

Kelvin temperature scale, 121

Kendall, Henry W., 16

Kepler, Johannes, 51 (ill.), 51–52, 187, 228

Kepler's Laws, 51–53

Ketterle, Wolfgang, 15, 286

Kilby, Jack St. Clair, 15

kilogram, definition of, 5

kilowatt-hour (kWh), 67, 252

kilowatts, 252

Kirchhoff, Robert, 279

kite experiment, Franklin's, 238, 246

KMS (meter-kilogram-second) system. See Metric system

Kobayashi, Makoto, 14
Koshiba, Masatoshi, 14
Kroemer, Herbert, 15
KTHI-TV tower, 94
Kuo, Shen, 261
Kusch, Polykarp, 19
kWh (kilowatt-hour), 67, 252

L

Lamb, Willis Eugene, 19
laminar flow, 108
Lamm, Heinrich, 213
Landau, Lev Davidovich, 18
Laplace, Pierre-Simon, 66
Large Hadron Collider
   (LHC), 249, 296, 297 (ill.),
   309–10, 314–16
lasers, 286 (ill.), 286–88, 307
latitude, 27, 28 (ill.)
Laughlin, Robert B., 15
Lavoisier, Antoine-Laurent
   de, 66, 274–75
Lawrence, Ernest Orlando,
   20
laws, physics
   Coulomb's Law, 236–37
   difference between
      theories and, 52
   inverse-square law,
      173–74
   Kepler's Laws, 51–53
   Lorentz Force Law,
      266–68
   Newton's Laws of Motion,
      33–53, 35 (ill.), 55–56,
      59, 95, 110
   Ohm's Law, 249–51
   Snell's Law, 188, 208
   thermodynamics,
      131–32, 131–33 (ill.),
      135
LCD (liquid crystal display)
   devices, 195 (ill.), 195–96
le Monnier, Louis-Guilli-
   aume, 233
LED lights, 189, 189 (ill.)
Lederman, Leon M., 16
Lee, David M., 15
Lee, Tsung-Dao, 19
Leggett, Anthony J., 14

Leibniz, Gottfried Wilhelm,
   66
lemons, creating batteries
   out of, 246 (ill.)
Lenard, Philipp Eduard
   Anton, 22
lenses, study of, 209–13, 223
leptons, 312–15, 314 (ill.)
Leucippus of Miletus, 273
level, ability of water to stay,
   96
levers, 75–77, 76–77 (ill.)
Leyden jars, 237–39
LHC (Large Hadron Collid-
   er), 249, 296, 297 (ill.),
   309–10, 314–16
Libby, Willard F., 293
life, Earth's magnetic field's
   importance to, 265
lift, airplane wings' creation
   of, 110
lifting objects in water, 102
light, atoms' emission and
   absorption of, 279–81, 281
   (ill.), 283, 283 (ill.)
light, physics of, 187–230
light, properties of, 281–82
light bulbs, inventor of, 252
light quantum, 282
lightning, study of, 168,
   241–45
lightning rods, 245
lights, northern and southern, 271
light-years, 192
limits to noise pollution,
   establishment of, 184
line spectra, 279–80
Lippershey, Hans, 228
Lippmann, Gabriel, 22
liquid crystal display (LCD)
   devices, 195 (ill.), 195–96
liquid pressure, similarity to
   gas pressure, 99
liquids, evaporation of,
   127–28
"Little Boy" bomb, 302–3
live load, 94
load of a bridge, 94
location of lightning strikes,
   242–43
longest bridges, 91

longitude, 27, 28 (ill.)
longitudinal waves, 137, 138
   (ill.)
Lorentz, Hendrik Antoon,
   22, 266
Lorentz Force Law, 266–68
Los Alamos, weapon lab in,
   302–3, 306
loudness, difference between
   sound intensity and, 176
lowest possible temperature,
   123
lunar eclipses, 197, 199, 200
   (ill.), 201

M

Mach 1, 114–15, 167
machines, simple, 72–75,
   73–74 (ill.), 78, 81
Mackinac Bridge, 91
MAGLEV trains, 270
magnetic force, 47
magnetic resonance imaging
   (MRI), 249
magnetism, physics of,
   261–71
Magnus Force, 111–12
Maiman, Theodore, 287
makeup of the universe, 319
Manhattan Project, 13, 301
Marconi, Guglielmo, 22, 144
Marsden, Ernest, 278
Maskawa, Toshihide, 14
mass
   definition of, 40–41
   measurement of, 5
   of particles, 278, 314
      (ill.), 314–16
   relationship to
      acceleration, 34–36, 45
   role of in rotation, 58
matching, impedance,
   158–59
Mather, John C., 14
matter, dark, 319–20, 320
   (ill.)
matter, definition of, 273
matter, states of, 124–28
Maxwell, James Clerk, 136,
   143, 188, 190, 267–68

Maxwell's Equations, 143
measurements in physics
    brightness of light, 193
    distance, 27
    electricity, 236, 250
    energy, 67–68, 68 (ill.)
    focal length of lenses, 210
    force, 33
    length, 4
    mass, 5
    pressure, 96, 98, 100
    speed, 28–29
    speed of light, 190–92,
        191 (ill.)
    standards, 3–6
    temperature, 118–23
    time, 4–5
mechanics, study of, 2
media, index of refraction
    for, 208 (ill.), 208–9
media, light traveling
    through, 192–93, 193 (ill.)
media, sound traveling
    through, 166–67, 167 (ill.)
medical physics, 3
medicine, antimatter used
    in, 296
Meitner, Lisa, 299–300
Mendeleev, Dmitri, 289
mercury, thermometer, 120
metal, danger of inside
    microwaves, 153
metal detectors, 269, 269
    (ill.)
metals attracted to magnets,
    types of, 262–63
meter, measurement of a, 4
meter-kilogram-second
    (KMS) system. See Metric
    system
Metius, Jacob, 228
Metric Conversion Act, 4
metric system, 3–6, 4 (ill.)
Michelson, Albert Abraham,
    22–23, 190–91
microwaves, 146 (ill.),
    152–53
military uses of fission,
    300–301
Milky Way, 30 (ill.)

Millikan, Robert Andrews,
    21, 278
mirages, 209
mirror telescopes, segment-
    ed, 229–30
mirrors, study of, 204–7,
    229–30
molecular physics, 2
momentum, 55–59, 64,
    66–67
Monadnock Building, 92
moon, shadow of the, 199
Morse, Samuel S.B., 144,
    144 (ill.)
Morse Code, 144
Moseley, Henry, 289
Mossbauer, Rudolf Ludwig,
    18
motion, Brownian, 12
motion, rotational, 58
motion and its causes, 25–53
motors, difference between
    generators and, 268
Mott, Sir Nevill F., 17
Mottelson, Ben, 17
movement of fluids, 108–9
Mozart, Wolfgang Amadeus,
    158
MRI (magnetic resonance
    imaging), 249
Muller, K. Alexander, 16, 249
music, creation of through
    resonance, 158
music vs. noise, 184
musical acoustics, 180–84
musical instruments, stand-
    ing waves in, 155 (ill.),
    155–56

N

Nambu, Yoichiro, 14
National Ignition Facility,
    307
National Institute of Stan-
    dards and Technology
    (NIST), 5, 191
Natural Philosophy (Aristo-
    tle), 1
nearsightedness, 224, 225
    (ill.)
Neckham, Alexander, 261

Neel, Louis, 18
negative electrical charges,
    234, 239
neutral electrical charges,
    234
neutrinos, 312, 315–16
neutrons, 290, 309, 311–13
Newton, Sir Isaac
    comparison to Maxwell,
        143
    contributions to field,
        9–12
    description of
        momentum, 66
    ideas about light, 188,
        279
    illustration of, 11 (ill.)
    invention of the
        telescope, 228
    knowledge of sound
        media, 166
    laws of motion, 35–36,
        38–40, 52–53
    study of color, 216–17,
        221
Newton's Cradle, 64
Newton's Laws of Motion,
    33–53, 35 (ill.), 55–56, 59,
    95, 110
Newton's Theory of Colors,
    188
NEXRAD Doppler radar,
    162–63
NIST (National Institute of
    Standards and Technolo-
    gy), 5, 191
Nobel, Alfred B., 13
Nobel Prize
    antimatter, 295
    carbon dating, 293
    electrons, 278, 284
    lasers, 287
    neutrons, 289
    nuclear power, 295, 300,
        302
    quantum mechanics,
        285–86
    radio communications,
        144
    radioactivity, 292
    radiochemistry, 299
    speed of light, 190

superconductivity, 249
winners of the, 12–23
nocturnal animals, sight of, 225–26
noise pollution, 184–86
noise vs. music, 184
Nollet, Jean-Antoine, 233
normal force, 86
Norman, Robert, 265
northern lights, 271
Novoselov, Konstantin, 14
nuclear power, 2, 295, 299–308, 300 (ill.)
nucleus, properties of the, 289–90

O

ocean, color of, 220
ocean temperature, sound determining, 166
Oersted, Hans Christian, 266–67
Ohm, Georg Simon, 249
Ohm's Law, 249–51
Oliphant, Mark, 305
Omnibus Trade and Competitive Act of 1988, 4
one-way mirrors, 204
Onnes, Heike Kamerlingh, 248
opaque materials, 196
open circuits, 253–54
Oppenheimer, J. Robert, 301–2
The Optical Part of Astronomy (Kepler), 187
optics, study of, 2, 187, 202–3, 213 (ill.), 213–14
order of rainbow colors, 221
Osheroff, Douglas D., 15
Osiander, Andreas, 9
outlets, electrical, 257 (ill.), 257–59
overtones, 183
oxygen, discovery of, 275

P

pain, maximum decibel level without, 175

parallel circuits, 256–57
particle physics, 2, 281–82, 309–17
Pascal, Blaise, 95
Pascal's Principle, 95, 107
path of a thrown ball, 48–50
Paul, Wolfgang, 16
Paul III, Pope, 9
Pauli, Wolfgang, 20, 294
penumbra of a shadow, 198–99
Penzias, Arno A., 17
Peregrinus, Petrus, 261
period of waves, 139
Perl, Martin L., 15
Perrin, Jean Baptiste, 21
person's center of gravity, location of, 84
PET (Positron Emission Tomography), 296
Petronas Towers, 93
Philharmonic Hall, 179
Phillips, William D., 15
Philoponus, 29
phones, mechanics of cell, 151 (ill.), 151–52
photoelectric effect, 12–13
photograph red-eye, 227–28
photons, 282
physical optics, 202–3
pilots' breaking sound barrier, 114–15
pinhole cameras, mechanics of, 210, 211 (ill.)
pions, 313
pitch, difference between frequency and, 180
Planck, Max Karl Ernst Ludwig, 21
planes, inclined, 74–75
planets, surface temperature of, 123, 123 (ill.)
plasma physics, 2, 124 (ill.)
Plato, 8
playground, resonance in the, 156–57
plutonium, bombs containing, 301–3
Pluto's planetary status, 42
pneumatics, 108
Podolsky, Boris, 287

Poisson, Simeaon, 188
polarization of light, 194–96
poles, magnetic, 262, 262 (ill.), 270–71
Politzer, H. David, 14
pollution, noise, 184–86
Pont du Gard, 90
position, definition of, 25, 27
position, relationship to time, 43–44
positive electrical charges, 234, 239
Positron Emission Tomography (PET), 296
potential difference, 247
Powell, Cecil Frank, 19
power
    definition of, 68
    generators, 267 (ill.)
    lines, safety of animals on, 251, 251 (ill.)
    nuclear, 2, 295, 299–308, 300 (ill.)
    outputs, 70–72
    relation to volume, 185
    uses of electric, 252–53
precision, difference between accuracy and, 6
prefixes, metric, 5–6
pressure
    atmospheric, 49, 99–101
    blood, 98–99
    difference between force and, 48
    in ears, 97
    measuring, 96, 98, 100
    similarity between liquid and gas, 99
    water, 95–98, 96 (ill.)
Price, Derek de Solla, 80
Priestley, Joseph, 236, 274–75
primary colors, 217–18
Principle of Equivalence, 41
Principle of Relativity, 49
Principle of Superposition, 153–56, 154 (ill.)
printers, colors used in inkjet, 218 (ill.), 218–19
prisms, 188 (ill.), 217

probability as model for atom, 284

Prokhorov, Aleksandr Mikhailovich, 18

properties
elements, 290, 296 (ill.), 298 (ill.), 298–99
light, 281–82
magnets, 261–62, 262 (ill.)
nucleus, 289–90
waves, 138, 138 (ill.), 281–82, 284–85

protection, hearing, 175–76, 176 (ill.)

Proton Synchotron, 297 (ill.)

protons, 290, 309, 311–13, 315

psychoacoustics, 186

Ptolemy, Claudius, 9, 187

pulleys, 78–80, 79 (ill.)

Pulse Width Modulation (PWM), 147

pumps, vacuum, 277 (ill.)

Purcell, Edward Mills, 19

PWM (Pulse Width Modulation), 147

Pythagorean Theorem, 26

Q

quantum, light, 282

quantum mechanics, 285–87

quantum physics, 2

quantum teleportation, 318–19

quarks, 311–15, 314 (ill.)

qubits, 318

questions in physics, unanswered, 309–21

R

Rabi, Isidor Isaac, 20

radar, study of, 161–63

radar guns, Doppler Effect's use in, 160–61, 161 (ill.)

radiation, 196–97

radio astronomy, 163–64

radio communications, 144–46, 146 (ill.)

radio stations, reception distance of, 151

radioactivity, 291–93, 295

Rafale, 162

rainbows, 216–17, 221–22

Rainwater, James, 17

Ramsey, Norman F., 16

Rayleigh, Lord John William Strutt, 22, 220

rays emitted by radioactive materials, 291–92, 295

reactors, fusion, 306–8

real images, 206

rearview mirrors, mechanics of day/night, 205

reception distance of radio stations, 151

reception of electromagnetic waves, factors in, 145–46

recording, hearing oneself on a, 170

red shift, 160

red-eye, photograph, 227–28

reduction of noise pollution, 184–85

reflection, light, 202–4, 203 (ill.)

reflection, sound, 178

reflection, total internal, 212–13

Reflections on the Motive Power of Fire (Carnot), 132

reflector telescopes, 229, 229 (ill.)

refraction, light, 207–9

refractive index, 193

refractor telescopes, 228–29

refrigerator magnets, 263, 263 (ill.)

refrigerators, mechanics of, 134, 134 (ill.)

Rehm, Albert, 80

Reines, Frederick, 15

relative velocity, 32

relativity, 2, 13. See also General Theory of Relativity; Principle of Relativity; Special Theory of Relativity

repulsive electrical forces, 234

research, physics education, 2

resistance, electrical, 247–48, 250, 250 (ill.)

resistors, 248

resonance, 156–58

reverberation time, 177–78

reversal, mechanics of mirror, 204–5

On the Revolutions of the Celestial Spheres (Copernicus), 9

Richardson, Robert C., 15

Richardson, Sir Owen Willans, 21

Richter, Burton, 17

ringing in ears, cause of, 175

river, width of affecting flow, 109

rockets, acceleration of, 57, 58 (ill.)

rods, lightning, 245

rods, optical, 223

Rohrer, Heinrich, 16

Romer, Ole, 191

Rontgen, Wilhelm Conrad, 22

Roosevelt, Franklin D., 13, 300

Rosa, E.B., 191

Rosen, Nathan, 287

rotation of objects, 57–59, 59 (ill.)

Roy G. Biv, 215

Royal Society of London, 11

Rubbia, Carlo, 16

Rubin, Vera, 319

Rumford, Count (King George III), 66–67, 117

Ruska, Ernst, 16

Rutherford, Ernest, 278–80, 289, 291, 309

Rydberg, Janne, 281

Ryle, Sir Martin, 17

S

Sabine, Wallace Clement, 179

safety of animals on power lines, 251, 251 (ill.)

safety precautions for lightning strikes, 243–45

Salam, Abdus, 17, 295

satellites, motion of, 52–53
saturation, 219
The Sceptical Chymist (Boyle), 274
Schawlow, Arthur L., 16, 286
Scheele, Carl Wilhelm, 275
Scheiner, Christoph, 228
Schrieffer, John Robert, 17, 249
Schrodinger, Erwin, 20, 284–85
Schwartz, Melvin, 16
Schwinger, Julian, 18
scientific method, 9
screw as a simple machine, 75
sea breezes, 129
Sears Tower (Willis Tower), 93
second, measurement of a, 4–5
second law of thermodynamics, 132–33, 132–33 (ill.), 135
secondary colors, 217–18
secondary rainbows, 222
seeing, mechanics of. See Eyesight
segmented mirror telescopes, 229–30
Segre, Emilio Gino, 19
sensitivity, eye, 223
series circuits, 256–57
shadows, 197–202
Shakir, Ibm, 9
Shanghai World Finance Centre, 93
shape, lens, 223
shape of concert hall, optimal, 178–80
shapes, aerodynamic quality of, 112–13
shear, 89
shifts, color, 160
ships, buoyancy of, 104–5
shock waves, 113–14
Shockley, William, 19
shocks, static, 235–36
short circuits, 254
Shull, Clifford G., 15

SI (International System of Units), 3–4
side-view mirrors, appearance of objects viewed in, 207
Siegbahn, Karl Manne Georg, 16, 21
sight. See Eyesight
signals, difference between analog and digital, 147–48, 147–48 (ill.), 150
signals, transmission of using antennas, 145–46
simple machines, 72–75, 73–74 (ill.), 78, 81
sinking and floating, 101–7
sitting in chair, act of as static, 86
skiing, similarities between surfing and, 142
sky, color of, 219–20
skyscrapers, 92–93
Smith, George E., 14
Smoot, George C., 14
Snellius, Willibrord, 188, 208
Snell's Law, 188, 208
Soddy, Frederick, 278, 291
SOFAR (Sound Frequency and Ranging) Channel, 174
solar eclipses, 197–202, 200 (ill.)
solar system, sun as center of, 9
solid-state physics, 2
sonar, 171–72
sonic boom, 167–68, 168 (ill.)
sound, physics of, 165–86
sound, transmission of stereo, 150
sound barrier, 113–15, 114 (ill.), 167
Sound Frequency and Ranging (SOFAR) Channel, 174
sound waves, 165
sources of energy in the U.S., 69–70, 70 (ill.)
sources of sound, 165
southern lights, 271
space, magnetic fields in, 270–71

sparkle of diamonds, 212
Special Theory of Relativity, 30, 32, 39, 57
spectrum, definition of, 279–80
spectrum, electromagnetic, 143, 149–50
spectrum, sound, 183
speed, physics of, 28–30, 43–44, 46, 140–41
speed of light, 190–94
speed of sound, 165–68
spots in car windows, 194
SQUID (Superconducting QUantum Interference Device), 249
squirrels on power lines, safety of, 251, 251 (ill.)
Standard Model, 309, 314, 316–17
standards of measurement in physics, 3–6
standing waves, 155 (ill.), 155–56
Stark, Johannes, 21
stars, ability to see dependent on atmosphere, 208–9
states of matter, 124–28
static electricity, 231–37, 232 (ill.)
statics, 86–94, 104
statistical mechanics, 2
stealth planes, 161–62
Steinberger, Jack, 16
stereo sound, transmission of, 150
Stern, Otto, 20
Stoney, George Johnstone, 276
Stormer, Horst L., 15
Strassman, Fritz, 299–300
streamlines, 111, 111 (ill.)
string theory, 316–17, 317 (ill.)
strong force, 290–91
structure of atoms, 278–79, 279 (ill.)
structures, forces on, 89
structures, materials used in static, 87–88
structures, static, 90–94
structures, tallest, 93–94

Strutt, John. See Rayleigh, Lord John William Strutt
subfields of physics, 1–3
sublimation, 125
subtractive color mixing, 217–19
sun, temperature of the, 122–23
sun as center of solar system, 9
sundials, 5, 6 (ill.)
sunrises and sunsets, color of, 219–20
Superconducting QUantum Interference Device (SQUID), 249
superconductors, 247–50
superposition, 153–56, 154 (ill.)
supersonic flight, 115–16
supporting force, 86
surfing beaches, best, 141–42
Sushrata, 261
suspension bridges, 91, 92 (ill.)
sustainable energy, 72, 72 (ill.)
symbols, chemical, 275
symbols, electrical, 250
synthesizer, mechanics of the, 182, 182 (ill.)
Szilard, Leo, 300

T

Tacoma Narrows Bridge, 157–58
tallest structures, 93–94
Tamm, Igor Yevgenyevich, 19
Tatara Bridge, 92
Taylor, Richard E., 16
Taylor Jr., Joseph H., 15
TDMA (time division multiple access), 152
TDWR (Terminal Doppler Weather Radar), 163
technology, electromagnetic, 268–70
teleportation, quantum, 318–19

telescopes, mechanics of, 228–30
telescopes, radio, 163–64, 164 (ill.)
temperature, 118–23, 120–21 (ill.), 125, 130, 166, 168
Terminal Doppler Weather Radar (TDWR), 163
terminal velocity, 45–46, 46 (ill.)
Tesla, Nikola, 255
Thales of Miletus, 232–33, 261
theory, difference between law and, 52
Theory of Colors (Newton), 188
thermal energy, 64–66, 65 (ill.), 117–18, 128
thermal physics, 117–36
thermodynamics, 2, 130–36
thermographs, 122
thermometers, 118–20, 119 (ill.)
thermostats, 122, 122 (ill.)
third law of thermodynamics, 135
Thompson, Benjamin, 66, 117
Thompson, William (Lord Kelvin), 121
Thomson, Sir George Paget, 20
Thomson, Sir Joseph John, 22, 277–78, 291
three-dimensional vision, mechanics of, 226
three-prong outlets, 257 (ill.), 257–58
tidal waves, 142
tides, gravity's cause of, 42–43
time, speed and position's relationship to, 43–44
time, velocity's effect on, 30
time division multiple access (TDMA), 152
Ting, Samuel C.C., 17
tipping over objects, ease of, 84–85, 84–85 (ill.)
Tomonaga, Sin-Itiro, 18

tone, frequencies of a, 180–81
tones, difference, 183–84
tornadoes, early warning of, 172 (ill.), 172–73
A Toroidal LHC ApparatuS (ATLAS), 310
torque, 58
Torricelli, Evangelista, 100
torsional waves, 158
total energy of a system, 63, 64 (ill.)
total internal reflection, 212–13
total lunar eclipses, 200 (ill.), 201
total solar eclipses, 199m 200 (ill.), 201
towers, water, 97, 97 (ill.)
Townes, Charles H., 18, 286
traffic lights, "smart," 269–70
trains, MAGLEV, 270
transference of energy, 60–62, 61–63 (ill.), 64–65, 65 (ill.), 73–74 (ill.), 128
transformers, 159
translucent materials, 196
transparent materials, 196
transverse waves, 137, 138 (ill.)
Treatise on Electricity and Magnetism (Maxwell), 143
trees, safety of during lightning storms, 245
"Trinity" bomb, 303
Truman, Harry, 13, 283
Tsui, Daniel C., 15
tsunamis, 142
turbulent flow, 108
tweezer, laser, 287–88
two-slot outlets, 257 (ill.), 257–58

U

ultrasonics, 170–73, 171 (ill.)
ultrasound, 170–71
ultraviolent radiation, 196–97
umbra of a shadow, 198–99

unanswered questions in physics, 309–21

United States, energy in the, 69–70, 70 (ill.)

United States, longest bridge in the, 91

United States, measurement system of the, 4

United States, total solar eclipses in the, 201

United States, voltage system in the, 258

units of measurement. See Measurements in physics

universe, expansion of the, 321

universe, makeup of the, 319

uranium, 301–2

Urban VIII, Pope, 10

uses of energy in the U.S., 69–70, 70 (ill.)

V

vacuum pumps, 277 (ill.)

Van Allen Belts, 270–71

Van de Graaff generators, 239–41, 251

Van de Graaff, Robert Jemison, 239

van der Meer, Simon, 16

van der Waals, Johannes Diderik, 22

van Musschenbroek, Pieter, 237

van Vleck, John H., 17

Vatican, antimatter's relation to the, 296

vector, representing quantity of a, 25

vehicles. See Cars

velocity, 29–32, 44–46, 49–50, 137–38, 139 (ill.)

Veltman, Martinus J.G., 15

Venkataraman, Sir Chandrasekhara, 20

Verrazano Narrows Bridge, 91, 92 (ill.)

vision, mechanics of. See Eyesight

Vitruvius, 81

voice, wind instruments' similarity to human, 182

Volta, Alessandro, 246–47, 276

voltage, 247, 250 (ill.), 250–51, 258

voltaic pile, 247

volume, relationship to power, 185

von Eotvos, Baron, 40

von Guericke, Otto, 276–77

von Helmholtz, Hermann, 67, 276

von Kleist, Ewald Jurgen, 237

von Klitzing, Klaus, 16

von Laue, Max, 21

von Liebig, Justus, 204

von Mayer, Julius Robert, 66–67

W

Wald, George, 282

Walton, Ernest Thomas Sinton, 19

Warszawa Radio Tower, 93

water
ability to stay level, 96
clocks, 5
danger of operating electrical devices in, 258–59
freezing point of, 119–20
lifting objects in, 102
pressure, 95–98, 96 (ill.)
seeing under, 212–13, 224
submerging balloons in, 100
towers, 97, 97 (ill.)
using to model voltage and current, 247
waves, 140 (ill.), 140–42

Watson-Watt, Robert, 161

watt, definition of, 68–69, 252–53

Watt, James, 66, 69

waveform, sound spectrum relating to its, 183

wavelength, 138, 139 (ill.), 223

waves, physics of, 137–64

waves, properties of, 138, 138 (ill.), 281–82, 284–85

waves, shock, 113–14

Weakly Interacting Massive Particles (WIMPs), 320

weapons, fusion, 306

weapons, nuclear, 303–4

weapons containing uranium, 301

wedge as an inclined plane, 75

weightlessness, 46–47, 47 (ill.)

Weinberg, Steven, 17, 295

Westinghouse, George, 255

Wheeler, John, 42

wheels and axles, 76–78, 77 (ill.)

white light, 215–17

Wieman, Carl E., 15, 286

Wien, Wilhelm, 22

Wigner, Eugene P., 18

Wilczek, Frank, 14

Willis Tower (Sears Tower), 93

Wilson, Charles Thomson Rees, 21

Wilson, Kenneth G., 16

Wilson, Robert W., 17

WIMPs (Weakly Interacting Massive Particles), 320

wind, increase of in cities, 110

wind, speed of and wave type, 140, 140 (ill.)

wind, throwing a discus into the, 111–12

wind instruments, mechanics of, 181–82

winners, Nobel Prize in physics, 14–22

wires, grounding, 257 (ill.), 258

work, difference between energy and, 65–66

world, makeup of the, 273–88

World Health Organization, 184

World Trade Center, 93

Wright, Orville, 113

Wright, Wilbur, 113
Wright 1903 Flyer, 113

X, Y, Z

X rays, 285
Yang, Chen Ning, 19
Yeager, Chuck, 114–16
Young, Thomas, 188
Yukawa, Hideki, 19
Zeeman, Pieter, 22
Zernike, Frits (Frederik), 19
zeroth law of thermodynam-
    ics, 131
Zulu War, 202
Zwicky, Fred, 319